'This is an utterly engrossing book' *Foreign Affairs*

'Finely honed, consistently compelling … In the end this is a cautionary tale, a story of what happens when the dreams of humankind are hijacked by the darker aspirations of politics'
Publishers Weekly

'A swift, exciting history of the race to the moon, from Sputnik to "The Eagle has landed" … Cadbury's prose is heart-racing as she describes the individual missions, poignant as she acknowledges the loss of American and Soviet lives. First-rate research and reporting'
Kirkus Reviews (starred)

By the same author

The Feminisation of Nature

The Dinosaur Hunters: A True Story of Scientific Rivalry and the Discovery of the Prehistoric World

The Lost King of France: The Tragic Story of Marie-Antoinette's Favourite Son

Seven Wonders of the Industrial World

DEBORAH CADBURY

SPACE RACE

The Battle to Rule the Heavens

HARPER PERENNIAL
London, New York, Toronto and Sydney

Harper Perennial
An imprint of HarperCollins*Publishers*
77-85 Fulham Palace Road
Hammersmith
London W6 8JB

www.harperperennial.co.uk

This edition published by Harper Perennial 2006

3

First published in Great Britain by Fourth Estate in 2005

A catalogue record for this book is available from the British Library

ISBN-13 978-0-00-720994-1
ISBN-10 0-00-720994-0

Set in Minion and Helvetica

Printed and bound in Great Britain by Clays Ltd, St Ives plc

CONTENTS

LIST OF ILLUSTRATIONS

PLATES:

Yuri Gagarin became the first person in space on 12 April 1961. © Novosti London/Science Photo Library

Crowds pack the streets of Moscow to celebrate Gagarin's triumph. © Novosti London

Vostok 1 landing capsule. Gagarin himself ejected and parachuted to safety. © Novosti London/Science Photo Library

Yuri Gagarin with Chief Designer Sergei Korolev. © akg-images/Ullsteinbild

Wernher von Braun alongside his massive F-1 engines, which could consume 40,000 gallons of fuel a minute. © Science Photo Library

A section of the giant Saturn V rocket being transported to Cape Kennedy. © Hulton Archive/Getty Images

The Saturn V – over 36 stories high – was moved on a giant crawler to the launch pad. © NASA/Science & Society Picture Library

Apollo 8 sets out for the moon, 21 December 1968. © NASA/Roger Ressmeyer/Corbis

The Earth viewed from Apollo 8 orbiting the moon. Digital image © 1996 Corbis, original image courtesy of NASA/Corbis

16 July 1969: Apollo 11 lift-off at the Cape. © NASA/Science Photo Library

Buzz Aldrin walking on the moon. Digital image © 1996 Corbis, original image courtesy of NASA/Corbis

TEXT ILLUSTRATIONS:

Map of launch sites. © HLStudios, Long Hanborough

Commemorative stamp of Laika, the first living creature in space. © Hulton Archive/Getty Images

'Lighting-Up Time?' © Punch Cartoon Library

Drawing in *Pravda*, 'USSR September 1959'. © Photos12.com-Oasis

'Then, at 900,000 feet, you'll get the feeling that you MUST have a banana.' Cartoon © Emmwood/Daily Mail/Solo Syndication

Soviets with parade of aliens. © Punch Cartoon Library

PROLOGUE

As the two great superpowers, America and the Soviet Union, confronted each other during the Cold War, the race to the moon became a defining part of the struggle for global supremacy. Victory in this race meant more than just collecting moon rocks or planting flags on a barren wasteland. The development of missiles and rockets went hand in hand with the struggle to develop the capacity to deliver nuclear weapons, to spy on the enemy and to control space. Above all, the space race became an open contest between capitalism and communism. Victory was not just a matter of pride. National security and global stability were at stake.

The architects of this race were two extraordinary men destined to operate as rivals on two different continents at the height of the Cold War. Both were passionate about transforming their dreams of space travel into a reality yet both were cynically used and manipulated by their political paymasters as pawns in the wider conflict between the two superpowers. Both were men of their times but with visions that are timeless. Both were hampered by the legacy of a past which returned to haunt them, threatening to destroy the achievement of their dreams. One had collaborated with the Nazis to produce rockets in slave-labour camps during the Second World War. The other had been denounced as 'an enemy of the people', swept up in Stalin's purges and incarcerated in the Gulag in appalling conditions. Yet their ingenuity and vision would inspire the greatest race of the twentieth century: the race for the mastery of space.

For much of his life, the Russian Sergei Pavlovich Korolev was obliged to live in almost complete obscurity. Referred to as simply the 'Chief Designer', his name was obscured in the official records, never mentioned in the press and was virtually unknown to the public in his native country during his life. Such was the paranoia in the Soviet Union that this brilliant scientist might be assassinated by Western intelligence, he was shadowed constantly by his KGB 'aide'. When his bold exploits in space produced national celebrations in Red Square, he rarely appeared on the balcony beside the Soviet leaders and received none of the national acclaim for his achievements. Often working in harsh conditions deep within the Soviet Union, short of resources and at times challenged by jealous rivals, he pursued his quest relentlessly, with no regard for the enormous toll this took on his personal life. In the early years as Chief Designer of the Soviet Union's missile programme, Korolev understood that Stalin controlled his fate. Lavrenti Beria, Stalin's notorious Chief of Secret Police, was watching. False rumours, repeated failures or simply incurring displeasure could finish him at any moment. His family life destroyed by his long sentence in the Gulag, and with the loss of friends and colleagues during Stalin's purges, Korolev's future held no certainty. But now, with the release of classified information in Russia, for the first time the true story of this extraordinary man can at last be pieced together.

From his place in the shadows, Sergei Korolev was well aware of his rival in America, the charismatic Wernher von Braun. With his film-star good looks, his aristocratic manner, his brilliance in inspiring others, von Braun's smiling face often appeared in the American press and his ideas were studied closely by Korolev. Yet through all his glory years of success at NASA designing rockets that came to symbolize the might of America, von Braun carried a secret from his work as a Nazi during Hitler's Germany. During the Second World War, thousands of slave labourers had died of disease, starvation and neglect, or had been executed at the slightest whim of their SS guards while building the rockets that von Braun had designed to win the war for Nazi Germany. Sinister details of his assignment to save the Third Reich as Hitler's leading rocket engineer were classified after the war by the US authorities under the codename Project Paperclip – so called because a paperclip was allegedly attached to every file which was to be whitewashed. Von Braun's own secrets have only recently been unravelled.

These two men – Sergei Pavlovich Korolev in the Soviet Union and the former Nazi, Wernher von Braun in America – were both obsessed by the same vision of breaking the bounds of gravity and reaching the moon and beyond. 'In every century men were looking at the dark blue sky and dreaming,' Korolev told his wife. 'And now I'm close to the greatest dream of mankind.' Both found their ideas were way ahead of their time. When Sergei Korolev campaigned simply to speak publicly about launching the world's first satellite, 'a second moon', to the Academy of Artillery Sciences in 1948, he was repeatedly opposed, his ideas being dismissed as 'dangerous dreams'. Such notions had no place in Stalin's Soviet Union. As for von Braun, his vision of launching rockets and exploring the universe was considered so far-fetched in America even by the early 1950s that the only professionals who would take him seriously were those in the film industry.

In the 1950s, President Dwight D. Eisenhower feared that the Soviet Union would regard the development of rockets by America that were capable of putting men into space as a hostile military act. If men could be launched into space, so could spy satellites and nuclear warheads. The fragile peace between the Soviet Union and America could be blown apart. However, by 1957, much to the American public's consternation, the Soviet Union took the lead in space. Sputnik inspired terror – a Soviet satellite was flying over America. The US was in a race for survival, declared the *New York Times*. Away from the public's gaze, America's politicians and military elite panicked. They were horrified by the lead the Soviets had apparently developed in space technology.

As the Cold War escalated in the 1960s and the need for increasingly sophisticated weaponry grew, their ideas were no longer confined to the realms of science fiction. Both men endured enormous pressure from their political masters to win one of the most fiercely contested battles of the Cold War. Against this backdrop, the world struggled to come to terms with the constant threat of nuclear war. The Cuban Missile Crisis of 1962 brought the world to the very brink of disaster, the war in Vietnam raged on and the nuclear arms race threatened to spiral out of control. In the Soviet Union, Red Army troops were trained for nuclear

PART ONE

The Race for Secrets

'It will take 30 hours to get to the moon and 24 hours to clear Russian customs officials there ...'

Bob Hope, 1959

CHAPTER ONE

'The Black List'

In the mid-winter of 1945, the war in Europe had reached its final stages. Germany was crumbling under continued heavy Allied bombing. Cities were being obliterated, magnificent buildings returned to their original elements of so much stone, sand and lime. The massive Allied raids had demolished towns and cities on such a scale that Bomber Command was running out of significant targets. The attack on the Western Front was unrelenting, the dark shapes of Allied soldiers slowly advancing across occupied lands. The Rhine would soon be in Allied hands. From the east, with an unstoppable fury, the Soviets were approaching. In January 1945, the Red Army launched a massive offensive as 180 divisions overran Poland and East Prussia. Berlin was in their sights.

Right in the path of the advancing Soviets, at Peenemünde on the Baltic coast, lay a hidden village housing some five thousand scientists and their families. Discreetly obscured by dense forests at the northern tip of the island of Usedom, it was here that Hitler's 'wonder weapons' were being developed. The trees ended suddenly to reveal a chain-link fence and a series of checkpoints. At the local railway terminal a notice reminded passengers: 'What you see, what you hear, when you leave, leave it here.' Across a stretch of water known as the Peene River, a large village could be seen. It looked like an army barracks with regimented rows of well-built hostels. The sound and smell of the sea were never far away but remained invisible. About half a mile further on, hidden among the trees, was a scene from science fiction at the very cutting edge of technology, known as 'Rocket City'.

The world's largest rocket research facility was created by a young aristocrat named Wernher von Braun. At thirty-two, he was head of rocket development for the German army. A natural leader, he possessed the 'confidence and looks of a film star – and knew it', according to one contemporary account, although what people remembered most about him was his charm. He had a way of lifting the most ordinary of colleagues to a new appreciation of their worth. His organizational skills had turned Peenemünde into a modern annexe of German weaponry. However, very few people were allowed to see beyond the practical engineer, who dreamed not of destructive weaponry, but, improbably, of space. He was driven by the ambition of building a rocket that could achieve 'the dream of centuries: to break free of the earth's gravitational pull and go to the planets and beyond'. He envisaged space stations that would support whole colonies in space. 'In time,' he believed, 'it would be possible to go to the moon, by rocket it is only 100 hours away.' But in Hitler's Germany he was forced to keep such visions to himself. These were dreams for the future – a future that was increasingly in doubt.

Hitler had pinned his last desperate hopes of saving the Third Reich on von Braun's greatest achievement: a rocket known as the A-4. Even those working with von Braun were overawed on seeing this strange vehicle for the first time. In 1943, his technical assistant Dieter Huzel remembered being taken to a vast hangar which loomed above the trees. Once inside, the noise was deafening, a combination of overhead cranes, the whir of electric motors and the hiss of compressed gas. It took a second for Huzel's eyes to adjust to the strong shafts of sunlight, which cut across the hangar from windows high in the far wall. 'Suddenly I saw them – four fantastic shapes but a few feet away, strange and towering above us in the subdued light. They fitted the classic concept of a space ship, smooth and torpedo shaped…' Painted a dull olive-green, standing 46 feet tall and capable of flying more than two hundred miles, the A-4 was the most powerful rocket in the world. 'I just stood and stared, my mouth hanging open for an exclamation that never occurred. I could only think that they must be out of some science fiction film.'

Far removed from any fanciful notion of space exploration, for Hitler this rocket represented the ultimate weapon that could save the Third Reich and prove German superiority to the world. In July 1943, Wernher von Braun had been summoned to Wolfsschanze, the Führer's 'Wolf's Lair' in Rastenberg, East Prussia, to give a secret presentation. Walter Dornberger, the army general who ran rocket development at Peenemünde, had not seen Hitler since the beginning of the war and was 'shocked' at the change in him. The Führer entered the room looking aged and worn, stooping slightly as though carrying an invisible weight. Living in bunkers for much of the time had given his face the unnatural pallor of someone who spent his days in the dark. It was devoid of expression, seemingly uninterested in the proceedings, except for his eyes, which were worryingly alive, touching everything with quick glances.

Hitler's original response to the rocket had initially hampered von Braun's team. He had had a dream that no rocket could ever reach England, and simply refused to believe in the idea. Now, in the half-light, as he watched footage of the first successful launch of the A-4 shooting faster than the speed of sound over the Baltic Sea, his concentration became intense. With apparent satisfaction he took in an impression of the blond, blue-eyed von Braun, a perfect specimen of his 'master race' talking with uncontained enthusiasm as he outlined technical details of assembly, mobile launching facilities and testing. Here was the perfect terror weapon that could carry a 1-ton warhead, could be launched from any location and was undetectable on approach.

As von Braun finished his presentation, it was clear that Hitler had been greatly affected. His former listlessness vanished as he fired questions with increasing excitement. He had found the weapon that could win him the war. 'Why could I not believe in the success of your work?' he said to Dornberger. 'Europe and the rest of the world will be too small to contain a war with such weapons. Humanity will not be able to endure it.' His next demand was that the A-4 – soon renamed the V-2, or Vengeance Weapon 2 – should carry a warhead not of 1 ton, but 10 tons, and be mass-produced with output of rockets raised to 2000 a month. As though recharging that core of nervous energy that responded to ideas of destruction, he continued: 'What I want is annihilation – an annihilating effect.' Recklessly, he was to gamble his dwindling resources on experimental rocket science. 'What encouragement to the home front when we attack the English with it!' With its deadly warhead it would surely turn the tide of the war and ultimately allow a German victory.

Eighteen months later, this was looking less certain. With the Eastern Front collapsing, the Soviets liberated Warsaw on 17 January 1945. The Red Army swept across the country in little more than a week and reached the Oder. Berlin was only one hundred miles away. At Peenemünde, von Braun and his men could hear the steady barrage of the Soviet guns, increasing in volume as the wind changed, imparting a feeling of urgency. Outside, the streets teemed with the human flotsam and the debris of war as refugees fleeing the Red Army trudged through the freezing Baltic winter like some ragged army.

With the death of well over twenty-seven million Soviet citizens caught up in Hitler's war, the Russian appetite for revenge could not be satisfied. Writers such as Ilya Ehrenburg, who drafted Soviet propaganda broadcasts and articles, constantly urged retaliation against German civilians, unchecked by Stalin. 'We now understand that the Germans are not human,' he raged. 'Let us kill. If you do not kill a German, a German will kill you ...' As the Soviet army swept across Europe from the east, there were endless reports of brutality. In countless villages, houses were plundered and torched, civilians summarily shot. 'The tales of horror were as constant and unvarying as the unending stream of pitiful humanity,' wrote Dieter Huzel. 'Pillaging, burning, wanton killing – and worst of all rape and murder of old women, mothers to be and young women.' German civilians fled leaving villages abandoned with no attempt at defence against the Soviet onslaught.

Von Braun could hear the workforce of Peenemünde, the civilian engineers being trained in the use of rifles in the forlorn hope of defending the town before the great tidal wave of the Soviet army engulfed them. Any hope of trying to defend the town seemed futile – yet it was almost impossible to escape. The SS were trained in the brutal suppression of all opponents of the regime and were effectively in control of Germany. Roadblocks were being set up to catch deserters; even relatives of deserters, it was rumoured, could be sent to concentration camps. Some of von Braun's engineers who had openly expressed their doubts about defending Peenemünde were now dead, their bodies strung up by piano wire and hanging stiffly from trees in the main street. They bore placards that read: 'I was too cowardly to defend the homeland.' Their SS murderers had left their bodies hanging as a warning to others.

Confronted with this perilous situation, von Braun feared for the loss of his staff and his life's work. He was determined to find a way to evacuate his entire team to a relatively safer part of Germany, but the scale of such an operation could hardly be concealed from the SS. It would require the movement of thousands of workers as well as their families and truckloads of heavy equipment. There was also a treasure trove of documents: 65,000 technical drawings and blueprints alone had been required simply to bring the first V-2 from drawing board to test site. This included data from thousands of painstaking hours of trial and error testing, together with meticulous drawings of each component. There was nothing like this in the world. Whoever acquired these blueprints would inherit the cutting edge of rocket research. More important still, for von Braun the drawings of the A-4 – and its descendants the A-9 and A-10 – represented his vision of space flight. He had guarded these documents against sabotage, theft and air raids, and now he still hoped he could find a way of saving his work and his men.

He was confronted by several conflicting orders on how to cope with the advancing Red Army: from the local defence commander, the Gauleiter of Pomerania, the Ministry of Armaments, the army ordnance department in Berlin. Some directed von Braun and the rocket team to stay and fight to the death with the home guard 'in defence of the holy soil of Pomerania'. Others ordered immediate evacuation. 'I had ten orders on my desk,' von Braun admitted to his team. 'Five promised death by firing squad if we moved and five said I would be shot if we didn't move.'

• • •

For Wernher von Braun the operation at Peenemünde represented the culmination of a vision that had started in childhood. His first experiments in rocketry began when he was a boy of twelve. With his younger brother, Magnus, he had created a glorified go-cart powered by six enormous skyrocket fireworks. Zooming down Tiergarten Strasse in Berlin in a haphazard course, it had frightened all the summer-frocked housewives before crashing into a grocer's shop. His aristocratic parents, Baron Magnus von Braun and his wife Emmy, both from families that were very much guardians of tradition, were surprised at their son's appetite for science. In the hope that he would stop firing his rockets among the tenant farmers on the family estate in Silesia and follow more gentlemanly pursuits, they bought him a telescope. It had the opposite effect, merely inflaming his interest. With it he explored the moon and wondered at the stars, new worlds that, one day, could be explored by rockets.

'I was deeply disappointed by the sketchy information I could glean from my space gazing,' he wrote later, perhaps frustrated by the small size of his telescope. Then one day he came across an essay 'which showed how a certain Professor Hermann Oberth claimed it would be possible to *fly* to the moon and the planets on rockets. It seemed to me that this was a much better way of learning about our nearest planets than that offered by the telescope.' Born in Transylvania, Hermann Oberth had studied mathematics and physics before he set out his startling vision of space travel in 1923 in his book *The Rocket into Interplanetary Space*, in which he claimed it would one day be possible for rockets to carry men into space. Oberth's ideas were subsequently incorporated into a popular 1929 film by Fritz Lang, *Frau im Mond* (*Woman in the Moon*).

For the young von Braun – poring over Oberth's studies showing the possibility of launching a satellite into the earth's orbit or building a space station – it was a revelation. Here at last were the calculations that would allow humankind to defy gravity, and he was enraptured at the idea of flying to distant planets. Oberth described a 'recoil rocket' based on principles first defined by Isaac Newton in his third law of motion: for every action there must be a reaction of equal force but in the opposite direction. It is not unlike the firing of a bullet from a gun. When the trigger is squeezed, the bullet rushes out of the barrel creating a recoil which jolts the rifle butt back on the marksman's shoulder. In a similar way, a rocket is like the barrel of a gun – and the gases ejected from the back are like the bullet creating a recoil which propels the rocket upwards. The power of the rocket can be measured in tons of recoil – or thrust. To achieve liftoff, the thrust must exceed the weight of the rocket. And just as a shell rapidly reaches a certain speed and then coasts through a curved path towards its target, so a rocket needs a considerable initial speed before it is carried by its own momentum. For the young von Braun the message was clear: gain enough speed and space could be within his grasp. He just needed a powerful enough engine.

By the time he was nineteen, von Braun had joined the Society for Space Travel based at the optimistically entitled *Raketenflugplaz*, or 'Rocket Airport'. Here, in a disused army dump outside Berlin, a group of rocket enthusiasts met to experiment with rocket design. They developed small prototypes and attempted test firing. Oberth had argued that rather than solid-state fuels like gunpowder, which, once alight, burn uncontrollably, liquid fuels were the future. The flow of a liquid fuel to the engine can be turned on and off and regulated like a tap, allowing for more controlled combustion. He recommended alcohol or gasoline fuel combined with liquid oxygen as the oxidizer. Von Braun and the other enthusiasts experimented with these ideas in their first rocket known as the *Mirak*, or 'Minimum Rocket'.

The first flights were dogged by explosions or wildly erratic firing – a million miles from any dreams of space. Yet by 1932, von Braun, never one to lack confidence, and his colleagues were ready to demonstrate their test rocket at the army's artillery range at Kummersdorf, south of Berlin. Although the rocket rose just 100 feet before veering off course and crashing, von Braun's technical knowledge impressed Captain Walter Dornberger, who ran the rocket programme for the army. By the time he was twenty, while still a student at the University of Berlin, von Braun was recruited by the army and charged with building a rocket that was superior to the largest guns.

Von Braun began designs on his first rocket, known as the A-1. They were ready to test it in 1933 – the year Hitler came to power – but the liquid-fuelled engine blew up on launch. That same year, von Braun's father exchanged his senior government position as Minister of Agriculture for a quiet country life on his estate in Silesia and invited his son to join him. Wernher von Braun, however, was absorbed by plans for the more elaborate A-2, which successfully flew about one and half miles. By 1936, as Hitler's new, enlarged army marched to reoccupy the demilitarized Rhineland, von Braun was hard at work on the A-3 and then, even more ambitiously, on the A-4 – a rocket of such size and significance that secret new launch facilities were to be developed at Peenemünde. The army wanted a rocket that could travel 160 miles bearing a 1-ton warhead which would land within half a mile of the target. For von Braun it was the perfect opportunity: it would be the largest and most powerful rocket ever created. 'We were only interested in one thing – the exploration of space,' he claimed later. 'Our main concern was how to get the most out of the Golden Calf.'

Research for the A-4 proceeded slowly at first as von Braun's team tried to introduce major innovations. The engine, designed by the brilliant Dr Walter Thiel, was to incorporate several original features that would enable it to achieve more thrust. The fuel was ejected into the engine combustion chamber as a fine spray, which allowed it to mix better with the liquid oxygen. This improved mixing of the propellants gave more efficient combustion and reduced the risk of explosions. Thiel changed the design of the combustion chamber to give greater volume and incorporated a pre-chamber where the propellants were mixed: these innovations also facilitated a smoother burn. Pumps were used to direct the fuel to the engine at a faster rate to give yet more thrust. The result was an engine producing about 56,000 pounds of thrust, seventeen times more powerful than any previous design.

Apart from the new engine design, the shape of the rocket itself was more aerodynamic, with large fins for stability and rudders at the bottom of the fins for control. The guidance system was greatly improved. It was based on an inertial guidance system, a spinning gyro or wheel which could measure the position and acceleration of the rocket and then regulate guide vanes in the exhaust which could deflect the thrust to control direction. There was even a radio transmission system to communicate data to the ground – the first developments in telemetry. With a growing staff of several thousand at Peenemünde to help him accomplish this, in the space of a few short years von Braun had progressed from amateur enthusiast to the technical director of the largest rocket facility in the world. And as more funds conveniently flowed through, von Braun joined the Nazi Party in 1937. If *Kristallnacht* – 'the Night of Broken Glass', the smashing up of Jewish homes and businesses in November 1938 – did anything to make him question his party membership, his work was so absorbing that he did not act on it.

When Hitler invaded Poland in September 1939, research for the army assumed a new urgency. Despite high hopes, the initial testing on the A-4 in 1942 ended in midair explosions, but by 3 October they were at last rewarded. The rocket flew about 120 miles at a speed of 3500 mph, and, with an altitude of almost sixty miles, was the first to reach the fringes of space. This was an extraordinary achievement. At a celebration later that day, Dornberger triumphantly announced that rocket propulsion was indeed 'practical for space travel'. It was the dawn of a 'new era of space transportation'.

The film of this successful launch was what Dornberger and von Braun had shown Hitler in July 1943, but their success was short-lived. After months of testing, problems remained, the missile frequently exploding before it reached the target area. Under increasing pressure to deliver the Führer's wonder weapons, von Braun went to the new testing site in Poland in July 1944, intending to stand at the spot where he estimated the rocket should explode in order to see for himself what could be going wrong. The rocket was fired two hundred miles away from the test site at Blizna. This time, to his astonishment, he saw it was exactly on target, aiming for the very place where he was waiting. In panic he started to run, hoping he had the direction right and realizing that he was likely to be killed by his own expertise. 'I was hurled high in the sky by a thunderous explosion,' he remembered. Clearly the guidance system was beginning to work.

With this increasing success, the SS became more interested in the rocket programme. The head of the Gestapo and Reichsführer of the SS, Heinrich Himmler himself, began to take notice. After an army officer, Count Claus von Stauffenberg, failed in his attempt to assassinate Hitler on 20 July 1944, Himmler was able to consolidate his power, gaining control of the Home Army as well. Just over two weeks later, von Braun's friend and colleague of thirteen years, General Walter Dornberger, found himself replaced by the notorious SS officer Special Commissioner Hans Kammler.

Kammler, an engineer himself, had gained rapid promotion within the Third Reich, dedicating his ruthless intelligence to the promotion of his own career. With unswerving fanaticism, he had built concentration camps, including Auschwitz-Birkenau, and managed and controlled slave labour, relentlessly demanding heavy work from inmates who were close to starvation. He had systematically dismantled the Warsaw ghetto, destroying all evidence of German atrocities. His dedication to the cause won him Himmler's trust, who gave him full power over the rocket programme and proclaimed that 'my orders and his directions are to be obeyed'.

Von Braun and Dornberger, whose responsibilities were now greatly diminished, were appalled at this projection of Kammler to the top and recognized the danger that now stalked their enterprise. The man was a chimera: with mobile features and eyes full of expression went the coarse hands of a labourer. Despite the favourable first impressions he created, wrote Dornberger, like 'some hero of the Renaissance', he soon found that Kammler had an air of 'brutality, derision, disdain and overweaning pride'. He was intensely alert to any slight, alternating this with overconfidence. His amazing ego swallowed flattery whole, and he would hold centre stage, careful to surround himself with obvious inferiors, issuing orders, never taking suggestions, erupting into megalomania. Logical, ordered thought and careful forward planning were things of the past. The whole V-2 enterprise now spun in the mad chaotic orb of the ambitious Kammler's whim. He continued to work feverishly for a German victory, supervising the development of a secret factory buried underground in the Harz Mountains in central Germany near Nordhausen, known simply as 'Mittelwerk', or 'Central Works', where the A-4 was being built. By August 1944 more than a thousand missiles had been made and Kammler ordered that they were to be put into action at once.

Even as he was drafting the manual to issue to the military, von Braun could not quite dismiss his fanciful notions of space in his 'Top Secret' guide for the firing crews, a fact which doubtless contributed to Kammler's view of von Braun as 'too young and too childish … for the job':

> On this planet where you live
> In an age of guided missiles
> A sky ship in the universe

A long dream of mankind
May someday fascinate our century
But first you must master a new weapon…
(Extract from the V-2 manual, 1944)

On 8 September 1944, the V-2 bombardment of London began. The age of the ballistic missile arrived in Chiswick in west London as the first V-2 struck at 6.43 p.m. Travelling at four times the speed of sound, the V-2s soon created terror as they plunged silently on to their targets: the explosion was always heard first, followed by the dreaded sound of its approach. For long-suffering Londoners who had survived the Blitz and countless bombing raids, the massive explosion without the accompanying warning sound of the bomber had just the unnerving effect that Hitler had hoped for. Despite production difficulties during the autumn, attacks intensified. The Allies responded, bombing roads to the mobile launch sites on the Dutch coast and blocking fuel supplies. Yet even with the situation disintegrating in Germany, Kammler insisted on stepping up production of the V-2.

By January 1945, the Soviets were advancing rapidly across East Prussia but Kammler was slow to issue instructions to stop the team in Peenemünde from falling into enemy hands. In spite of the increased danger to the community at Peenemünde, von Braun could not readily escape without Kammler's approval. In doing so he would be perceived as a deserter and Hitler had made clear his intolerance of deserters on the wireless:

> I expect every German to do his duty to the last and take upon himself every sacrifice he will be asked to make. I expect every able bodied German to fight with complete disregard for his personal safety … I expect all women and girls to continue supporting this struggle with utmost fanaticism …

The very next day, on 31 January, the crisis at Peenemünde came to a head. From Nordhausen, 250 miles south from where the V-2 rockets were manufactured, an order finally came from Kammler to evacuate Peenemünde. The engineers were to carry on with their production at Mittelwerk, well away from the Allied armies. Within hours of

Kammler's order, however, an equally specific command arrived from the army chief instructing von Braun's team to defend the area to the last. All men at Peenemünde were directed to take up arms and join the 'people's army' in defending the Fatherland against the Soviets.

Von Braun summoned his senior staff into his office. He looked at each serious face around the table. 'Kammler has just ordered the relocation of all the most important defence projects into central Germany.' His expression was grim. 'This is an order not a proposal.' Silence followed this statement. No one wanted to be first with a dangerous objection. All knew it was suicidal not to follow SS orders. And no one relished the prospect of falling into the hands of the Soviets.

Von Braun had already raised the possibility of surrendering to the Americans with senior members of his team. Some months earlier, when it had become clear that Hitler would not win his war, they had discussed it discreetly while walking on the beach, the wind quickly dispersing their treasonable words into the vast Nordic sky. They had all agreed: only the Americans could fund a space programme. Huge sums of money would be needed to pierce the stratosphere with a rocket and travel to the moon, then the planets; but it could be done. Speaking quietly, von Braun reminded them of their own plan to get to America, impossible though it seemed at the time. 'Gentlemen,' he continued, 'our choices are limited but following Kammler's order would seem to give us the best chance of putting ourselves in the way of the Americans. Who else thinks so?' Almost silently, each man in the room signalled his assent. The pact was made. They would follow Kammler's orders, but their secret goal was America.

Von Braun proceeded with great thoroughness. Decisions had to be made about who would go, and in what order. Department heads were ordered to report back in two hours on how many people they would need to move, the loading space required and, most importantly, what would be left for the Soviets. It was a massive operation to plan, hampered by the fact that they could only travel at night as Allied planes hunted down and bombed anything that moved by road or rail in daylight hours. 'We will go as an organisation,' insisted von Braun. 'We will carry our administration and structure straight across Germany. This will not be a rout.' He decided that the safest way to cross Germany was to look as though they were on an authentic SS mission.

After days of organization, they commandeered cars, trucks, trains, ships and barges to sail from Peenemünde harbour, mules with carts, not to mention a snowplough and fifty select horses purchased from a wealthy woman fleeing East Prussia. With the state of near-collapse in Germany, the SS were controlling major routes and had set up roadblocks everywhere. Foreseeing difficulties, von Braun had labels printed which implied that the convoy was under the direct protection of Himmler, and part of Dornberger's agency, BZBV. However, a mistake in printing made nonsense of the large initials which emerged as the incomprehensible letters VZBV. There was no time to re-order the printing. Von Braun took a gamble and decided to use the labels anyway, on the grounds that any kind of official-looking label was better than none. VZBV would represent a top-secret agency, reporting to Himmler himself. And these letters were plastered all over the trucks and trains. If their ploy was uncovered they could face being sent back to Peenemünde or even being charged as deserters.

By mid-February the enormous task of dismantling Peenemünde had been completed and the first train carrying more than five hundred people moved slowly away from the approaching Soviets. The families travelled in freight cars in uncomfortable conditions, sleeping on straw. There was room for a few cows on the train, so that the children could have fresh milk – but there was no food for the cows. They ate the straw, and the beds slowly disappeared.

Leading the first convoy of trucks by road, von Braun ran into an army roadblock. Unknown to him, civilian traffic was forbidden to travel by road and an extremely efficient army major would not allow him through. The major was adamant but so was von Braun, who recalled that 'they faced each other like bulldogs' while precious time was slipping away. The whole expedition hung in the balance while the major considered ringing the army group commander at Peenemünde for confirmation that the exodus was legitimate. If this happened von Braun knew that they were in trouble, as the Gauleiter would make no such confirmation. They would be sent back to fight the Soviets or even arbitrarily shot as traitorous escapees.

Thinking on his feet, von Braun now rose to the occasion. He explained that VZBV was the abbreviation for Vorhaben zur

Besonderen Verwendung – Project for Special Dispositions – and that perhaps the major would care to continue his dispute with Himmler personally as the convoy was under his direct command. Upon very brief reflection, the major decided to forego that particular experience and the convoy was allowed through.

The journey had hardly begun. It seemed that danger was waiting round every corner. The 250 miles ahead to Mittelwerk would take many days as they could only travel at night, slowly, under dimmed headlights, stopping when low-flying fighters appeared. Fighters strafed the roads and railway lines. There were many roadblocks and if the convoy ran into a crack troop of SS, von Braun doubted they could survive. Their ridiculous VZBV stickers would fool nobody.

And as von Braun counted the dangers, he was unaware that he himself was a wanted man: both American and Soviet intelligence had him in their sights and were busy hunting him down.

• • •

On 3 February 1945, a young American engineer, Major Robert Staver, was sent from US Army Ordnance to London to find out everything possible about the secret German weapon. He lost no time in contacting British intelligence who dryly noted his air of confident enthusiasm and the strong signals he conveyed that he was not someone to be tied up in red tape. He was a man of action who could be relied upon to get things done. Indeed, his enthusiasm overrode any fear he might have felt on first close encounter with the V-2, which happened when he visited the American Embassy in Grosvenor Square. Staver claims he soon observed the destructive power of the V-2 at first hand: he was in a meeting when, without warning, a powerful blast flung both him and his senior officer across the room.

Staver was astonished at the silent, deadly approach of the V-2. He could hear the crash of falling debris, and from a window he observed 'a big round cloud of smoke where a V-2 had exploded overhead'. Later, at his hotel near Marble Arch, he was thrown violently out of bed when a V-2 landed at Speaker's Corner in Hyde Park. The amateur orators were silenced; more than sixty people died. In spite of protective covering, thousands of windowpanes in nearby buildings were shattered and Staver watched 'the drapes of his window stand straight out from the wall'.

Staver was impressed. There was nothing in the US armoury to compare with the V-2. He had been sent by Colonel Gervais Trichel, US Army Ordnance, chief of the rocket branch, who was concerned that America was twenty years behind the times. Trichel had already established a firing range at White Sands, New Mexico, with a view to creating a US rocket programme that could benefit from the German expertise. To develop the technology, he wanted to examine a V-2, obtain all possible technical data and interrogate the men who had designed the weapon; the blueprints alone would help advance the US armoury. He contacted Colonel Holgar Toftoy at the US Ordnance Department in Paris to ask him to track down V-2s that could be brought over to White Sands. Staver was to acquire the technological secrets in this field that would put America firmly in the lead.

Staver soon found British intelligence was generous with its information and had pieced together quite a story. The scientific intelligence officer at Bletchley Park, R. V. Jones, had been aware of the possibility of a German rocket programme since the beginning of the war, when an anonymous report outlining the development of experimental new weapons including rockets was posted to the British Naval Attaché in Oslo in Norway. At the time, leading officials had doubted the authenticity of the 'Oslo report' and believed it had been a hoax planted by the Germans. However, in March 1943, a secret recording between two captured German generals revealed that a rocket programme did indeed exist. Unaware that their conversation was being recorded, one of the generals speculated over what could have caused the delays in attacking London. He was reassured by the other general: 'wait until next year and the fun will start … There's no limit to the range.'

That same month, the Polish Intelligence Service gave an idea of the scale of the research. They passed on reports of an SS camp almost 180 miles south of Warsaw at Blizna. It was suspected that the site was being used for missile test firings. British aerial reconnaissance soon revealed craters a few miles from the camp and other signs of equipment and machinery similar to that at Peenemünde. Polish agents attempted to reach some crashed missiles before the Germans in the hope of finding out more about the technology, and on 20 May they discovered a complete, unexploded rocket by the River Bug some eighty miles from Warsaw. Hearing the Germans approach, they pushed it into the river and watched from cover while the Germans searched for but failed to find it. Later, they dismantled it and, in grave danger of discovery, cycled through enemy lines to take key components to an airfield, en route to England.

By June, Peenemünde itself had been infiltrated. A student from Luxembourg sent by the Germans to work on the secret rocket site smuggled out letters to his family describing the huge rocket and the ominous sound it made, like 'a squadron at low altitude'. That same month, R. V. Jones was studying photographs of Peenemünde when he suddenly spotted on a railway truck 'something that could be a whitish cylinder about 35 feet long and five or so feet in diameter, with a bluntish nose and fins at the other end'. It was direct evidence. 'I experienced the kind of pulse of elation that you get when after hours of casting you realise that a salmon has taken your line,' he said.

As news of the scale of the problem reached Churchill, instructions went out to Bomber Command. Over two nights in mid-August 1944, six hundred aircraft set out to destroy the site and target key personnel in the camps and sleeping quarters. Without warning, the camp became a maelstrom, vibrating to the whistle and crash of bombs. Dr Walter Thiel, von Braun's brilliant engine designer, died with his wife and four children, but most of those who perished that night were foreign labourers rather than the key German staff. The raid also failed to do as much damage as was hoped to key installations and launching facilities.

By early September, there was such confidence in London among senior officials that Hitler's secret weapon had been destroyed that this was even announced at a press conference. The very next day, the first V-2 struck London. R. V. Jones recognized the chilling significance of the V-2 programme. The V-2, he had written in a report as early as 1944, might be 'a feasible weapon to deliver a uranium bomb, should such a bomb become practical. It would be almost hopeless to counter by attacks on the ground, because the increased range would allow an almost unlimited choice of firing site … production would probably take place underground.'

British intelligence learned that the bombing of Peenemünde had not stopped production, and that the Germans were still manufacturing the missile in the vast underground site at Mittelwerk. They also knew the hidden factory was buried in the side of the Harz Mountains and consequently was impossible to destroy by bombing.

Major Staver's file grew thick. He knew where the work on guided missiles was carried out in Europe. He understood the significance of Peenemünde and what was being manufactured in the secret factory near Nordhausen. And he had a growing number of names of men he needed to trace for their expertise before the Soviets got to them. The Combined Allied Intelligence Report for August 1944 identifies a number of key personnel including a 'Dr Hans von Braun'. By January the Allied report was updated and listed his name correctly. Dr Wernher von Braun had moved to the top of Staver's black list.

• • •

Unknown to Major Robert Staver, in the Soviet Union the NKVD, or People's Commissariat of Internal Affairs – the predecessor of the KGB – had been on the trail of Wernher von Braun since 1935.

When Adolf Hitler first came to power, Soviet intelligence had been put on high alert. Vasily Zarubin, one of the NKVD's most successful agents, went to Germany along with his glamorous wife and colleague, Lisa. There they met a German spying for the Soviet Union, 'Agent Breitenbach', who worked for the German police under the name of Willy Lehman. Encouraged by Zarubin, Breitenbach provided much information, some of which even reached Stalin. Although these reports are still classified by the KGB, according to the German historian Matthias Uhl one letter sent to Breitenbach from the NKVD reveals the extent of Soviet interest in von Braun. They asked Breitenbach to find out more about von Braun including where he lived and worked and even whether there was a chance of infiltrating his laboratory.

When Zarubin left Germany in 1937, Agent Breitenbach continued to provide reports on the German rocket programme. He was so successful as an agent that on 19 June 1941 he was able to warn the Soviet administration that Hitler intended to attack the USSR on 22 June at 3 a.m. He was out by just half an hour – but it was his last report. In 1942, the Gestapo exposed a Soviet spying ring, Rote Kapelle, or Red Orchestra, and Breitenbach got caught up in the same sweep. Hitler himself took a personal interest in the trial of the fourteen leaders of Red Orchestra. The eleven members of the ring who faced the death penalty were executed in the most barbaric way: they were thrust on to meat hooks and left to bleed to death, hanging by the throat. Whether Breitenbach met the same fate is not known: on Himmler's instructions, details of his execution that December were never released.

After the death of Agent Breitenbach, Stalin was anxious to receive more information on the V-2 through the NKVD and military counterintelligence – but this proved difficult. Consequently, in early 1944 a team of specialists was recruited to develop Soviet rocket technology and investigate German progress; this was headed by Major General Lev Mikhailovich Gaidukov, a senior military figure known to the men simply as 'Desk General'. The team was established in a specially designated scientific research institute, NII-1, in Likhobory, on the outskirts of Moscow.

As concerns escalated over the programme at Peenemünde, in the summer of 1944 Soviet intelligence officers devised another way of spying on Nazi rocket research. Putting their faith in German prisoners of war, who were equipped with radios and false documents and parachuted into Pomerania, close to Peenemünde, they were to report back to the Soviets. One prisoner of war, Lieutenant Brandt, duly filed several reports before he was picked up by German intelligence services and shot.

On 13 July, Winston Churchill too advised Stalin of the secret German technology. At the time, Soviet forces in Poland were closing in on the test facilities at Blizna, where Polish agents had previously uncovered the German missile launching site. Recognizing that the weapons were 'a serious threat for London', Churchill asked if British experts could examine the site, but Stalin was in no hurry to help the British representatives. He delayed while a Soviet team of experts was hastily dispatched to the front line in Poland. Eventually, Stalin agreed that the British could examine Blizna but again delayed access. Five weeks elapsed while the experts were held up in Tehran, first for failing to meet the Soviets' visa requirements, then with dysentery. They were further delayed by a curious and no doubt spurious detour to Poland via Moscow. When they finally reached Blizna in mid-September, they sent a signal back to London describing the success of their mission.

They claimed to have found various V-2 components which were crated up to be sent to England. On arrival, however, the crates were opened to reveal old aircraft engines, the Soviets having seized the crucial parts for themselves. The British team were far too naïve, reported R. V. Jones, for the 'wiles of the Soviets'.

Stalin was all too aware that any allegiances between the Allies were temporary, necessary while Hitler was the common foe. After the war political differences would isolate the Soviet Union, which he knew had often presented itself as an inviting target for petty European dictators. He needed the new weapons technology. Components recovered from the remarkable new missile, including a rocket engine, were hastily dispatched to NII-1 for exhaustive analysis. Initially, only senior party members were allowed to investigate, but eventually 'sense took over and the engineers were allowed in', recalled one scientist at the institute, thirty-two-year-old Boris Chertok. His expertise lay in rocket plane engines, radar and precision instruments.

Chertok entered the large hall and at first all he could see were the legs of a man who was right underneath the rocket engine: 'His head was somewhere inside.'

'What is this?' Chertok enquired.

'This is what cannot be,' came the immediate reply. It was his colleague, a rocket engine specialist named Alexei Isayev.

To Isayev and Chertok, studying what remained of Dr Walter Thiel's V-2 engine in Moscow, it seemed almost unbelievable that in wartime conditions the Germans had developed such a bold and innovative design. By testing various components, such as fuel pumps, and taking careful measurements of the engine's combustion chamber and the smooth curvature of the exhaust nozzle, the Soviets gradually deduced that this extraordinary vehicle was not powered by traditional fuels like kerosene, but by alcohol and liquid oxygen. To their amazement they calculated the rocket itself must weigh over 12 tons and had a thrust of around 20 tons. 'We were shocked,' said Chertok.

Stalin realized the Soviets were behind in many technologies, including rocketry. He was eager to import as much German military industry as possible and to step up his search for the secrets of the missile technology. Early in 1945, he set up a special committee under the chairmanship of Georgi Malenkov, a member of his inner circle

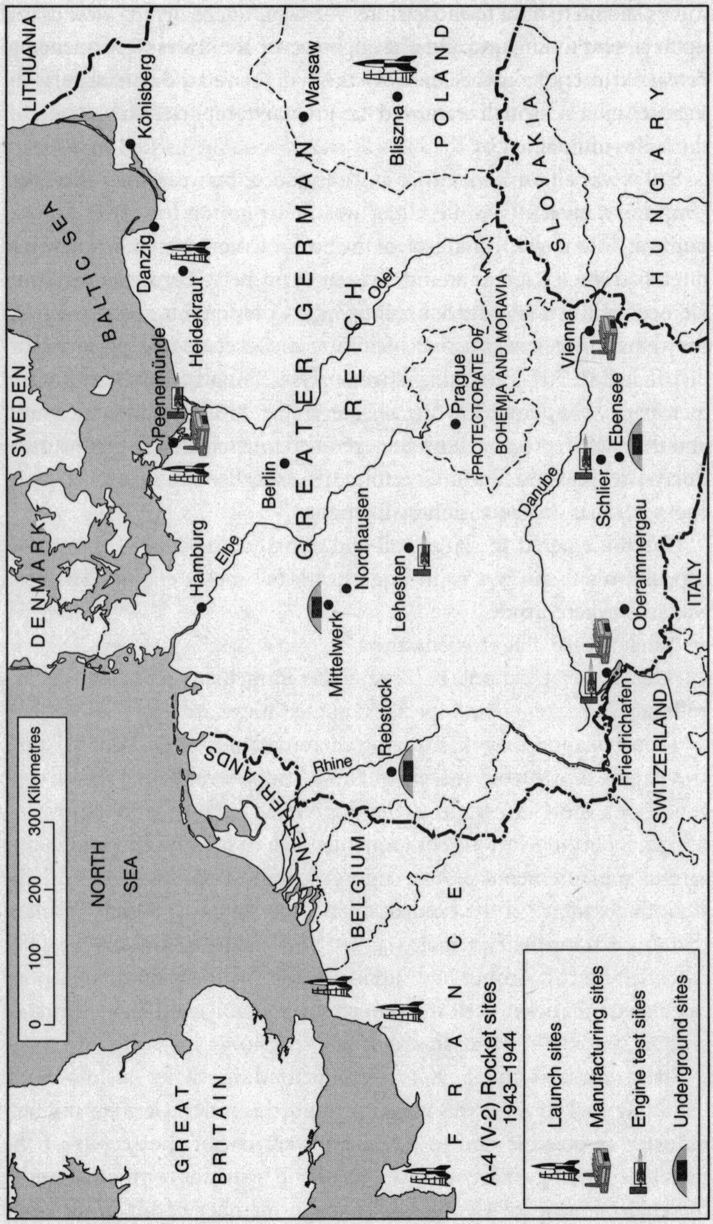
A-4 (V-2) Rocket sites
1943–1944
Launch sites
Manufacturing sites
Engine test sites
Underground sites
0
100
200
300 Kilometres
GREAT BRITAIN
NORTH SEA
BALTIC SEA
SWEDEN
DENMARK
LITHUANIA
NETHERLANDS
BELGIUM
FRANCE
SWITZERLAND
ITALY
HUNGARY
SLOVAKIA
POLAND
GREATER GERMAN REICH
PROTECTORATE OF BOHEMIA AND MORAVIA
Königsberg
Danzig
Heidekraut
Peenemünde
Hamburg
Elbe
Berlin
Warsaw
Bliszna
Oder
Prague
Vienna
Danube
Ebensee
Schlier
Oberammergau
Friedrichafen
Rebstock
Rhine
Mittelwerk
Nordhausen
Lehesten

who is estimated to have been responsible for many thousands of deaths during Stalin's purges. The engineers at NII-1 were instructed by Malenkov to try to gather data on the V-2 – and to do so speedily – because Stalin personally required the information. Malenkov's team of experts, accompanied by the NKVD, were to follow the Red Army into Eastern Europe and hunt down both the secrets of the V-2 – and the men who made it.

And so it was that Wernher von Braun, pursuing his vision of space travel, had attracted the interest of world leaders. While Hitler saw him as the saviour of a last-ditch 'final redoubt', von Braun topped the list of every Allied intelligence service. Stalin, with the despot's expectation of the immediate fulfilment of an order, was ready with roubles. British intelligence, gentlemen of course, tempering their ardour with good manners, were politely giving carefully plotted information to the Americans. Meanwhile, Staver, embodying all the brash confidence of the New World, began the chase in earnest.

CHAPTER TWO

'The Germans have set up a giant grill'

When von Braun's team finally reunited in central Germany in early March, they were absorbed into villages in the Nordhausen–Bleicherode area near Mittelwerk where the V-2s were made. This immense underground metropolis had been designed by Kammler to produce almost one thousand V-2s a month under the general manager, Dr Georg Rickhey, and the director of V-2 production, Arthur Rudolph, a committed member of the Nazi Party. Towards the end of the war, it was one of the largest operational industrial complexes in Germany.

Von Braun was reunited with his younger brother Magnus, whom he had installed as a manager at the factory to oversee the production of gyroscopes for the V-2 guidance and control system. They were instructed to continue with research and production, even though Germany's fall looked imminent. Although plagued by scarcities and despite the absurdity of the order, von Braun told his team: 'For now, let's do what we can to get our people settled and back into operation.' It was essential to give a strong signal to watchful SS eyes that he was loyal to the Führer and dedicated to producing V-2s for the 'ultimate victory'. Any hint that he might feel otherwise, any small gesture of defeat, could be noted and reported to Himmler, with the direst of consequences.

On 17 March, von Braun made his way to the Ministry of Armaments in Berlin to raise more funds for V-2 production. Travelling at night to avoid Allied planes, he soon fell asleep – as did the driver. The car crashed into a wall, somersaulted over it and down a railway embankment on to the track. In spite of a broken arm and shoulder, von Braun managed to pull his unconscious driver free before the car burst into flames. When he awoke in hospital, he found that he had suffered a head wound that needed stitches, broken his arm in two places and badly smashed his shoulder. Although the injury was complicated, he soon discharged himself from hospital. Wrapped up like a mummy, swathed in bandages and plaster casts, his broken left arm permanently held out before him, von Braun was gratefully installed in one of the grander houses in Bleicherode to recuperate.

This imposing two-storey house was called the Villa Frank and had once belonged to local Jewish cotton-mill owner who had been deported from Germany by Hitler. On 23 March, in the lavishly furnished accommodation, a party was held for von Braun's thirty-third birthday. All his old friends and colleagues were gathered, men he had worked with for years. According to his technical assistant, Dieter Huzel, everyone attempted to maintain 'at least a façade of normal activity', but the atmosphere was nonetheless subdued. Uncertainty hung in the air like the cigarette smoke that lingered in the beautiful rooms. The soft-carpeted opulence could not blot out the war; defeat seemed imminent.

The Third Reich was doomed, crumbling into the dust of history as the victorious Allies moved inexorably across Germany; the Soviets thundering across the country from the east, the American and British armies moving in to meet them from the west. The US army had crossed the Rhine between Mainz and Mannheim and soon they would be at the Elbe in the eastern half of Germany, shaking hands with the Soviets and comparing stories. On one occasion, Huzel recalled hearing the 'ever present rumble of distant cannon fire grow in loudness until it was a roar'. He looked up, momentarily blinded by the sun, and was shocked to see 'hundreds upon hundreds of Allied bombers, surrounded by the tiny points of their fighter escorts'. He was aghast 'at the immensity of the formation' and stared open-mouthed. Berlin was bombed night and day in terrible retribution. For mile after mile the stony skeleton shapes of once majestic buildings stood open to the sky.

At the very centre, enclosed in his tomb-like bunker, Hitler was still there, shouting at his generals, issuing directives. On 19 March, he had issued a decree that anything of value to the enemy was to be demolished:

DEMOLITIONS ON REICH TERRITORY
'Nero Decree'

> Every opportunity must be taken of inflicting, directly or indirectly, the utmost lasting damage on the striking power of the enemy ... anything ... of value within Reich territory, which could be in any way used by the enemy immediately or within the foreseeable future ... will be destroyed.

Hitler would not listen to dissenting voices and was convinced the German people were unworthy of him. 'If the war is lost,' he ranted, 'the nation will perish ... it will be better to destroy these things ourselves because this nation will have proved to be the weaker one ... Only the inferior will remain, for the good ones have all been killed.' According to Hitler's decree, nothing that could be turned against Germany was to remain; his enemies would acquire only 'scorched earth'.

Von Braun and Dornberger, who was now installed nearby in Gad Sachasa, quietly discussed Hitler's directive. They were still guarding 14 tons of precious blueprints. If these were destroyed by the SS, or if they themselves were killed, years of groundbreaking work in their field would be lost. They thus took the fateful decision to go against Hitler's express orders and at least hide the documents. They knew exactly what their fate would be if Kammler discovered their plan – but Kammler was distracted. He had been promoted yet again and now carried the impressive title of 'Special Commissioner for breaking the Air Terror', an impossible, Canute-like task. The Luftwaffe was impotent. Allied bombers ruled the skies and at this late stage of the war nothing could stop wave after wave of bombers blackening the heavens. The ominous drone of their engines filled the air and the bombs continued to fall like a deadly hailstorm.

Whatever steps Kammler took in an attempt to manufacture yet more German aircraft, the Allied planes were always above him. No one around him dared to murmur even a word of opposition. Kammler had special powers; if any Luftwaffe generals failed to carry out his orders they could be shot. Kammler's temper was tuned to breaking point. 'He was on the move day and night,' observed Dornberger.

> Conferences were called for 1 o'clock in the morning somewhere in the Harz Mountains, or we would meet at midnight somewhere on the autobahn and then, after a brief exchange of views, drive back to work again. We were prey to terrific nervous tension. Irritable and overworked as we were, we didn't mince words. Kammler, if he got impatient and wanted to drive on, would wake the slumbering officers of his suite with a burst from his machine gun. 'No need for them to sleep,' Kammler would say. 'I can't either.'

Secretly, von Braun picked two trusted men, Dieter Huzel and Bernhard Tessman, and ordered them to collect all blueprints and plans and find a safe hiding place for them, such as a disused mine. They represented a small, treasured seed of possibility which, if von Braun lived, he would reclaim one day. The A-4, the V-2 prototype, was for von Braun just one of a family of rockets – for which the latest designs embodied the future of space travel. The A-9, which had winged surfaces and included room for a pilot in the cockpit, was designed to travel several thousand miles and was for him the first step towards a space ship. The A-10 was a booster that would enable the A-9 to span even greater distances. These documents were irreplaceable and represented von Braun's future and his hopes that space exploration could one day be pursued without limitations. Their safety was paramount. In spite of the enormous risk involved in disobeying Hitler's orders, he was determined to save them.

Equipped with three trucks and trailers, along with false papers from von Braun which appeared to validate his mission, Huzel set off with Tessman to find a suitable site. This was not easy: abandoned mines were suddenly at a premium. Huzel was soon at breaking point, struggling to avoid endless low-flying enemy aircraft which plagued even minor roads and worried by the menacing proximity of enemy guns, when to his great relief he discovered a forgotten and untenanted mine. Down a long passageway he found a wonderfully dry room behind a locked door. To his relief, a nearby SS post did not detain him or question his mission, but warned him that American troops were entering the next village. This was the spur Huzel needed and he and Tessman worked through the night to hide the documents.

Meanwhile, in Nordhausen the usually dashing figure of von Braun had been obscured under a bizarre arrangement of plaster casts and bandages. Although recuperating in the sumptuous splendour of the Villa Frank, his arm would not heal and it caused him some pain. It had not been properly set and further treatment was needed if he was to avoid the risk of amputation. There was no time to seek medical help, however, with the Allies closing in. It was difficult to know just how near they were since no real news was transmitted on German radio. In its absence, rumour took on a convincing reality – the American army were sited just twenty miles away. Von Braun and Dornberger were hoping that they could persuade the Americans to employ the rocket team for further research in space. Quite apart from their own expertise, the unique hoard of documents would be a useful bargaining tool. Yet, surrounded by the SS, there was no chance of surrendering. It was unclear what might happen if they tried to escape.

Dornberger had informed von Braun how edgy and unpredictable Kammler had become. Although he still cut an authoritative figure in his black leather SS military greatcoat as he emerged imperiously from a handsome German car with a flurry of attendant guards, the leonine features of the general were now taking on a drawn and worried look. Dornberger had overheard Kammler instructing his Chief of Staff, SS Major Starck, to remain at all times 20 feet behind him, and, if they found themselves in an uncompromising position from which there was no escape, Starck was to shoot him in the back of the head. Frantically trying to prop up the last remnants of the regime, the 'berserk warrior', as Dornberger now called him, had every reason to be nervous as a high-ranking Nazi officer with a very long list of war crimes behind his meteoric rise in rank. If the Americans caught him he would almost certainly be hanged. He was as good as dead already – except that he had a strategy. The time had come to implement a plan that would protect him.

Kammler was aware that in his hands he had a tantalizing prize to offer the Americans in exchange for his life: the rocket team. A manageable number – five hundred of the top men under close SS guard – would be enough to impart a feeling of security. And if he failed to strike a bargain, it would not take long to shoot them as Hitler had ordered, rather than let them fall into enemy hands. On 1 April, therefore, Kammler ordered that von Braun and five hundred of the leading scientists were to move immediately to the Bavarian Alps. Hitler and the best SS divisions would be there, Kammler explained, employing the natural fortifications of the landscape from which to mount a victorious attack on the enemy.

Kammler's train offered an atmosphere of luxury entirely divorced from the war, with wagons-lits, and a dining car offering excellent cuisine and choice of wines in which the general increasingly indulged on his not infrequent trips. Yet as von Braun's team departed on their tortuously slow four-hundred-mile journey south to the Alps chaperoned by a hundred SS guards, it was difficult to disguise the feeling that they were being imprisoned. At least one scientist, the talented young engineer Helmut Gröttrup, was beginning to question von Braun's leadership. At Peenemünde, Gröttrup had worked as deputy to the director of Guidance, Control and Telemetry. With the situation in Germany increasingly dangerous and unpredictable, he was worried about leaving his young wife, Irmgardt, and their two children. In addition, he was aware of the persistent rumours 'that prisoners on the train were to be killed on account of their knowledge of secret information'.

'I don't even know where he will be taken or when he will return,' wrote his wife Irmgardt in her diary, despairing that her whole family 'was scattered somewhere around the burning world … In the evenings I sit and wait until late at night … As we said goodbye it all became so empty and dreadful … They keep coming to take more of our group away. Will we ever see one another again? It astounds me that man endures so much suffering …'

Helmut Gröttrup's train left at midnight on 6 April yet with so many lines damaged or destroyed, the speed rarely exceeded 3 mph. Just before Munich, he seized his opportunity to escape. Despite the presence of the SS guards, he jumped from the train and managed to hide. Moving only at night, he began the long walk back towards Nordhausen while the remainder of von Braun's leading scientists continued south towards the Bavarian Alps for Hitler's alleged last stand.

• • •

In early April, a few days after von Braun left the Nordhausen area, John Gallione of the US 104th Army Infantry Division set out to find a concentration camp that was rumoured to be in the vicinity, providing labour to make the Nazis' V-2 rockets. The 104th Infantry Division had swept almost two hundred miles from the Rhine in a relentless push to the east. 'We felt lucky to have made it this far in the war,' said Gallione, 'but something was driving me on to find the prisoners. I just had a gut feeling that something horrible was going on …' He decided the best way to find a camp was to follow the railway line. The only way the Nazis could sneak large numbers of people into a camp, he reasoned, was by train.

He followed the track for miles until, to his horror, he came across a carriage full of dead bodies. 'The smell was horrible. Where I was standing, I could see a hidden tunnel coming out of the side of a mountain.' While he was investigating the tunnel and the bodies, trying to find any papers that would help him identify the nationality of the dead, he was spotted by an SS guard. 'I was shot at and took cover … we shot back and forth for a while, but he seemed to let me go. He was in a hurry to get out of there …'

John Gallione went for help and found a soldier with a Jeep. They retraced the route back to the train and slowly became aware of a foul smell: a strong, sickly, bad smell, impossible to identify with any certainty but undoubtedly a precursor of trouble. They found a gate into what seemed to be a vast enclosure and broke the lock. The night was almost over and in the half-light of dawn they became aware of shadowy forms, not quite human, moving about. Caught in the headlights of the Jeep, skeleton figures could be glimpsed, apparently wrapped in skin. They were moving painfully slowly among piles of the dead who were grey in colour with limbs like sticks. Picking his way towards them through the bodies was one of the living skeletons, his face devoid of recognizably human features, his skin drawn tight over the bony head, his eyes those of a very old man. He gesticulated and tried to explain. 'There are people in here,' he cried out, 'desperate for help.'

This glimpse of a ghoulish, subhuman scene was too much for the terrified driver. He reversed the Jeep and kept on reversing until they were well away from the camp. 'That's how frightened we were that we might be captured,' explained John Gallione. 'We did not know what was going on in there and we did not want to end up like the people we had seen.' Thoroughly unnerved, the two men raced back to headquarters and explained that they had seen the inmates of some kind of concentration camp. They immediately radioed for help from other divisions in the area. By 11 April, a massive rescue operation was underway involving tanks, medical teams and the Red Cross.

The place they had chanced upon was Camp Dora, which had been built by Kammler to house the men who provided slave labour for the manufacture of the V-2. John Gallione found the courage to return and tried to describe the scene. 'There were dead bodies piled up. The smell was so bad, like nothing you can imagine. The people were so happy to see us; they were tugging our clothes, feeling our uniforms between their fingers like they were gold. They just wanted to touch us, thanking God over and over again. They looked like the walking dead; skin and bones – that's all. Some of them were so weak they didn't even get to see their own rescue. They died on the way.'

Nothing had prepared the combat soldiers and medical teams for the grim sight that they uncovered. This was as close to hell as it was possible to get. 'The gun and the pursuit of the enemy was dropped,' reported one medic from the 104th Division. 'All hands were turned to the overwhelming task of saving the living if at all possible.' Another medic, David Malachowsky, heard machine-gun fire. To his horror, as he approached a rise in the land he saw 'the SS … frantically trying to finish the job. They had a bunch of prisoners lined up against the fence and were gunning them down.' On the hill he could see the huge cavern-like entrance to the factory. Six thousand emaciated bodies covered the ground far into the distance. Halfway up the hill in a low building, the furnaces of the crematorium were still smouldering; the doors open, awaiting more human fuel. 'They had been shovelling people in and burning them up.'

Before long the US army came across yet another concentration camp in nearby Nordhausen. More slaves for the Mittelwerk V-2 factory were housed in rancid-smelling buildings; the dead and living shared wooden bunks, the difference between them hardly discernible, those still alive, immobile, waiting patiently for death. Sergeant Ragene Farris of the 329th Medical Battalion, 104th Infantry Division, was approached by a Frenchman who was anxious to get immediate help to his fellow French prisoners in the dark cellars of another building. 'It was like stepping into the Dark Ages to walk into one of these cellar cells and seek out the living,' reported Sergeant Farris. 'One French boy was huddled up against a dead comrade, as if to keep warm, having no concept that the friend had been dead two or three days and unable to move his limbs.' There were so many bodies, but his eyes were drawn to one girl in particular: 'I would say she was about 17 years old. She lay where she had fallen, gangrened and naked. In my own thoughts I choked up – couldn't quite understand how and why war could do these things.'

The camp had been bombed and in one bomb crater lay about twenty bodies. 'We pulled about 3 or 4 feebly struggling living ones from the bottom of the pile,' continued Farris. 'They had been struggling for 5 or 6 days to get out, but the weight of the other bodies on them had been too much for their starved emaciated frames.' One man feebly staggered to attention for the US medics, 'and tried to salute us as tears slowly trickled down his cheeks. Too weak to walk, this man was genuinely moved to pay tribute to those who were helping him – showing him the first kind act in years.' Night and day the army of workers buried the dead and did what they could for the dying. Few of the inmates could stand or walk, but those who could reached out for the hands of their liberators, their pale faces washed with tears – and the hardened, battle-weary men of the 104th cried too.

After the bombing of Peenemünde in 1943, the Nazi hierarchy had been determined to form an impregnable factory where the V-2s could be mass-produced. Kammler, Himmler's right-hand man, was keenly aware of Hitler's insistence on the immediate production of five thousand rockets 'to force England to her knees'. The SS had taken prisoners from Buchenwald concentration camp at Weimar to a disused mine in the Harz Mountains in central Germany. Out of the solid rock of the mine and without modern tools, the prisoners had laboured to build a vast underground factory. They forged two tunnels, each two miles long, connected by a series of intersecting tunnels. The main assembly line ran down one of the tunnels while supplies and subassemblies came through from the parallel tunnel. Once the prisoners had built the production line, they manned it and began producing

V-2 rockets. At first the prisoners lived in the tunnels, but, as production increased, concentration camps were created at Dora and Nordhausen to house them.

Their daily food ration was well below subsistence level, about 1000 calories a day. There were no washing facilities or latrines in the mine; a few buckets were provided for use as toilets. In the absence of drinking water, the labourers were compelled to drink from the puddles formed by water oozing from the walls. Disease and skin infections were prevalent; dysentery, gangrene and starvation offered the most commonplace exit from this hell on earth. Encouragement, in the form of a beating, was often administered to those too weak to work. The SS had worked out with chilling precision the most cost-effective way of working the prisoners. In the hot and humid tunnels, and with the minimum of food, a fit man could usually give six months' work before he died of starvation. Around 160 workers a day died, but that was irrelevant to the camp administration; they could always be replaced. Now and then, if a little more 'encouragement' was required, there would be a public hanging, the victims selected for random, arbitrary reasons. At least twenty thousand slaves died creating Mittelwerk and building the V-2.

Everyone at the camp was required to watch a hanging. Twelve prisoners could be hanged at the same time by a man operating a crane. The prisoners had wooden gags in their mouths to stop them screaming. Their hands were tied behind their backs and they were hanged by a length of wire. They were left hanging for days, sometimes dying slowly. Most of the bodies lost their trousers and shoes, and puddles of urine and faeces covered the floor. Since the ropes were long, the bodies swung gently about 5 feet above the floor; if the corpses pushed against each other, they would spin round. Those walking by received bumps from knees and tibia soaked in urine. One day as many as fifty-seven were hanged.

The Americans were horrified and determined to know who was responsible. The evidence was difficult to obtain, the information confusing. The remaining Germans shifted the blame and admitted nothing. The names of Dr Georg Rickhey and Arthur Rudolph were mentioned, but neither man could be found. Eventually the finger of guilt was pointed at the director of production, Albin Sawatzki, who claimed he had been promoted to this position only a month earlier. On 14 April, he gave a deposition while in American custody. He confessed that there had been public hangings and that in his capacity as overseer he had often kicked workers in order to make them work harder. He blamed the SS administration who set long hours and high production quotas and claimed that he had often complained about the executions. Justice for all the unnamed horrors perpetrated at Mittelwerk was difficult to obtain but the rumour spread that Sawatzki met a violent and untimely end at the hands of American soldiers who supplied their own more definitive judgement.

Even with the liberation of the camps, it was not quite over for all the prisoners from Dora. Shortly before the Americans arrived, the SS had sent many prisoners who were still fit enough to walk on death marches in order to destroy the evidence. One group left on 5 April; the SS rounded them up and herded them into a wagon train. They travelled for five days, some dying on the way. This was followed by a further day's march, during which the prisoners dared to hope that the Allies would find them in time. Surely the SS would have killed them by now if that were their plan?

On 13 April, they reached an isolated barn near Gardelegan, where some thousand prisoners were ordered inside. 'Straw is scattered all about everywhere,' wrote Yves Béon, a French survivor of Dora. 'The door closes behind them, but, for Christ sake, what a smell of gasoline! No need to paint the picture; the prisoners understand everything. The Germans have set up a giant grill. It's the end.' The SS threw open the door and flung in burning torches. As some of the prisoners rushed forward on to the flames to protect the others, the SS opened fire on them. The Americans later discovered 1016 bodies.

• • •

As news spread that the Americans had secured Mittelwerk, Major Robert Staver from US Army Ordnance set out to investigate the site. After gathering as much information from the British as possible, he had joined Colonel Holgar Toftoy of US Ordnance in Paris, who was equally anxious to obtain the secrets of the V-2 programme. Colonel Gervais Trichel had asked him to recover one hundred V-2s that could be transported to the testing site at White Sands, New Mexico, for research and test firing. Above all, he wanted to find Wernher von Braun. 'Get the Germans,' Toftoy said to Staver, leaving him in no doubt that his mission was a top priority. 'Find the German scientists who are years ahead of us and can teach us from their success.' Equipped with a detailed map of the area acquired from British intelligence, Staver was determined to be first at Mittelwerk to see what could be requisitioned.

Staver avoided the soldiers and the Red Cross who were dealing with the living dead from the infamous factory; he was primarily interested in the machinery. He did see about two hundred bodies piled up, he later admitted, but he was preoccupied with his orders and stopped for nothing. He walked the length of the two miles of tunnels alone and in silence. Laid out before him systematically was the assembly line, the shiny new parts placed in order: nuts and bolts, intricate shapes, bits of unidentifiable metal laid out in rows. The complicated machinery was taking shape, one part interlocking with another then another, until, finally, he saw the massive shell of the mighty V-2, ready for transportation, almost as though invisible hands had assembled it as he had walked the line.

Staver was impressed, remaining oblivious to the lingering stench and his ominous surroundings until he came upon a cup still warm, left by someone in a hurry to depart. He later learned that many German SS hid in the tunnels for weeks, yet at this stage nothing could affect his exhilaration at his find and the knowledge that he was the first to make such a discovery. He realized that there was enough material here to meet Colonel Toftoy's request for one hundred V-2s and that it must be crated straight away, ready for transport. He had no intention of sharing anything with British intelligence.

Staver was disappointed not to find von Braun at the site, but there nonetheless lingered a tantalizing hint of the man's presence. The extraordinary V-2 was evidence of his scientific brilliance. He was too late, he realized, as he walked around the Villa Frank, now in American hands. Staver saw that it still bore traces of von Braun's recent occupation. It was impossible to know where the enigmatic von Braun might be.

And now there was a very real question to be asked. Von Braun and

CHAPTER THREE

'Kill those German swine'

Oberammergau was a quaint resort town in the Bavarian Alps full of picturesque steep-pitched chalets. Snow lay on the pine-covered upper slopes. The air was like champagne. The deep valleys guarded a profound silence. It was hard to imagine that there was a war raging around them. Yet when Wernher von Braun arrived on 4 April, he found the situation confirmed his worst fears. This was hardly terrain in which to make a last stand and there were no discernible attempts to do so – no marshalling of resources, no attempt at defence. And no sign of Kammler either, but the SS were present in large numbers and von Braun soon learned that they would guard the camp for which the scientists were destined. This was surrounded by barbed wire and they were to be kept as prisoners.

It was unnerving waiting for Kammler; no one knew quite what his intentions were. Several days passed before von Braun was summoned. Without warning, Kammler had arrived with his Chief of Staff, SS Major Starck. The general kept him waiting and from an adjacent room in the hotel von Braun could hear fragments of their conversation and the sound of wine being poured generously into glasses. The mood was convivial, accompanied by much laughter; the discussion centred on ways of quietly vanishing – of performing the perfect disappearing act on the Allies. Major Starck suggested hiding in a nearby abbey famous for its liqueur. Kammler roared with laughter. Yes, how convenient; one minute a flamboyant Nazi general, the next a faceless monk with only intoxicating liquor to tempt him. He would burn his Nazi uniform and disappear.

Eventually an SS guard took von Braun to see Kammler and Starck. The two men were unaware that von Braun had overheard them. Kammler was smiling; Major Starck's pistol was prominently displayed. Kammler was enjoying the role of genial host, his lean features relaxed, but his eyes still wary in spite of the empty wine bottles. He was solicitous, exuding his particular blend of charm with a convincing display of concern for von Braun's arm. Von Braun provided the reassuring news that his men were still involved in research for the Führer. The general was feeling benign for he had collected yet another important post: to his distinguished list of titles had been added 'General Commissioner for Turbojet Fighters'. This required him to be away for a while, he explained, but von Braun was not to worry: the SS guards would protect the scientists. While he was away they would be under the wing of SS Major Kummer. They should carry on with the research; victory was in sight. 'Heil Hitlers' and heel clicking signalled that the interview was over.

It soon became clear that Kammler really had left the area. Like a thief in the night, he had mysteriously disappeared under cover of darkness. No one could inform von Braun of his whereabouts. The meeting with Kammler had confirmed to von Braun that his team of scientists were indeed in a vulnerable position. They were unarmed, virtually under lock and key and easily controlled. There were too many SS guards about. Even if they could escape, five hundred scientists wandering about the snowy hillsides would be hard to conceal. Von Braun decided to gamble on a plan to outmanoeuvre SS Major Kummer. He hoped to persuade the major to disperse the scientists legitimately – and on his own orders.

Von Braun enlisted the support of a long-standing colleague from Peenemünde, Dr Ernst Steinhoff. The two men visited Major Kummer together, ostensibly to congratulate him on his new appointment. They expressed their delight that Kummer was looking after them; it was a great blessing. After a while, von Braun apparently began to confide in the major. Even though the Americans would obviously never take the Alpine redoubt, he was worried about the destruction a US bomber might inflict until the new turbojets were in production. Supposing the camp took a direct hit? It could be difficult to explain to General Kammler that *all* the top scientists were dead. A whole generation would be wiped out. There would be no more V-2s.

This was clearly not something that had occurred to the major. The room was silent as he pictured the carnage and pondered the immensity of the dilemma. While the two men waited for his reply, as though on cue there came the roar of a pack of US P-47 Thunderbolts strafing the valley, conveniently demonstrating the problem they faced. This made a difference. The major saw their point, but did not see how he could help. Von Braun and Steinhoff were ready with the answer: why not billet the men in the surrounding villages? Then, no matter how intensive a raid might be, the scientists would be in less danger.

Once again Kummer saw the point but was still unable to help as there was no petrol for the vehicles needed for such an operation. And once again the major was trumped. Steinhoff had the magic recipe for a home-made fuel. The major had a little petrol; the scientists had plenty of rocket fuel, alcohol and liquid oxygen. If these three were combined there would be more than enough fuel for such an undertaking. There was silence again while the major pondered Steinhoff's suggestion, a long silence that augured defeat for the strategy. Then right on cue the P-47 Thunderbolts came roaring back up the valley, splintering the silence. This time the major was convinced.

The scientists and engineers now moved freely out of the camp. Rooms were found for everyone in the nearby mountain villages. Von Braun had won a temporary victory, but they still had not shaken off the presence of the SS. Everywhere they faced the dark uniforms, always creating that sting of dread, that fear not easily suppressed. The future was still unknown. Any kind of unpredictable madness could overtake them all.

As von Braun and his brother Magnus moved into a house in the village of Weilheim, just south of Oberammergau, it was becoming obvious that von Braun's arm, broken in the car accident, was still not healing. Unless he got professional medical treatment soon, amputation could still not be ruled out. It was clear he could no longer postpone the fifty-mile journey over the mountains to a hospital at Sonthofen. The surgeon there was a man noted for his skill in dealing with the complications caused by climbing accidents, but the prognosis was not good. Von Braun had neglected what was now a serious injury. With great care the plaster was removed and the aching limb and shattered shoulder examined. His arm was rebroken and set again without anaesthetic and he was put in traction. He was to lie completely still. The surgeon promised to look in after a week, when he had the time.

Von Braun had no option but to wait and hope his arm healed before the advancing enemy arrived. He was unsure just how near they were and did not know that on 16 April the Soviets, who had been massing two and half million troops around Berlin for more than two weeks, had at last attacked and were ready to seize their long-awaited prize. By 24 April, they had encircled the city and there was hand-to-hand fighting in the suburbs. In his memory every Soviet soldier carried images of the rape and devastation of his homeland: the burnt villages, bodies frozen in the snow, the dead and dying on the streets of Leningrad as the besieged citizens slowly starved to death. Now they had arrived in the streets of Berlin and would exact a heavy payment. Some 325,000 Germans were to die, and more than 100,000 women were raped. It is estimated that almost 10,000 committed suicide.

At the heart of Berlin lay the Reich Chancellery, now reduced to a mere shell of its former grandeur, its polished floors and huge chandeliers ground to dust by the endless bombing. On 20 April, in the bunker beneath it, 50 feet underground where no daylight fell and the sounds of war were muffled, Hitler celebrated his fifty-sixth birthday. A wreck of a man in the early stages of Parkinson's disease, his hands were shaking, blank eyes reviewing a busy inner world of triumph, unable to accept the inexorable march of the enemy across the glorious Third Reich that he had prophesied would last for a thousand years. He drank champagne with his generals and pondered moving south out of the dark shelter to the sunlight of his Alpine redoubt. His generals were given their orders to mount a counterattack. No one dared tell his Führer that there were few men left to fight, there would be no counterattack, that he was defended now in his sanctuary by sixteen-year-old boys from the Hitler Youth. One by one, leading figures in the Third Reich found an excuse to leave the doomed city.

The Allies were closing in, tightening the noose, the British and Canadians on a wide front in the north and west, the Soviets wiping out resistance in the east. In the south, the Allied armies were pushing their way up the length of Italy. The US First Army was less than a hundred miles from Berlin at the Elbe in central Germany. To the south-west, the French had crossed the Danube and were driving towards the Bavarian Mountains not far from Oberammergau. German soldiers tired of fighting a losing battle were surrendering in their thousands. For the Allies there was an almost palpable feel of victory in the air.

Allied intelligence reported the possibility of a redoubt in the mountains of southern Germany, anywhere from Oberammergau to Sonthofen. It was known that German scientists were involved in developing new weapons, possibly even an atomic bomb. Sporadic rumours that German scientists had succeeded in mounting poison-gas or even atomic warheads on to a long-range rocket could not be completely dismissed. If such a weapon were fired into the oncoming armies, the devastation caused could alter the course of the war even at this late stage. The US Seventh Army was directed towards the Bavarian Alps to investigate the redoubt.

In his bed in hospital, von Braun lay absolutely still, as he had been directed. Overhead, American bombers were mounting an attack. Suddenly, quite helpless and unable to move, he found himself in the thick of war. The bombs fell near the hospital and other patients were moved to the cellar. Von Braun could not leave for his condition would not allow it. He was left alone in the ward, listening to the barrage.

The days passed slowly. In his narrow hospital bed he longed to know the war news. German radio offered no clues. Had Berlin fallen? At any moment, while he lay immobile, he thought the SS could enter the ward. Losing all track of time, he woke one day from a deep sleep to see the feared dark shape bending over him. Whether it was Soviet, British, American or SS was not clear. A quick, instinctive fear told him it was the end.

The stranger said that he had orders to take him to a mountain rendezvous. Ignoring questions about who had sent him, the man urged von Braun to accompany him immediately – otherwise he would become a prisoner of the French who were poised to enter Sonthofen. Von Braun protested that he could not travel in his present condition but the soldier was insistent and they set off in an ambulance.

The mountain roads were full of twists and turns. As they slowly negotiated the convoluted ribbon bends, the bare mountain slopes rising to the high peaks gave an impression of a world untouched by war. Finally they arrived in a village, lost in the clouds, seemingly sitting on the very edge of habitable existence. The country below them was spread out in the sun like a map. Then coming towards him von Braun saw with enormous relief General Dornberger, his brother Magnus and several friends from Peenemünde. He soon became aware, too, of the black uniforms of the SD soldiers – the security service of the SS. Even here, in this remote hideaway, it was impossible to get away from the Nazis. His first thought on seeing them was to associate their presence with Kammler, but Dornberger assured him that Kammler was not with them. He had not been seen for a week.

To avoid the SS, Dornberger had moved the men into a hotel in the isolated village of Oberjoch but he had not been able to rid himself of a detachment of SD troops. He had arrived with a hundred of his own soldiers and was trying not to think about what would happen if the SD decided to follow one of Hitler's rather more unpleasant directives. The quiet resort village with its steep-pitched houses seemed an unlikely venue for a shoot-out yet the soldiers and the SD detachment outnumbered civilians. Von Braun found it difficult to relax in their presence and Dornberger was edgy, anticipating trouble. Dornberger decided to find out from the SD major in charge what his instructions were.

Acting on the assumption that a little alcohol would help loosen tongues, he invited the major to his room for a friendly drink. The major was only too happy to accept. Dornberger had provided an exceedingly good wine and, by the third bottle, the major, now the worse for wear, had been gently relieved of his information. What exactly was he doing here, Dornberger wanted to know. The major replied that he was protecting scientists from the French and the Americans. Dornberger expressed his surprise that the major and his thirty men were expected to take on the French and the US battalions. 'If we fail,' the major revealed, 'my orders are to shoot you all.' 'And have you considered what the Americans will do to you?' Dornberger replied. 'If you shoot a group of innocent civilians you will be hanged as a war criminal, probably immediately.' The major was not a happy man; he had seen the inevitability of this outcome himself but no other option presented itself. Survival for the major, now in tears, seemed an unobtainable luxury.

Dornberger had the perfect scheme that would save them all. The major, still overcome with emotion, refilled his glass and listened, more in hope than expectation. Dornberger's plan was simple. The major and his thirty men were invited to lose their SD identity completely, to burn their uniforms and hand over their weapons to Dornberger, who would supply ordinary German army uniforms that would complete the metamorphosis. When the US arrived, they would all become prisoners of war with a life ahead of them. As SD soldiers they would almost inevitably be shot.

The major was impressed by such a luminous understanding of the problem. He pulled himself together and agreed wholeheartedly to the plan. Next day the uniforms were burned, the guns were handed in and thirty more ordinary Wehrmacht soldiers joined Dornberger's one hundred. The easing of the tension that had plagued the little village was felt immediately. The major was seen to smile, as was Dornberger, fully aware that SS tattoos were not so easily removed.

The days thus passed unnoticed. They were all in limbo, waiting, divorced momentarily from the real world of war. They played cards or chess, drank quietly through the night or sat in the late spring sunshine and talked companionably. The silence of the mountains acted on them like a calming drug. 'About us towered the snow-capped Allgau Mountains,' Dornberger wrote in his diary, 'their peaks glittering in the sunlight under the clear blue sky. Far below us it was already spring. The hill pastures were a bright green. Even on our high mountain pass the first flowers were thrusting buds through the melting snow. It was so infinitely peaceful here. Had the last few years been nothing but a bad dream?'

The spell was finally broken by the sounds of human voices and traffic in the valley road below. At first it was insignificant: people calling, a line of carts, possessions piled high. Then the trickle increased, trucks appeared, and lorries, the stream turning into a never-ending flow of people on the move, of refugees and army convoys and all the noise and paraphernalia of war.

There was no news, but rumours emerged like ghouls on Halloween. Munich had fallen to the American Seventh Army which was about to make an all-out attack on the Allgau foothills. It was suspected that the Nazis were gathering there for the redoubt. In the west, the French could almost be seen in the blue distance, and a company of French colonial soldiers would soon be in Oberjoch. The final end to the war in Europe was close. Everyone in the hotel was sure of that, but the threat of violence was undiminished. One could almost smell the scent of peace and understand what it would be like to wake in the morning without fear. But at the moment there was still nowhere to hide.

By the end of April, the once handsome city of Berlin was a pile of rubble, a vast tomb. A smell of death hung in the air. So many thousands had been killed in the Allied raids and were buried by the rubble where they fell. The soldiers of the Soviet army had fought relentlessly and were now within a few hundred yards of the Reich Chancellery which hid the deep bunker where the architect of this monumental catastrophe still enjoyed his freedom.

Hitler's little world was shrinking. The Soviets would soon arrive. He could just hear the faint music of war: the staccato sound of machine guns, the sharp crack of rifle fire echoing in the streets; a fitting requiem for the many dead. The Soviets lost 300,000 men clawing their way through the streets of Berlin to get to Hitler, the prize they wanted to take back to Stalin.

Hitler had received news that Himmler had tried to surrender to the Allies. Momentarily his corpse-like demeanour became animated as he gave one of his more demonic performances, screaming at this betrayal by his former right-hand man. But there was little strength left in him now and he soon lapsed into listlessness. Buried 50 feet below ground, in the heart of the deep bunker, so grave-like and joyless, Hitler married his devoted blonde mistress. Eva Braun wore his favourite dress, with the décolleté neckline and roses over the bodice. A simple woman, her composure did her credit. She smiled although the mood was sombre. Goebbels and his wife were witnesses and a few old generals were there too. It was time for the bride and groom to be alone. Then, behind the heavy padded door where custom decreed long-held marriage rites should begin with kisses and affection, in a solitary, unwitnessed, very private moment, he offered her the kiss of death. Historians believe she swallowed a cyanide pill and that he did too. He then raised a pistol to his head, just to make sure. Marriage to Hitler meant so much to Eva

Braun that she was prepared to accept this ignominious, shuffling exit from the maelstrom he had created.

On 1 May, in the hotel in Oberjoch, Bruckner's 7th Symphony was playing on the wireless, trying to compete with the babble of laughter and conversation in the lounge. Suddenly the music was interrupted by the sound of an announcer's voice. Silence descended as though responding to an unheard command. Into the stillness came the beautifully articulated, disembodied voice from the wireless:

> Our Führer, Adolf Hitler, fighting to his last breath against Bolshevism, fell for Germany this afternoon in his operational headquarters in the Reich Chancellery.

The room was still quiet as everyone waited for a fuller explanation but no details about the end of the war were forthcoming. The announcer continued with the news that Grand Admiral Karl Doenitz, Commander in Chief of the Navy, was the new Führer and would continue Hitler's war. In reality the conflict was over but as yet there was no peace. The country was in a vacuum. The situation was without precedent; anything could happen. Von Braun considered the position of the scientists to be more dangerous than ever.

He summoned the team to a private meeting and outlined their position. The French First Army was known to be approaching; they might arrive before the Americans. Worse, it was rumoured that a detachment of French Colonial Moroccan troops was creating havoc in a nearby valley. Most sinister of all were the SS who were in the area in large numbers. Extremists among them, ever-loyal to Hitler, shot anyone they thought would surrender to the Allies. Equally, the scientists could all be shot in cold blood to prevent their knowledge falling into the hands of the enemy. The scientists were unarmed. There was nowhere to hide and von Braun's ability to act was hampered by the cumbersome plaster around his torso and arm. The question was how to engineer ending up in American hands?

They formulated a plan that they hoped would not draw attention to themselves and entrusted von Braun's young brother, Magnus, whose English was good, to carry it out. On 2 May, as sunlight struggled to break through the early morning mist, he left the hotel on a bicycle, conscious that their future depended on him. His hopes centred on meeting the Americans rather than a welcoming party led by Hans Kammler.

In late April, the American Seventh Army, after taking Munich, had proceeded south with all speed to the edge of the Bavarian Alps expecting to meet heavy resistance from the rumoured Alpine redoubt. Their 44th Infantry Division had passed quite near Oberjoch on its way south, meeting little hostility. On the morning of 2 May, one of the anti-tank units was easing its way along the isolated mountain roads near Oberjoch, on the lookout for Nazi resistance, when a figure appeared out of the mist.

The soldiers debated whether or not to shoot. Could this be preliminary to an ambush? The sun's rays cut wide beams through the pines bordering the steep-sided road, still thick with winter snow. The misty figure became clearer, a young blond man on a bike cycling frantically and waving madly. Private Fred Schneiker from Wisconsin narrowed his eyes. Was the figure waving something white? He did not have time to work it out. The young man was suddenly among them, his face split by a broad grin that would not go away once he learned they were American even though he was ordered to put his hands up. His first words to Schneiker made the private doubt the young man's sanity. He needed to see Ike (Eisenhower), he said. Then he revealed that he was a messenger representing a group of rocket scientists who were staying at a hotel nearby. They were responsible for making the V-2 and they wanted to surrender. Would the private come with his tanks and fellow soldiers and rescue them? They were in grave danger from the SS Obergruppenführer Hans Kammler, who was in the mountains somewhere nearby and had orders to shoot the scientists rather than allow them to be captured by the Allies.

Private Schneiker, who knew little about rocketry, sensed the situation was taking on a bizarre quality as they stood deep in enemy territory, talking about rocket scientists, conscious that fanatical Nazis were in the area. 'I think you're nuts,' he replied, 'but we'll investigate.' After searching Magnus, they took him back to their headquarters in the nearby town of Reutte and delivered him to the Divisional Counter Intelligence Corps. The American officers listened as the young man in the grey leather coat told his story with an eagerness that became more vehement as he watched the growing disbelief in the eyes of his interrogators. They were unsure how to proceed but after a discussion they gave him 'safe passes' and instructions to came back with the rocket scientists.

Magnus returned to the hotel in the early afternoon and explained the situation. Losing no time, von Braun commandeered three cars. Ten of them, including von Braun himself, Magnus, Dornberger and his Chief of Staff, Huzel and Tessman, were to make the momentous journey along the winding roads; each man was conscious that he was taking a step into the unknown. It was dark as they arrived at the Counter Intelligence Corps in Reutte. They were questioned briefly by Lieutenant Charles Stewart, who was in charge of them overnight, then given comfortable rooms. For von Braun and his core team, the apparently insurmountable barrier of getting from Nazi-ridden Germany to cosy quarters under American protection had finally been accomplished. Nothing could hurt them now, it seemed.

But as they lay asleep upstairs, confident that at last they had reached safety, a Polish kitchen hand, hearing that Germans were sleeping the sleep of the innocent in feather-bedded luxury under American protection, decided to take retribution into his own hands. He crept upstairs, gun at the ready. There was only one thought in his mind: he would 'kill those German swine'. In the semi-darkness, there was the unmistakable sound of a safety catch being released on a pistol.

Fortunately for the sleeping scientists, he was intercepted by the observant Lieutenant Stewart who had gone downstairs to investigate suspicious noises. The scientists never knew how close they had come to death that night – ironically the first time they believed themselves to be safe. Over the following days, US counterintelligence found most of the remaining scientists around Oberammergau. With von Braun and Dornberger they were taken to Garmisch-Partenkirchen, a ski resort in Bavaria, for questioning. Housed together in a large building under guard, their future now depended on their perceived value to America.

Two days later, on 4 May, Helmut Gröttrup, the one man who was quite certain that America was not his preferred destination, at last arrived back in the Nordhausen area. He had escaped the SS when he jumped from Kammler's private train and laboriously walked the long journey home to the haven of his wife and family. Irmgardt Gröttrup was overjoyed. 'The sun is shining and Helle [Helmut] has returned,' she wrote in her diary. 'He was beaten. His arms and legs look dreadful, covered in bruises. What he has gone through – how he escaped the SS. He should be dead!'

But their trials were far from over. Unlike von Braun and most of the other scientists, the Gröttrups wanted to stay in Germany. The idea of being forcibly taken to Russia or America had no appeal, but it was difficult to see how their skills in rocket science would be accommodated by a Germany completely bankrupt and divided up among the Allies. As they struggled to eke out a living 'like peasants', Irmgardt noted ominously, 'I cannot sleep. I keep thinking that someone is coming to take me away.'

CHAPTER FOUR

'First of all, to the moon'

In the week after Hitler's death events moved quickly. German armies north of the Bavarian Alps surrendered and the war in Europe came to an end on 8 May 1945 when Germany surrendered unconditionally to the Allies. The scourge of Nazism had finally been eliminated. Six years of war gave way to victorious celebrations to mark the end of the worst turmoil Europe had ever known. More than fifty million Europeans had died; there were thirty million refugees and huge areas had been laid waste.

Now, with the end of the Third Reich, the bells rang out across Europe. Blackout curtains were torn down and people marvelled to see streetlights again. In Paris, the crowds took to the streets for a huge party. Allied soldiers, pelted with flowers, became heroes. Wine appeared from hiding places, saved for this day. In London, crowds packed the city to a point where movement was impossible. They roared with joy to see Churchill give his V for Victory sign standing beside the royal family on the balcony at Buckingham Palace.

In Germany, government did not exist. At a conference held in Yalta by the Allies in February 1945 in an air of optimism, the division of a defeated Germany had been decided. Churchill, Roosevelt and Stalin had agreed on the zones that each country's forces would occupy and manage for the foreseeable future. Communication of every kind was in ruins. Food was like gold dust; cigarettes were currency; bartering became the order of the day. The SS quickly disappeared and those

German soldiers that could quietly walked home, praying that home was still standing.

At Mittelwerk in central Germany, Major Robert Staver was obsessed with trying to recover as much as possible for the Americans and suddenly found that time was not on his side. At the end of the war, the position of the Allied armies bore no relation to the zones agreed at Yalta. The US armies had progressed more swiftly than was thought possible. A large section of Germany, some four hundred miles long and 120 miles wide, which had been captured by the Americans, was now to be ceded to the Soviets. Nordhausen and Mittelwerk were in the allotted Soviet zone and due to be taken over by them on 1 June. If Staver was to stop the vital secrets of the V-2 falling into Soviet hands, he had to act fast.

He had the support of Colonel Toftoy in Paris who immediately authorized an operation to recover as much as possible from the underground factory: this was called *Special Mission V-2.* A recovery expert, Major William Bromley, was sent out to Mittelwerk to retrieve everything of value from the mine: any drawings, assembly jigs and components to create a hundred V-2s. Hundreds of tons of equipment were to be sent by train to Antwerp, Belgium, ready for shipping to White Sands, New Mexico. Bromley set about hiring and requisitioning anything on wheels and found employment for the recently freed workers packing hundreds of tons of parts. He achieved his objective, removing almost everything of value within days, leaving very little for the British or the Soviets.

Meanwhile, on 8 May, Dr Richard Porter, leader of General Electric in Paris under contract to US Army Ordnance, flew out from London to start interviewing the scientists who had surrendered. The Americans wanted to know as much as possible about the V-2, but von Braun and Dornberger had already briefed everyone 'not to be too chatty'. They were to be spare with information until they saw what sort of opportunities might arise with the Americans or British. 'Our primary concern was stability and continuity,' von Braun admitted later. 'We were interested in continuing our work, not just being squeezed like a lemon and then discarded.' They wanted to see just what they could get out of the situation – surely a three-year contract from the Americans at the very least?

The interviews were not recorded and very little documentation remains to show just what ground was covered. Rather than investigating possible war crimes and the use of slave labour at Mittelwerk, it appears that the overwhelming concern of the interrogators at this stage was to try to establish what value these men might have in America and whether they would be cooperative. According to the scientists themselves, the response from the interviewers was 'disappointing'. There were far too many interrogators working for different interest groups and they often showed little understanding of the science. Von Braun answered questions at some length on jet motors and V-2 oxygen vent valves, only to find himself answering the very same questions from a different interviewer a week later. 'They didn't know what to ask us,' Dornberger said later. 'It was like they were talking Chinese to us.'

Hoping to provide enticing glimpses of the future if the rocket team were able to continue their work, on 15 May von Braun wrote an eight-page report which he hoped would assure the Americans that the future of rocketry promised 'revolutionary consequences' for civilization. Interned in Garmisch-Partenkirchen, the country all around them in ruins, von Braun set out his dreams for the future of space exploration. Rocket aircraft, space telescopes, spy satellites, the lengthening of daylight hours with giant space mirrors to reflect light, even interplanetary travel – all this was confidently promised:

Survey of Previous Liquid Rocket Development in Germany and Future Prospects

We are convinced that a complete mastery of the art of rockets will change conditions in the world in much the same way as did the mastery of aeronautics ... In the more distant future, the development of rockets offers, in our opinion, the following possibilities:

1. Development of long-range commercial planes ... The flight duration of a fast rocket aircraft going from Europe to America would be approximately 40 minutes ...

2. Construction of multi-staged piloted rockets, which would reach a maximum speed of over 4.7 miles per second outside earth's atmosphere. At such speeds the rocket would not return to

earth, as gravity and centrifugal force would balance each other out. In such a case, the rocket would fly along a gravitational trajectory, without any power, around the earth in the same way as the moon ... The whole of the earth's surface could be continuously observed from such a rocket. The crew would be equipped with very powerful telescopes and be able to observe even small objects, such as ships, icebergs, troop movements etc ...

3. It would be possible later on to build a station ... as an 'observational platform' outside the earth ... The work would be done by men who would float in space by means of rocket propulsion units, the nozzles of which they would point in the required direction.

4. According to a proposal by the German scientist, Professor Oberth, an observation station ... could be equipped with an enormous mirror consisting of a huge net of steel wire onto which metal foils could be suspended ... This would enable large towns to get sunlight during the evening hours ... The weather, too, could be influenced by systematic concentration of the suns rays onto certain regions ...

5. When the art of rockets is developed further it will be possible to go to other planets, first of all, to the moon ... To conclude, we think after what has been said above that a well-planned development of the art of rockets will have revolutionary consequences ...

Von Braun, 15.5.45

Recognizing that all this might seem a little far-fetched in 1945, he pointed out that any prophecy of the development of aviation made in 1895 for the next fifty years would have seemed 'at least as fantastic'.

In Washington, the authorities were aware that the horror of the concentration camps had caused great public concern. There were difficult questions to deal with before Hitler's rocket team could be brought into the country. Why did von Braun join the Nazi Party and later the SS? Was he aware of what went on inside the concentration camps and to what extent was he himself guilty of perpetuating the inhumane conditions there?

While there is no record of the extent to which these issues were explored at the time, it is evident from an affidavit von Braun gave in 1947 that, when questioned, he claimed he was under pressure to join the Nazi Party. His work was attracting more and more attention at higher levels and he said he could not continue without membership: 'My refusal to join the party would have meant that I would have to abandon the work of my life. Therefore I decided to join. My membership did not involve any political activity.'

Interestingly, some of his colleagues, notably Dornberger himself, did not feel obliged to become a member of the Nazi Party. There is also the question of von Braun's membership of the SS, which he joined in May 1940. He was promoted every year, reaching the rank of Sturmbannführer, or major, on 28 June 1943. Von Braun later claimed that Himmler himself pressured his leading rocket scientist to join. Early in 1940 an SS colonel, SS-Standartenführer Mueller, had arrived at Peenemünde in order to persuade him:

> Reichsfuehrer SS Himmler had sent him with the order to urge me to join the SS. I immediately called on my military superior for many years, Major General Dornberger. He informed me that … if I wanted to continue our mutual work I had no alternative but to join … After having received two letters of exhortation from Mueller, I finally wrote him my consent. Two weeks later I received a letter showing that Reichsfuehrer SS Himmler had approved my request for joining the SS and had appointed me Untersturmfuehrer …

The extent to which von Braun was under pressure to join the SS, or whether he simply recognized the expediency of joining and went along with it, may never be fully resolved. In his defence, von Braun has maintained that he himself was left in no doubt that he had to obey his SS masters. Indeed, he was even arrested by the Gestapo, who made it clear that Hitler's leading rocket scientist was expected to cooperate in full with the regime. In February 1944, von Braun had been summoned to meet Himmler at his headquarters in East Prussia. Himmler pressured von Braun to join his staff rather than report to the army. 'Time is running out,' Himmler had said. 'I hope you realize that your A4 has ceased to be a toy and that the entire German people eagerly await the mystery weapon …'

Three weeks later, at 3 a.m., the early morning hour favoured by the Gestapo, von Braun had been arrested and taken to the SS jail in Stettin. He was given no explanation but placed in solitary confinement to ponder the fact that those arrested by Himmler seldom lasted long. Arrested with him were Helmut Gröttrup and Klaus Riedel.

When he heard what happened, Dornberger was incensed. 'I could not believe my ears,' he wrote. 'Von Braun, my best man, with whom I had worked in the closest collaboration for years, whose whole soul and energy, whose indefatigable toil by night and day was devoted to the V-2! It was incredible!' Dornberger was equally baffled at the arrest of Gröttrup and Riedel.

He demanded to know the charge from Field Marshal Keitel, who represented army interests to Hitler. Their crime, he was informed, was sabotage. At a party the three men had been overheard saying that they were not really interested in weapons of war: space was what fascinated them. It was suspected that progress on the V-2 had been neglected in favour of their real interest and for this disloyalty they could pay with their lives. Dornberger argued strenuously for their release but Keitel had no inclination to bother Hitler and invite his rage with the information that, if von Braun was shot, the V-2 programme could be in jeopardy. Neither did he wish to interfere with Himmler's orders. Dornberger was on his own. There was nothing to be done. He decided that, whatever the risk, he would try to see Himmler personally.

Two weeks passed before von Braun was taken to a courtroom, where it became clear that he would be tried by the SS. He was accused of undermining progress on the V-2 and it was asserted that he had every intention of fleeing to England with valuable rocket data, travelling in a small private plane he kept. Although he had no such plan, this was impossible to disprove. The court had a way of making supposition appear to be fact, so that von Braun later claimed he felt he would be lucky to get away with being shot. A much worse punishment was probably intended for him. He had given up hope when Dornberger entered the room demanding to see the SS officer in charge.

For the previous two weeks, Dornberger had been busy pleading von Braun's case with the SS elite in Berlin, who informed him that he too was not beyond suspicion. By swearing an oath that 'the arrested men were indispensable to the programme', he obtained three months' grace for them, to be reviewed every three months. Von Braun was released and left with Dornberger, intensely relieved but now acutely aware of the extent of SS power and careful not to provoke it. Incidents like these impressed on the rocket scientists that it was dangerous to express opposition of any kind to the SS. However, records of the arrest have not been located to verify their account of events.

Neither von Braun nor Dornberger admitted to having had any say in the conception, design or running of Mittelwerk and the accompanying concentration camps. At the time, the evidence pointed conclusively to the fact that this had been under Kammler's jurisdiction as head of the SS construction department and, by 1943, also in charge of the rocket programme. He had run the enterprise with the help of the Totenkopfverbande SS – the 'Death's Head' units. Constantly driven by Hitler's demand for five thousand rockets to fire on London, Kammler was untroubled by conscience. 'No matter the number of human victims, the work must be executed and finished in the shortest possible time,' he had said. If von Braun or Dornberger had wanted a more humane approach, they understood all too well that they would be throwing their lives away.

With limited knowledge about quite what had happened at the V-2 factory, and under pressure to resolve the situation to America's advantage with the continuing war in the Pacific, the moral issue did not appear to have the same urgency that it has had for later generations. Even so, the question of what to do with the scientists was to launch a massive row within the US government. The glorious future funded by American wealth that von Braun had envisaged was not quite as certain as he would have wished and more evidence was yet to emerge about his own role at Mittelwerk.

• • •

Meanwhile, throughout May, Robert Staver and William Bromley were making progress in the race to seize anything of value from the rocket factory before the Soviets. Not satisfied just with securing the actual V-2s, Staver became convinced that documentation and blueprints must exist that would show how to construct the rockets. During his interviews with those who had been left in the area after Kammler had headed the five hundred senior staff south, he came across a former Peenemünde engineer near Nordhausen, Karl Otto Fleischer, who aroused his suspicions. Staver thought he detected a hint of anxiety in Fleischer's behaviour when he asked about blueprints. Fleischer had indeed been given an approximate idea of where the papers were hidden by Dieter Huzel and was put on the spot. He was prepared to introduce Staver to some of the leading scientists who were still living quietly in the area but would do no more. The whereabouts of any blueprints remained a mystery.

As he gathered more information, Staver decided that the time had come to approach Fleischer again. Calling Fleischer's bluff, he told him that von Braun, Dornberger – even Kammler – were interned at Garmisch-Partenkirchen, and had confirmed that the blueprints were buried in a local mine and that Fleischer alone knew how to locate them. Staver watched carefully. Would his lie work? Again he detected nervousness in Fleischer, who played for time. Fleischer was a worried man. Overnight, he considered his dilemma. If he gave away crucially important information to Staver, against von Braun's wishes, he would be in deep trouble. And if he did not cooperate with Staver, apparently at von Braun's request, he was still in deep trouble. He alone would be the man who had betrayed his country and given valuable information to the enemy.

Next morning there was a message for Staver. Would he meet Fleischer in a village called Haynrode? The directions led Staver through dark alleyways to a priest's house. Inside, Fleischer stood, dishevelled and unshaven, looking like a man under sentence of death. He admitted that he did know something. It was clear he was fearful of the implications of what he was doing and that he had been up most of night struggling with his conscience, helped by the priest.

Slowly, he gave up the information he had. Staver listened, showing no sign of the excitement he was feeling. The papers were hidden in a mine – an old, disused mine near the village of Doren about thirty miles away. If Staver would supply petrol, he would try to locate the documents. They would find the mine more quickly, Fleischer claimed, if he was not accompanied by an American. Once Fleischer had set off, suitably equipped with the permissions, petrol and passes allowing him to be out after curfew, Staver looked at a map of the area only to discover that there was no such village called Doren. He wondered if he would ever see Fleischer again. He was called away for two days on a military inspection. During that time, Fleischer's mission was constantly on his mind.

Fleischer, however, was now committed. Enquiring at every village, calling at every abandoned mine – often half-buried, with their entrances overgrown with foliage so as to refute all evidence of previous activity – Fleischer hunted for the documents. He could not find them, and, worse, like Staver, he discovered there was no village called Doren. There was one called Dornten, however, and he felt sure the mine would be there. The caretaker lived nearby, but could not help. He was an old man and obstinately insisted that there was nothing of value in the mine, certainly no important papers. Fleischer persisted. The old man looked blank. Fleischer threatened him, saying that he was acting for important generals: Dornberger and Kammler. Was one old man to defy the orders of those running Germany's rocket programme?

Miraculously, the old man's memory returned as he started to describe how in early April lorries had trundled through the darkness delivering boxes. They had been stowed in the mine, then dynamite had been used to block the entrance to the gallery. Fleischer could see that removing the papers would not be easy. An enormous pile of stone and rubble 30 feet deep sealed off access to the gallery.

Staver was delighted when he heard the news. The top German scientists were in the American zone. A hundred V-2s were seemingly in American hands and now he had discovered arguably the most wanted documents in Germany. However, his satisfaction at finding the papers was diminished considerably when he learned that a decision at highest levels had changed the boundaries of control between American and British zones. In less than a week, the mine with its hidden treasure would be under British administration. Suddenly, there was no time to lose. Because there was no way of communicating to Ordnance Headquarters Technical Division in the Champs-Elysées for the appropriate authorization, a short trip to Paris was unavoidable. Meanwhile, he instructed staff from the US Technical Intelligence Team to work with Fleischer at clearing the mine.

Staver was back the next day having organized the men and 10-ton trucks he would need to remove the large containers. Unaccountably, progress had been slow in his absence; the gallery was still blocked and the British were due to take over administration of the region on 27 May. It then transpired that a small detachment of British officers had arrived at the mine while he was away and had organized an inspection. They had claimed to be looking for weapons that the retreating Germans might have hidden and began a meticulous search. Staver's team had decided that the best way of dealing with the problem was to pose as geologists and precious time had been wasted while they pretended to look for iron ore samples, which were convincingly graded and packed by a team of miners until the British departed.

With just three days left in which to clear the pile of rubble blocking the gallery, crate the documents and transport them to Nordhausen, Staver now devoted his energy to hurrying the project along. On the afternoon of 26 May, the 15 tons of documents were finally packed and awaiting transport, but there was no sign of the 10-ton lorries. Enquiries revealed that they were not even in the area. Working well into the night, Staver finally acquired six two-and-a-half ton trucks and early in the morning, with the British arrival just hours away, guided them to the mine. The boxes loaded, the convoy made the return journey back over the Harz Mountains into the new American zone just as the British moved in to claim their new area, which, unknown to them, was now a much less valuable one.

By the end of May, Staver felt he had made real progress. Arrangements were made for the blueprints to be sent to the USA, to the evaluation centre in Maryland, and Bromley had crated up enough parts to assemble a hundred V-2 rockets, which were on their way to Antwerp for shipping to White Sands, New Mexico. Staver's prime concern now was to ensure that the team of top German rocket scientists would also reach American soil. On 22 May, anxious that the urgency of the situation would be recognized in Washington, he had drafted a secret cable to Colonel Trichel in the Pentagon:

> Have in custody over 400 top research development personnel of Peenemünde, developed V-2 ... The thinking of the scientific directors of this group is 25 years ahead of US ... Have begun development of A10 to have thrust about 220,000 pounds. Latest version of this rocket should permit launching from Europe to US ... Recommend that 100 of very best men of this research organisation be evacuated to US ... Also recommend evacuation of all material drawings and documents belonging to this group to aid work in the US ... Immediate action recommended to prevent loss of whole or part of this group to other interested agencies. Urgently request reply as soon as possible.

So far Staver had had the field to himself, but now the war was over things would not remain that way. He had failed to share anything with British intelligence, despite the fact that they had been so generous with information at the beginning. Once the British and the Soviets realized what was going on, they would surely do everything in their power to acquire the German scientists for themselves. The Soviets had revealed a considerable interest in the V-2 when they had arrived on 26 May for an inspection of Mittelwerk. They had also begun appealing on the radio for a workforce of scientists and technicians, promising generous salaries and housing. They were especially interested in the scientific leaders – they offered von Braun a bonus of 50,000 Reichsmarks – and it was obvious that the Soviets were planning to continue work on ballistic missiles using the German scientists. What especially worried Staver was the fact that, on 1 June, not only Mittelwerk but the majority of German scientists from Peenemünde, including some four thousand experienced technicians, would find themselves in the Soviet zone.

When Staver returned to his office after transporting the blueprints to Nordhausen there was still no reply to his cable to the Pentagon pointing out the need to take a nucleus of top German scientists to America. There was widespread antagonism in both the State and War Departments to the idea of inviting them into the country, and, in the words of one general, 'treating them as honoured guests'. On 28 May, Robert Patterson, Under-Secretary of War, issued a secret memo regarding the German scientists:

> These men are our enemies and it must be assumed that they are capable of sabotaging our war effort. Bringing them to this country raises delicate questions, including the possible strong resentment of the American public, who might misunderstand the purpose in bringing them here … Taking such a step with out consultation with our allies, including the Soviets, might lead to complications …

Despite these reservations, Patterson was anxious to exploit any source that could potentially help in the war against Japan. While Washington vacillated, Toftoy compromised, instructing Staver to move as many Germans as possible to an area under US control. Staver discovered he had only three days left in which to find, interview, persuade and transport thousands of Germans and their families before the Soviets arrived to claim their hard-won territory.

Staver immediately set out on what seemed an impossible task. He started by enlisting the help of colleagues, but not everyone saw Staver's point of view. 'I don't care if the Soviets get all of those Krauts. I say good riddance' was the answer to one request for help. The greatest problem was asking men to leave their homes when he had no real incentives to offer them. Inviting people to walk out of their lives and step on a train with no kind of hope in front of them except that of being in the American zone seemed a ludicrous demand. It was beginning to look as though little could be achieved. Then he had a stroke of luck, which completely changed the picture.

To Staver's good fortune, some obscure order in lofty administrative circles set back the date of the Soviet takeover to 21 June. The best people to help, Staver decided, would be the Germans themselves. He immediately made arrangements for the leading scientists to be transported back to Nordhausen and assigned to searching out the most essential staff in the surrounding villages. Staver himself personally escorted von Braun along with Dr Richard Porter. Winning people over was difficult and, as the days passed, the number who would ultimately be on the train to the west was not easy to ascertain. When a cottage door was opened in a sleepy village, and the proposition put to a particularly indispensable technical expert, the response was frequently: 'What kind of treatment can we expect from the US for our assistance as compared to the approaches made by the Soviets?' 'The Soviets,' the German engineers said, 'were baiting them with homes, research facilities and special consideration for continuing their missile developments.' For others, however, the presence of von Braun was sufficient; his immediately recognizable figure, still in the cumbersome plaster cast, was persuasive enough. Von Braun himself had high hopes that America would be the Promised Land and he was unceasing in his efforts in persuading the men to join him.

It was thought that about a thousand would avail themselves of the chance to go to the American zone, but getting an unknown number of men, their families and possessions to the train in Nordhausen presented considerable logistical problems. There was not enough transport to take that number of people from the outlying villages on the day of departure. After a week of much begging and many small miracles, the morning of 20 June saw the gathering of three hundred assorted vehicles in a variety of conditions ready to play their part in the great escape. In each vehicle was a German assigned to collect one individual or family. The message delivered to all was the same: 'You have a quarter of an hour to make up your mind. Stay here and face the Soviets or pack up what you can and go with us.' Most made the doorstep decision to change their lives and accepted the American offer.

By noon, a crowd of nearly one thousand of the original Peenemünde team and their families were assembled at the station, but there was no sign of a train. There were some carriages in a siding, but no engine. The Soviets were due in less than twenty-four hours. The hot afternoon wore on. Enquiries yielded no information. Tempers became frayed, people were nervous, tension rose to critical levels as an increasing number of displaced persons joined the crowd. 'I was close to being a mental case waiting at the station,' recalled one observer. 'Every time a German would say "Ruski" I would jump ten feet.' Just as it was looking as though everyone would have to be transported into the American zone by car, at last the engine arrived.

The scientists and their families hurriedly boarded the fifty carriages provided. This, however, proved complicated, as, inevitably, German civilians and people displaced by war were also fighting for places on the train. When order was restored, the train moved south-west on a journey that took its passengers to the west banks of the River Werra in the American zone. It was a mere forty miles but the distance that was to separate their lives from those of their colleagues left behind was immeasurable.

There was one more outstanding task for Staver, once again to be undertaken against the clock. Before leaving for the Bavarian Alps in April, Dornberger had taken the precaution of hiding his own private documents and blueprints. With their exact whereabouts unknown and with less than twelve hours to find them before the Soviets arrived, Staver found himself hunting down a large-scale map of the area. Some hours later, with a small detachment of soldiers, some digging tools, an invaluable mine detector and a borrowed map, the search began in what was thought to be a likely area. A few more hours of nail-biting suspense and the mine detector located the metal boxes. With a certain feeling of satisfaction, Staver, knowing he had the last piece of the jigsaw, took the papers – all 250 pounds of them – through the lonely byroads to the US zone.

Staver had made good use of his few short few weeks in Mittelwerk. Von Braun's team were in the American zone, he had the 65,000 irreplaceable documents and a hundred V-2s were on their way to the United States. He had outmanoeuvred the British and the Soviets. Despite the Soviet's superior intelligence, Staver, with an entrepreneurial American touch, had run rings around them. The Americans had won the first lap in the race for the secrets of the V-2. Staver congratulated himself on having acquired everything worth collecting. Little did he know he was dangerously discounting Soviet persistence and ingenuity.

PART TWO

The Race for Supremacy

'I am convicted of a crime which I have never committed … The charges are all false and made up. I have never been a member of any anti-Soviet organization …'

SERGEI PAVLOVICH KOROLEV from Kolyma in Siberia, one of Stalin's notorious Gulag camps, 1938

'I was incredibly happy to be in the presence of Comrade Stalin …'

SERGEI PAVLOVICH KOROLEV, 1947

CHAPTER FIVE

'We've not got the right Germans'

When the Soviet army entered Peenemünde on 5 May, the sound of infantry boots echoed through derelict work sheds, research buildings and test stands. Soviet intelligence, hoping to inherit a vast fortune of German technology, found that the retreating Germans had blown up about 75 per cent of the site and removed anything of value, including precious documentation. None of the caretaker staff who had been left behind were able to impart any scientific information, yet the Soviets were in no doubt about the purpose of the site and saw it as a matter of increasing urgency to recover the men responsible for producing such advanced weaponry.

Within less than a week, the State Defence Committee set out plans to coordinate the activities of the army and the NKVD in the search for German weapons technologies. The ruthless secret police officer Colonel General Ivan Aleksandrovich Serov was put in charge of an NKVD operation to assess the value of any German specialists who could be found and to procure them for the Soviets. Serov had joined the NKVD in 1939 and been rapidly promoted during the war following the successful execution of a series of brutal campaigns in Soviet-occupied territories which had met the approval of his NKVD superiors, among them the notorious Soviet Chief of Secret Police, Lavrenti Pavlovich Beria. Serov's first task had been the 'expulsion of the anti-Soviet elements from the Baltic States', which he had accomplished by arresting families that fell under suspicion and segregating men, women and children before sending them to camps. Such campaigns of terror were repeated when he was appointed head of the NKVD for the Ukraine, working closely with Nikita Khrushchev, then Ukrainian Party First Secretary. Later in the war, Serov was responsible for the mass deportation of Crimean Tatars, Chechens and Kalmyks and during 1944 he established an NKVD presence in Poland and supervised the crushing of the Polish resistance. For these 'victories', Serov was awarded the title 'Hero of the Soviet Union' in May 1945, and given sweeping powers across Soviet-occupied territories. He was entrusted with applying his methods to ensure that German advances in atomic weapons and rocketry did not slip through the net.

However, in June, when the NKVD followed the infantry into Mittelwerk as the Americans departed, it was soon apparent that the Americans had lost no time in taking all the important trophies from this facility as well. At first sight, the Soviets had inherited nothing – no fully built V-2s, no documentation, no senior experts. Stalin's fury was uncontainable. 'This is absolutely intolerable,' he said to Serov. 'We defeated Nazi armies; we occupied Berlin and Peenemünde, but the Americans got the rocket engineers. What could be more revolting and more inexcusable? How and why was this allowed to happen?' Serov was left in no doubt that he had to rectify the situation.

As head of the NKVD in the Soviet-occupied zone in Germany, Serov ordered his men to investigate the factories and technological institutes that had fallen under his control to find out what could be salvaged. At Mittelwerk, former prisoners of war and people held in camps for repatriation were put to work clearing the site and searching for anything of value. In the absence of any technical information, Serov needed Soviet specialists in rocket technology to make any headway. This was difficult since Stalin had effectively destroyed the Soviet rocket programme before the war. During the purges of the 1930s, many of the senior figures in the field had been denounced and shot or imprisoned, and consequently many of the new recruits at the scientific research institute NII-1 were young and relatively inexperienced. Under the guidance of Major General Lev Mikhailovich Gaidukov, Chief of the Interdepartmental Technical Commission in Germany, engineers with suitable experience were found and sent out from Moscow. One of the first to arrive was Boris Yevseyevich Chertok, who was promoted to major in the Red Army and wrote an engaging account of his experiences in Germany after the war.

Boris Chertok reached Nordhausen on the evening of 14 July 1945 and joined forces with his colleague Alexei Mikhailovich Isayev. They had already made preliminary studies of von Braun's V-2 as various parts of the German missile had fallen into Soviet hands. In the morning, their first step was to examine the site at Mittelwerk. Outside Camp Dora a straggling line of people offered to help. A former Soviet officer stepped forward, curiously attired in assorted items of US army uniform. With great charm, he introduced himself as Lieutenant Shmargun and explained that he had been an inmate of Camp Dora and was the man most qualified to show them around. Noticing that he did not look as thin as the rest, Chertok and his colleagues wondered if he was an American spy and asked how he had managed to stay alive. Shmargun replied that the Germans had had to keep some people alive to dispose of the corpses of the dead and he was sure he would prove indispensable once again as he had contacts among the Americans. 'I know the places where the SS hid the secret components of the V-2 and the Americans have not found them,' he said. 'We, the prisoners, knew many things.' Chertok and Isayev decided that if SMERSH 'passed' Shmargun, it was in their interests to keep him.

SMERSH, the Ninth Section of the NKVD's Special Division, the 'Section for Terror and Diversion', was also operating in occupied Germany under Serov's control. SMERSH, an abbreviation for *Smert Shpionam*, or 'Death to Spies', had been formed in 1941 and operated on foreign territory as military counterintelligence. It aimed to remove all opposition and danger to the Soviet regime by liquidating anyone considered a threat. Victims who fell under suspicion could be dispatched ruthlessly; often they simply disappeared without trial. People of any nationality, not just defectors, but anyone engaged in activity that undermined the defence of the motherland, could be considered a threat. Serov was anxious to stop all Allied attempts at espionage on the Soviet rocket programme.

Despite his concern that Lieutenant Shmargun was a spy, Chertok soon found he proved his value when he took them on a detailed tour of Mittelwerk. It was a hot, bright July day but inside the rancid-smelling tunnels it was dark; the Americans had broken much of the lighting before they left. The Soviet engineers had witnessed industrial endeavours on a giant scale but had seen nothing like the huge complex at Mittelwerk, 200 yards underground with its seemingly endless forty-two miles of tunnels. Guiding them skilfully around abandoned equipment and inspection pits, Shmargun pointed to a bridge crane high in the shadows. There were still nooses on the beam where he claimed as many as sixty suspected saboteurs could be hanged at one time when the crane operator pressed a button. 'This was done in front of all the "striped people"' – the prisoners – he added. Shmargun, now warming to his task, described the horrors until Chertok felt the ghostly presence of those unfortunate phantoms around him in the darkness. The crematorium was visited; they were shown where the prisoners had to rake the ashes. Chertok had had enough. 'The unseen ash was hammering at our hearts and temples,' he wrote.

Chertok and Isayev had an altogether better experience when they spoke to the local burgomaster about accommodation. 'The Villa Frank would be accommodation worthy of your status,' he assured them. 'Von Braun used to live there.' A short drive up a cobbled road to the top of a hill took them to a great mansion. An elderly woman with a key was found and the massive cast-iron double doors opened on to a scene of magnificence that stunned the Soviet visitors. They took in the resplendent hall; the library with polished black wood shelves; the imposing drawing room with its view of the large garden with magnolias and scented roses around a fountain. Reached by a marble staircase were impressive bedrooms, their floors laid with thick carpets, the walls covered with pictures and mirrors in gilded frames. The largest bedroom, wrote Chertok, 'had a mahogany bed, big enough to sleep four people. Snow-white down-filled covers instead of blankets. And the ceiling. The ceiling is a mirror! You can look at yourself while lingering in bed.' As they eyed the opulence, for Isayev the magnificent bed was one temptation too far. Without pausing to remove his boots or his dusty uniform, he dived into the crisp linen and the sea of soft pillows, a contented smile on his tanned face as he retrieved a crumpled packet of Belamor, his favourite cigarettes, and declared: 'You know, Boris, being in this damned den of the fascist beast isn't so bad.'

It was clearly now the turn of the Soviets to sample the delights of the Villa Frank. However, the burgomaster and caretaker, concerned at their prolonged absence, soon came to find them and, seeing Isayev – who was 'dirty and dusty'– in the pristine bed, enquired politely: 'Is Mr Officer not well? Should we bring the doctor?' If Chertok and Isayev had any doubts at all that enjoying such luxury might be construed as anti-communist, they smothered such thoughts as they hit on the delightful idea of turning the Villa Frank into something of a club – a head-quarters for their group. It would be called the 'Institute Rabe' – in German short for 'Rocket Manufacture and Development'.

They established workshops and laboratories for the Institute Rabe in a large building which had formerly housed an electric power station in Nordhausen. Chertok, as one of the self-appointed leaders of the institute, was given an office, which, by Soviet standards, was grandiose, complete with an impressive row of telephones for internal and external use – even one for calling Berlin. Soon a young typist was employed, Fraulein Ursula, and an interpreter, Lyalya. A committee was formed with Chertok assuming command. They decided the state of affairs was not quite as bleak as first thought. Some blueprints for parts of the V-2 had already been found hidden at Mittelwerk and these were to be copied and developed. Germans could be employed for this task from the still plentiful number of scientists and technicians of lesser grades available in the area who were anxious for work. In addition, they aimed to recover and evaluate as many remaining V-2 parts, sub-assemblies and tools as possible. Chertok and his men were joined by more colleagues from NII-1, including the twenty-eight-year-old Vasily Pavlovich Mishin, who soon gained a reputation for being particularly assiduous in deducing details of the German missile from the fragments they could find. With painstaking attention to detail, they would dismantle every component they retrieved and had technical drawings made by the Germans. Piece by piece, the Soviets aimed to recreate the missing blueprints of von Braun's rocket.

As the Soviet team set about trying to understand German technology, according to Chertok's memoirs they began to attract interest from the Soviet Union. On one occasion a group newly arrived from Moscow joined Chertok and Isayev and wore them out with searching questions, from two professors in particular. At the end of the day, when Chertok and Isayev thought they would have peace at last in the opulence of the Villa Frank, they still could not get rid of the two professors, who 'closed in on us with an enormous hunger for information'. Isayev, who was fond of practical jokes, decided it was time to get rid of them – and he had just the plan.

The unsuspecting academics were invited to make a social call later that evening. Once the party was relaxing with a few drinks, Isayev casually announced that the timing of their presence was perfect because any moment now they expected the arrival of an English intelligence officer who knew the whereabouts of some of von Braun's secret documents; better still, he knew where von Braun himself could be found. The rocket team were rumoured to be held in the town of Witzenhausen almost forty miles south-west of Nordhausen. Would the comrades from Moscow help them to kidnap von Braun?

Suddenly there was a tap on the window. A face appeared in the darkness. Isayev, playing the part, produced a gun. Shmargun then appeared in his motley American uniform and gave an impressive performance as 'the agent'. A heated discussion in German followed which, when 'translated' to the astonished visitors, revealed that the Soviet professors were known about and, furthermore, were on somebody's 'wanted list'. They would be wise to leave. The professors did not hesitate. They left for Berlin the following day.

Later that night, Isayev and Chertok found their way to the local café, set up in a bomb shelter; popular and heavy with cigarette smoke, it sold beer and black-market schnapps. Although it was in the Soviet zone, American soldiers and their girlfriends still drank there. As soon as Chertok and Isayev entered, an American officer called to the bar and in no time mugs of beer were placed in front of them. The singer, a brunette, looking voluptuous in gypsy costume, gave Isayev – whom she could see bore the rank of lieutenant-colonel – a kiss and said: 'Finally, the Soviets have come. What do you want me to sing for you?' They were enjoying the party atmosphere when Chertok noticed that another Soviet officer had joined their table. He gave Chertok a big Russian bear hug and said quietly, 'I am from SMERSH and I want you and Lieutenant-Colonel Isayev to come to my headquarters tomorrow morning.'

It was with some trepidation the next morning that they found their way to SMERSH headquarters. Chertok was apprehensive, assuming that their treatment of the visiting professors had been noticed and that they were about to be sent back to the Soviet Union. Isayev was more bullish, advocating a policy of attack and admit nothing. They were shown into the office where the headquarters chief and the SMERSH officer from the previous evening were waiting, looking serious. With a sense of relief they found the conversation quickly turned to ways of recovering German technology and had nothing to do with the treatment of the two professors. The SMERSH officer warned that the US Special Forces were carrying out a full-scale operation to capture German specialists in the area. However, they were in competition with SMERSH agents who were themselves in the process of 'cleaning up' the capture of German specialists, checking on any Soviets engaged in this field and finding all remaining German rocket technology.

At this point, Chertok thought it prudent to reveal that they had a very innovative piece of machinery in their care, something they had found, with Shmargun's help, hidden under blankets in Camp Dora. This was a gyrostabilizer platform that was used in a V-2 rocket, consisting of an assembly of spinning wheels or gyros that would be finely balanced on the top of the rocket. The gyros would sense the movement of the rocket and send signals to the control system to keep it on the proper trajectory. This would be achieved by the use of aerodynamic flaps on the bottom of the rocket's fins and by graphite vanes that deflected the rocket exhaust gases. The gyroscopic unit that Chertok had acquired had, apparently, been hidden by the Germans and then found by the inmates of Camp Dora who had decided to save it for the Soviets. SMERSH was clearly pleased with this offering and discussions began on joint operations. The organization was able to reassure them that Shmargun was trustworthy and agreed to find accommodation and employment for the German specialists Chertok had so far acquired.

By August, the Institute Rabe had developed into a considerable enterprise. Laboratories had been set up to study gyro appliances, steering engines, electrical apparatus and radio equipment. Success in attracting German technicians and specialists had swelled the staff, the 'star' being Kurt Magnus, a top engineer in the field of gyroscopy. To give the Soviet enterprise a greater chance of success, they still hoped to capture von Braun, but he was closely guarded in the US zone and for the moment unobtainable, legitimately or otherwise.

For Soviet intelligence the enterprise was soon to take on renewed urgency. On 6 August, the Americans exploded an atomic bomb above Hiroshima in Japan. Stalin had been warned by Soviet intelligence of the development of nuclear weapons, but now the appalling power of the Americans' new weapon was all too apparent. Exploding with a force equivalent to 12,500 tons of TNT, almost the entire city of Hiroshima vanished, its citizens vaporized. Three days later, a second bomb of even greater power was unleashed on Nagasaki. Many thousands were killed, simply dissolved in the mega blast, the imprint of their bodies reduced to mere shadows on the ground. The injured, those who had not been at the centre of the explosion, haunted the city. Gulping in radiation, they soon became horribly disfigured. Their eyes melted and their skin peeled as a slower death consumed them.

Stalin knew there was no time to lose. The Soviet Union had no protection against such terrorizing weaponry. The Second World War had crushed the country – more than twenty-seven million had lost their lives and countless cities had been destroyed. Industry and agriculture were in ruins. Now there was a new peril for the Soviet people to consider as the full horror of the atomic age was unleashed. Serov's men had already been trying to track down uranium supplies and any technology related to the atomic weapons as well as rocketry. Now Beria himself was put in charge of a Soviet nuclear programme.

Meanwhile, high-ranking party officials soon came on a tour of inspection of the Institute Rabe. Major General Lev Gaidukov was far-sighted enough to begin to question what might happen if atomic technology and missile technology could be combined. The Americans had both atomic weapons and missile specialists; the Soviet Union had neither. Yet it was not inconceivable that these two technologies, put together, could create a weapon of even more deadly power. The immediate issue was how to advance their attempts at reconstructing a complete V-2. They needed more senior German specialists. In conversation with Chertok, Gaidukov pointed out that the Soviet aviation specialists appeared to know more than the Germans and that Chertok 'had not got the right Germans' – the Americans had them.

Since the NKVD considered it too risky to kidnap von Braun, Chertok decided on his own secret mission, Operation Ost (East), whose purpose it was to get by any means possible the real rocket scientists from Peenemünde. He issued instructions to his colleague Senior Lieutenant Vasily Kharchev to penetrate the US zone with effective agents and get some 'real' Germans before it was too late and they were taken to America. Kharchev was suitably supplied with cognac, butter and 'all sorts of dainties' including many dozens of wristwatches to serve as bribes. Special papers allowed him to cross the border between the two zones. Between learning English and German and bribing Americans, Kharchev hardly slept. When news came through that von Braun would be leaving in early September for America, it became apparent that action had to be taken immediately.

Kharchev soon met with success. One of his agents discovered that the wife of a senior German specialist who was in the American zone wanted to talk to the Soviets. A meeting was duly arranged by the border. An agitated, fair-haired young woman and her son soon emerged from bushes at the edge of the country lane. 'In case of trouble I will explain that we were walking and we got lost,' she said quickly. The woman was Irmgardt Gröttrup, wife of Helmut Gröttrup, von Braun's deputy in guidance and control. She wanted to know what the Soviets had to offer.

According to Chertok, Irmgardt Gröttrup let the Soviets know 'that it was she who made all the decisions' for the family. Both she and Helmut secretly detested the Nazi regime and had at one stage been arrested by the Gestapo, who were suspicious about their loyalties. The Americans looked little better to Irmgardt Gröttrup. She found them 'rude' with their chewing gum and the way they put their feet on the table. The leading rocket experts – including her husband – had been 'grabbed', she said, housed at Witzenhausen and subjected to lengthy interrogation, only to find the contracts offered by the Americans held no appeal. Helmut Gröttrup was required to leave Germany for America alone – without his wife and children. In addition, 'it was a contract terminable by one signatory only: the US army'.

In 1945, Germany was a dangerous place. Irmgardt saw herself as the person best equipped to lead her family through the difficult times. They were determined to stay together and suspicious of what both the Americans and the Soviets had to offer, so Irmgardt set out to get the best deal. Putting the pressure on Chertok, she explained that time was running out. The Americans intended to ship the rocket scientists to

White Sands in less than two weeks. If they wanted Helmut, they had better make a generous offer.

Accordingly, a Soviet agent crossed the River Werra into the US zone to open negotiations. Before long, Helmut Gröttrup himself slipped across the river into the Soviet zone under cover of darkness. He was immediately offered a suitable post, that of running 'Bureau Gröttrup', responsible for guided missile development. 'An incredibly risky walk across the green border flanked by Soviet commanders; a villa in Bleicherode, good pay, food from the Red Army. Are we heading for trouble?' Irmgardt wrote in her diary. Within three days, the Gröttrups and their young family moved into the Soviet zone where they were treated generously. A grand house was requisitioned for them, the unfortunate owner sent packing. A car was provided, servants and, most importantly, extra food rations.

Chertok decided that 'he had made the right choice with Gröttrup', but that Irmgardt was more of a handful. Relishing her elevation to wife of the top man, 'Frau Gröttrup turned out not to be as shy as we first thought,' observed Chertok. She bought two cows for the children, acquired two rather fine horses and a stable and demanded a Soviet chaperone when out riding. Even she may have realized she had gone too far when she checked the provisions in the Soviet canteen and took it upon herself to sack the cook for stealing food. However, she was also learning Russian, Chertok noted with satisfaction, and rode up to the Villa Frank on her motorbike to play Liszt, Beethoven and Tchaikovsky on the piano. Evidently Frau Gröttrup aroused mixed feelings among the Soviets, who nicknamed her the 'German Amazon'. One officer who was asked to help with the riding was offended. 'My wife survived in Leningrad's blockade. She's very sick now,' he exploded. 'And I have to take care of German mares. Go to hell, all of you!!'

Meanwhile, Kharchev, bolstered by this success, felt confident that the time had come to pursue von Braun himself. It was a last-ditch attempt. Despite the fact that military intelligence advised against such a risky mission, Chertok gave his approval. Kharchev felt that if he was going to come back with von Braun, he needed an impressive consignment of goods with which to buy his way in, and he duly set off to the border suitably equipped with enough wristwatches on his arm to open a shop, various 'gewgaws' he deemed would see him through awkward moments with an overzealous 'Yank' and the one bottle of genuine Moscow vodka to be found in Germany. The watches found new owners at the first checkpoint and such generosity inspired the American guards to give him a lift in a Jeep. He wanted to go to Witzenhausen, Kharchev said innocently, knowing full well that this was where von Braun was being held. He was duly taken to the American commandant's office, who in turn directed him to the town major. It was then that Kharchev's well-thought-out plan began to take an unexpected turn.

'I am taken to a large bedroom,' he confided later to Chertok and Isayev. 'On a wide bed, like the one in Villa Frank, lies the town major; also a very beautiful woman and lying in between them was a huge German shepherd dog!' The US major, not at all taken aback, 'flung off his bedclothes, shoos off the dog and suggests I get in bed with him, adding: "Anything for a Soviet officer and our neighbour at the border."'

Chertok and Isayev pressed Kharchev for more revealing details. 'Did you get in? Did the girl throw her bedclothes off too?' but he blushed and the answers were never satisfactorily revealed. He would only say that a great deal of whisky was consumed as he tried to persuade the American major to share the German specialists; after all, the rocket scientists were war trophies. The major, however, was unable to see Kharchev's position on this in spite of the whisky. He would only explain that von Braun was a criminal and was under heavy guard; and with that he got back into bed, while Kharchev was escorted back to the border.

When Gröttrup heard the story, he laughed and said von Braun would never come over to the Soviet side. He was a brilliant scientist and something of a genius, but he was also a baron and a member of the Nazi Party. Chertok wanted to know more and was invited to the Gröttrups' house where, he noted, Irmgardt, true to form, appeared to be one of the few women in Germany serving real coffee with whipped cream. Gröttrup explained that all his life von Braun had been enchanted with the idea of space travel but that 'life had made him use his talent for warfare'. While at Peenemünde, he and Dornberger had had to watch their every word, guard their every thought. Should the Gestapo have suspected they had any ambition regarding space travel, von Braun would have been considered a traitor. And even though he had a most attractive ability to gather talented people around him and bring them into his charmed circle, no one could have saved him. Although under close guard, while he remained in Germany he was not safe and would not be until he reached America.

As they gathered more expertise around them, the Soviets were anxious to assemble and test von Braun's V-2 rocket. Under the command of the Soviet Interdepartmental Technical Commission, the Institute Rabe grew to control several scattered departments and sites. It had a factory for manufacturing electrical equipment about ten miles from Nordhausen and absorbed a factory in Berlin, which developed missile control systems. General Gaidukov organized a framework for production using a new production plant, preferring to avoid the ghoulish Mittelwerk with its long shadowy tunnels and the penetratingly foul smell which was impossible to erase with even the strongest disinfectant. They set the Germans to work at the new plant, organizing all the paperwork on production and assembly of the V-2.

The Soviets counted themselves fortunate to have Gröttrup, but unless they captured von Braun there was a feeling in the Soviet camp that a real leader who would take them forward was still missing. Gaidukov recognized it was essential to bring their own expertise from the Soviet Union. He needed a man who had the ability to inspire and lead with specialist knowledge of rocket science: someone, in fact, who could match von Braun.

His enquiries soon revealed that such a man existed, one of the stars of the former Soviet rocket programme, much admired by his former colleagues, but whose name no one dared mention. He had been in the Gulag, condemned as a traitor and an enemy of the state. For those who had the confidence to acknowledge him, even though he was as yet unpardoned, he did have a name, however: Sergei Pavlovich Korolev.

CHAPTER SIX

'I am not guilty'

To the Commissar of Internal Affairs: 29 June 1938
Nikolai Yezhov, chm of the NKVD

I was arrested by the NKVD for anti-soviet activity, so I have decided to give you my statement ... I was part of an anti-Soviet enemy organization that infiltrated the Scientific Research Institute No. 3 ... This enemy activity included wrecking the work of the Institute No. 3. The Institute was involved in important defence work and we were involved in sabotaging this. We engaged in useless tasks for the Institute and created antagonism between the different groups. We set about disrupting and destroying the industrial base of the Institute, its shop floors and laboratories ... We were involved in wrecking many projects and expelling specialists and communists from the Institute. I am going to provide you with detailed information about our enemy Trotskyist organization ... I feel repentant and am ready to be changed. I will no longer do all the bad things I have done. I am asking for an opportunity to become clean again and be a free citizen of the country SERGEI PAVLOVICH KOROLEV

Sergei Pavlovich Korolev remembered the evening of 27 June 1938, the day of his arrest, with the clarity that only terror can etch upon the mind. He and his young wife, Ksenia, and their three-year-old daughter, Natasha, lived in Moscow, in a sixth-floor apartment at 28 Konyushkovskaya Street. It was nine o'clock at night. Korolev was home first, waiting for Ksenia. Suddenly he heard the sound of her hurrying footsteps outside; she had run up the six flights of stairs and burst in to tell him that there were men downstairs making enquiries, searching for someone. She was terrified for him. Recently his friend and colleague Valentin Petrovich Glushko had been arrested.

Stalin's years in power had been characterized by fearsome purges as he ruled through terror. In 1931, he had engineered a law that brought accused 'enemies of the people' to trial within ten days of their arrest. If found guilty, their execution was immediate, with no appeal. By the late 1930s, millions had been dealt with in this way, and the reign of terror was unstoppable. As in Germany with the Gestapo, the terrifying knock on the door in the middle of the night would inevitably lead to brutal ways of obtaining a confession. Often there was no trial, although for many of those arrested there was a sentence of death. For the fortunate, those who were not arbitrarily shot, a long stay in the Gulag system, a network of labour camps run by 'the Chief Directorate for Corrective Labour Camps', invariably destroyed something spontaneous and vital in those who survived. The regime was so harsh that at least 10 per cent of prisoners died annually in the Gulags spread across the Soviet Union.

Korolev saw Ksenia's anxiety and held her close. 'They must have come for me this time,' he said quietly. Korolev had recently bought a new record. He put it on and they sat holding hands, saying nothing, listening to the music as the dusk turned to night.

When the knock on the door came, three men from the NKVD entered, showing paperwork which authorized them to make a search and appeared to demonstrate that Korolev was an active member of a sabotage organization. Through the long hours of the night they confined Korolev and Ksenia to the sofa while they hunted for incriminating evidence; clothes were flung from the wardrobe, underclothes were examined, papers and books were scrutinized and they even looked through the crockery. The music had stopped long ago and the needle made a rasping noise as the record continued turning on the gramophone. Ksenia saw one of the men pick up her husband's malachite cuff links and put them in his pocket. She sat in silence, not daring to speak.

At six in the morning, the door of the study was sealed. Korolev was arrested and instructed to 'collect his things'. At first, Ksenia did not understand what 'things' meant, but then she realized Sergei was going to prison. According to Korolev's biographer, Yaroslav Golovanov, that was when she 'became really scared; scared for Sergei, for Natasha and for herself. Scared about her life to come.' She picked up a toothbrush, soap and underclothes from the floor and packed them into a small suitcase. The bells from the trams could be heard outside in the street. Sergei put on an old leather coat and scarf and embraced her for the last time. He looked at her purposefully and said with a reassuring steadiness that belied their predicament: 'You do know, don't you, that I am not guilty?' Then he was gone. She was not permitted to follow. From the window she watched as her husband was led into a car. With a sense of shock she saw the car pull away.

Korolev was accused of being a member of a counterrevolutionary organization and of committing acts of sabotage. He was interrogated, beaten and threatened that harm would come to his family unless he 'confessed' to the fabricated charges of his colleagues. Two of his senior colleagues, Ivan Kleimenov and Georgi Langemak, had already been forced to admit to false accusations against both Sergei Korolev and Valentin Glushko. Then they were shot. When Glushko was himself arrested, he too denounced Korolev under torture. Several other engineers were also being investigated, and such was the atmosphere of fear and paranoia created by the NKVD that a number of colleagues at the institute felt under pressure to write letters confirming Korolev's 'wrecking' or 'disruptive' activities. The absurdity of the charges is highlighted by the science historian Asif Siddiqi, who found that one of the claims was that Korolev had destroyed a rocket plane, yet the very same plane 'sat quite intact in the hangar of the institute headquarters'. It is a measure of Korolev's desperation at this stage that, within two days of his arrest, he signed his confession to the Commissar of Internal Affairs on 29 June, admitting to his 'crimes', all allegedly carried out with a view to stopping the rearmament of the Red Army with new kinds of weapons.

Korolev had no trial, but he did collect a ten-year sentence in the Gulag. His first destination was the notorious Kolyma camp on the fringes of the Arctic Circle in eastern Siberia. While in a transit prison for a few months, despite his 'confession' he repeatedly protested his innocence, writing letters to anyone who would listen, hopeful that his cries for help would be heard: 'I have never carried out any wrecking activities. I have never been a member of any anti-Soviet wrecking organization, nor have I ever heard or known about such an organization … I have always been loyal in every way to the general line of the party, to Soviet rule and to the Soviet Motherland.' His mother, Maria Nikolaevna Balanina, was beside herself, and sent many letters to Stalin himself. 'I implore you to save my only son, a young and talented rocket engineer and pilot,' she pleaded. Her words fell on deaf ears.

Within a few months, the awful reality of a ten-year sentence for an unknown crime had taken its toll. Korolev confided to Ksenia:

> I am so very tired of life. Almost everything that used to give me pleasure in life has gone. It probably hurts you to hear that I have lost all interest in life … I can see no end to my dreadful situation … I will be a stain on your and Natasha's life. I don't even know if we'll be able to live together again, or rather whether I can and should live with you. I am afraid to speak and think about it.

Of all the Gulag camps, Kolyma, a gold-mine camp in eastern Siberia, had the worst reputation. The summers were so brief that the ground barely had time to thaw. Winter days were shrouded in darkness in spite of thick snow. The cold and frost were so terrible they burned the skin on contact. For most, life at Kolyma was brutal and short; thousands of prisoners died each month.

The sequence of events that had led Korolev so inexorably to this hopeless point in his life had begun with an innocent childhood interest in the wonder of space flight – with little thought that such dreams might ultimately cost him his life.

• • •

Sergei Pavlovich Korolev was born on 12 January 1907, near Kiev. A happy child, he was adored by his mother, Maria Nikolaevna. 'He liked my stories,' she recalled years later. They 'flew together on a fairy-tale carpet' out into space and he would look out enthralled, 'wide-eyed into the sky'. However, she was intent on further education and joined her sister, eventually training to be a teacher. Leaving her husband, she placed Sergei with her parents. His grandparents gave him everything a child could need but in the quiet, undisturbed routine of an elderly household he was often lonely.

When he was six, his grandparents took him to see the celebrated pilot Sergei Utochkin fly a plane at the local fair and there his enduring interest in flight was born. After his mother remarried, they moved to Odessa where military seaplanes were stationed. At sixteen, Sergei was determined to see the planes at close quarters and would swim across the bay to watch the mechanics at work. It was not long before he was rewarded with his first flight as a passenger, watching the people on the ground become reduced to dots as he soared above them through the brilliant white clouds.

At seventeen, he was studying hard, designing a glider in his spare time and falling in love with his classmate, the pretty but unobtainable Ksenia Vincentini. He had grown strong and athletic, relishing exercise, his body physically expressing his rugged determination. But what people remembered most about Sergei Korolev was his lively intelligence, his dark eyes, absorbed, taking in information. He proposed to Ksenia who refused him: she wanted to study. Korolev was not easily dismissed, however, never doubting eventual success. His confidence, strength of will and single-minded tenacity would not allow him to give up on anything.

Korolev's fascination with aviation and aerodynamics was firmly established and he studied first at the Kiev Polytechnical Institute and later at the Moscow Higher Technical School. Within a few months of graduating, he was recruited as an aeronautical engineer at Moscow's renowned Central Aerohydrodynamics Institute. During his years of study he had continued to pursue Ksenia, and in August 1931 they were married in a modest civil ceremony. They were both twenty-four years old. Hours after they were married, Ksenia returned to work.

That same year Korolev began to develop a serious interest in space travel and soon formed a series of professional relationships that would prove crucial throughout his career. He was inspired by one of his colleagues, Fridrikh Tsander, who gave up his job to devote himself to creating an amateur rocket society in Moscow. Their society came to be known (in Russian) as the GIRD – Group for the Investigation of

Reactive Motion – and they aimed to build a successful rocket engine. Korolev and Tsander sometimes worked late into the evenings from a wine cellar in a back street in Moscow. These were the years during which Stalin was imposing collective farming on the Soviet Union with dire consequences; food was short; money even shorter. But making money was not the point of their enterprise. Korolev was soon spending all his spare time at the GIRD and the group were attracting to its circle intellectuals such as Mikhail Klavdiyevich Tikhonravov, who shared Korolev's passion for aeronautics and rocket design and became a close friend. 'Onwards to Mars, onwards to Mars!' was their idealistic greeting.

To a man they had been inspired by the ideas of Konstantin Tsiolkovsky, a Russian village schoolteacher who had risen to become the father of Soviet space science. Writing at the turn of the century, well before Hermann Oberth in Transylvania or Robert Goddard in America – well before aviation was even established, in fact – Tsiolkovsky brilliantly anticipated the era of space exploration. He was the first to recognize that rockets would be the best means of travelling into space and to translate the fundamental principles of physics into detailed calculations. Ever since Isaac Newton had first defined the laws of gravity in the seventeenth century, the relationships between the celestial bodies had been understood. Newton showed how gravitational pull holds the moon in orbit around the earth and the planets around the sun. His calculations revealed that the strength of gravitational pull is dependent on the mass of a body and decreases with the square of the distance from its centre: the effect of gravity diminishes the further an object flies from the centre of the earth.

Tsiolkovsky applied these principles to space travel, showing how, in theory, rockets could be used to escape the earth's gravity. His calculations showed that because gravity is strong near the surface of the earth, a large amount of thrust is needed to launch the rocket, but once that rocket has gained enough speed and distance from the earth, the engines can stop burning and the vehicle will coast. He theorized that multistaged rockets would be able to reach far out into space. Each stage would be powered by an engine or set of engines, allowing the speed of the rocket to be stepped up drastically. As each stage burned all of its fuel, it would be jettisoned so that the vehicle became progressively lighter and more efficient at higher altitudes. Well before Oberth, who had influenced von Braun, he showed that liquid, rather than solid fuels, would allow controlled combustion and was particularly interested in new fuels such as liquefied hydrogen and oxygen. His ideas were published towards the end of his life and were to influence a generation of future rocket enthusiasts.

Tsander inspired the group to try to turn these ideas into reality, but it was a daily battle against lack of funds and ill health. His own sudden death from typhus in March 1933 was a loss deeply felt by Korolev. He became the natural leader of the group and within a few months, working with Tikhonravov, they successfully launched their first liquid-fuelled rocket, which travelled all of 1300 feet. With this success, they began to attract the attention of other groups, notably a military research laboratory in Leningrad where a small group led by Valentin Glushko was studying liquid-fuelled rocket engines. Like Korolev, Glushko was equally enthralled by the possibility of travelling to space. He had even written to Konstantin Tsiolkovsky describing his all-consuming interest: 'This is my ideal and my life's goal. I want to devote my life to this great cause.'

A far-sighted military leader named Marshal Mikhail Tukhachevsky recognized the value in uniting the two groups, and in the autumn of 1933 succeeded in creating a new scientific institute, known (in Russian) as RNII: Reactive Scientific Research Institute. Korolev, with his natural leadership skills, was appointed deputy director – only to be demoted for falling out with his superior over the direction of research. For several years Korolev, Tikhonravov and Glushko were all working together on different aspects of missile development, Korolev on long-range military missiles and Glushko on liquid-propellant engines. Although their research was focused on military needs, Korolev still found time to write papers, among them 'Missile Flight into the Stratosphere' and other space-related research, including the design of a winged rocket. Korolev was happy and his career progressed rapidly. In 1935, he and Ksenia had a daughter, Natasha, and the following year they were able to get their own apartment.

Stalin's purges were to bring an abrupt end to all this. The under-developed Soviet Union of the early thirties was to be dragged into a bright new communist twentieth century. Five-year plans enforced radical changes in agriculture. Peasants were made to give up their land into state ownership. Twenty-five million were forced off the land into industry, which would now be rejuvenated and contribute to the making of a great Soviet Union. Any opposition to such overwhelming change was stamped out mercilessly. Stalin's suspicious nature saw enemies in his own shadow. In a handful of years, millions of innocent citizens died, arrested on trumped-up charges, confessing under torture and often shot within hours or days of confessing. Punishment and death became an industry. As the crisis deepened no one was safe. 'Enemies' were denounced everywhere, in the party and in the Red Army. As the 'Great Terror' got underway, Tukhachevsky and eight other colleagues were arrested for treason and shot, including the institute director, Kleimenov, and his deputy, Langemak. Tukhachevsky's mother, sister and brothers were also rounded up and shot.

It was impossible to escape the fallout. Over a period of a few weeks in early 1938, the finger of suspicion began to point at Valentin Glushko. Rumours that he might not be safe grew rapidly into poisonous allegations that he had had dealings with 'enemies of the people' and could no longer be trusted with military secrets. At a critical meeting in March 1938, almost all his colleagues denounced him – although Korolev continued to maintain publicly that Glushko could not possibly be guilty. Despite this, on 23 March Glushko was arrested.

A few weeks later, the NKVD knocked on Korolev's door.

• • •

For the prisoners at Maldyak camp, part of the Kolyma network in eastern Siberia, the day began at 4 a.m. and ended at 8 p.m. when the men would fight to crowd around the stoves which warmed their tents, the steam rising up from their ragged, stinking garments. The diet of cabbage soup and a small ration of bread was barely enough to sustain them. The prisoners were always hungry; they talked of food, they dreamed of food. A brutal system was employed whereby criminals wielded authority as petty functionaries, keeping order in the tents. It was all but impossible to get news from the outside world. Nonetheless, Korolev knew perfectly well who was alleged to have denounced him, as a later letter he wrote to Stalin himself makes clear:

> Kleimenov, Langemak and Glushko gave testimonies about my alleged membership of anti-Soviet organizations. This is a despicable lie. There are no facts to prove it, nor can there be ... Without examining my case properly the military board sentenced me to ten years' imprisonment. I was sent to Kolyma ... My personal circumstances are so despicable and dreadful that I have been forced to ask you for help ...
>
> KOROLEV letter to STALIN, 13.7.40

Korolev also knew of the fate of Marshal Tukhachevsky's relatives and wondered what had happened to Ksenia and Natasha. It was routine for a prisoner's family to be punished; even friends and acquaintances were not safe. He hoped his rapid confession may have spared them. A distance so great, seeming to encompass more than a mere physical extent, now separated him from his former life. In his new, freezing world of mindless labour, where surviving each day was a miracle, he had no way of knowing if they had been spared, only the cold certainty that he could not see them. Even if they survived, they might not dare associate with him.

Within a few weeks of his arrival, Korolev was unrecognizable from his former self. The regular beatings meant he had lost many teeth and his gums were swollen and bleeding. Malnutrition had given him scurvy. He could barely walk, his legs were so swollen. He had a broken jaw and a huge scar on his head. It seemed he had been forgotten, left to die in his own private hell, to become a faceless statistic in Stalin's purges. And yet some small seed of hope refused to die. He could not give up writing letters asking for his case to be reviewed.

'I am convicted of a crime which I've never committed,' he wrote to the Chief Prosecutor of the USSR:

> The charges are all false and made up. I have never been a member of any anti-Soviet organization, never committed sabotage and never heard of anything like that ... For 15 months I am kept away from my favourite work, which filled my entire life. I dreamed of creating unique supersonic high-altitude jet planes for the USSR, which would be the most powerful means of defence and an

> unrivalled weapon. I am asking you to revise my case and clear my name from the weighty accusations as I am not guilty.

Then one day a new prisoner, Mikhail Alexandrovich Usachev, former head of the Moscow Aviation Plant, arrived at the Gulag. He was tall and strong, an amateur boxer, with an imperious manner and an unwillingness to be bossed around by the 'head' criminal in the camp. A battle of wills and physical power soon settled the situation in Usachev's favour, putting an end to the criminal's favoured status. As his first act of obeisance, the head of the criminals showed Usachev round the camp. In one of the tents he pointed to a pile of dirty rags: 'a king, a weakling, one of yours,' he said. To his horror, Usachev recognized Korolev, who was 'extremely weak, sick and almost breathless'. Usachev discovered that the leader of the criminals had deliberately set out to starve Korolev because he obstinately refused to bow to his will.

Korolev was taken to the camp doctor and given potato juice to combat his scurvy. Once he was a little stronger he returned to work, but this renewed strength brought its own problems. On one occasion Korolev noticed an old man too frail to push his barrow. The 'criminal' on guard hit him, taking perverse pleasure in watching him collapse. Too weak to stand, the old man lay helpless on the frozen ground. Unable to contain his anger, Korolev hit the guard. The crowd watched silently, expecting the worst. Incredibly, Usachev's authority prevailed and Korolev went unpunished.

One cold November morning in 1939, an official came for Korolev. His first thought was that he would now face punishment for striking the guard, and, fearing the worst, he said his goodbyes to everyone. One of the men gave him his worn good coat. Korolev made his way to the camp commander to hear instead the news that he had been praying for for months. He was leaving the Gulag and returning to Moscow where his case would be reviewed. A convoluted series of circumstances had led to this stroke of good fortune. The head of the NKVD, Nikolai Yezhov, had been arrested and replaced by Lavrenti Pavlovich Beria. Beria had one of the most brutal reputations within Stalin's inner circle and was responsible for the deaths of tens of thousands during the purges and yet he was the one to authorize a review of Korolev's case. The charge against Korolev was reduced and a retrial ordered.

There was no transport available for Korolev to get to Moscow, so he tried to hitch a lift to the town of Magadan, about one hundred miles away. The lorry driver demanded his coat as payment but when they reached Magadan the last boat had left. There would not be another until the spring. He had no choice but to stay there for the winter. He looked around. Thick snow lay on the ground, brilliant in the moonlight. It was very cold, around 50 degrees C below zero. His clothing was thin. He had not eaten for two days and was desperate for food. And then, he said, some sort of miracle occurred. As he walked though the snow, he saw a loaf of black bread by a well. It was still warm. He thought he was hallucinating. 'I went up to the well, saw the loaf and shut my eyes. I realized if when I opened my eyes the loaf is not there – I am dead.' But the loaf was still there. He ate until he was full, and then stole back into the army camp, secretly hid under a bed and slept. In the morning when he woke, he found his clothes were frozen to the floor.

His health deteriorated as he slumped through the winter, doing odd jobs, barely keeping body and soul together. The spring came at last and he reached the mainland and boarded the Moscow train, but was taken off at Khabarovsk, too ill to travel, probably dying of scurvy. Korolev recounts that out of nowhere an old man appeared who took care of him, massaging his bleeding gums with herbs. He sat under a tree in the warm spring sun and could feel his vitality returning. When he opened his eyes he saw a butterfly, a fragile thing, alive and beautiful, and realized that he too, miraculously, was still alive.

On his return to Moscow, his case was re-opened and his sentence reduced from ten to eight years in the Gulag; but rather than being sent back to the labour camp his skills were to be put to use in a *sharaga* prison. This was a big improvement as the *sharagas*, special prisons for technical workers, were run on more humane lines than the manual labour camps. In 1940 he was sent to the Central Design Bureau 29 in Moscow, where he was to work with fellow scientists. For Korolev this was the difference between life and death, as the diet was vastly improved and working hours were from 9 a.m. to 6 p.m. After more than a year working on aspects of bomber design, he was moved to Omsk in Siberia and finally to Kazan, some four hundred miles from Moscow. Here he specialized once again in rockets and designed liquid-fuelled rocket boosters for the Pe-2 dive bomber.

Occasionally there were visits from his wife and daughter, but as the years passed a silence and a distance had grown between them. He was released in August 1944 after six years in prison, a very different man, but still he was not allowed to leave Kazan. The war now required him to stay on and work. He persisted in trying to revive his relationship with his wife. 'All that happened to me follows me like a shadow, reminding me of the past and sometimes, even against my will, I am remembering things that seem long forgotten,' he told her in his first letter upon release:

> I can imagine how you must have suffered and what you felt. How many tears have been cried during all this time? I am afraid to look back and remember about it … I called you on Sunday but got no answer. I sent you a telegram; I will be calling on Thursday 24 at 10.00. I am anxious to talk to you and just listen to you dearest. I have not heard your voice for three years now …

When he finally heard from Ksenia, her letter did not provide the reassurance he was looking for. He replied:

> I've got your last letter and I should say that I was shocked with its sad and hopeless words. I am very sad at you calling me your 'so far, closest friend' … What else do you want my dear, tired and therefore unjust friend?

Korolev wanted to see his wife but was also afraid to do so, afraid to acknowledge the distance that had grown between them. In prison he had constantly dreamed of the day he would return home, imagining every last detail, the warmth of his wife's greeting. In November 1944, when he finally obtained permission for a short visit to Moscow to see his family, everything was different. His wife and daughter were not at home. Some pipes had burst and they had gone to stay with his mother.

He had imagined his return so clearly that he knew the words he was going to say by heart. He knew that when the door opened he 'would see an unfamiliar little girl who he had not seen for years. She certainly wouldn't recognize him and would be a little afraid, and when he took her in his arms she wouldn't like it and try to strain away. When he kissed her she would run to her mother.' Now, as he made his way to his mother's apartment, he knew it would all be different.

As Korolev turned into the courtyard, according to his biographer, Yaroslav Golovanov, 'his heart was hammering: he had been away for seven years. He walked upstairs to the flat and suddenly winced, his hand recognized the handrail. He had never thought about it, never even recollected the staircase where he had been running up and down in his youth. He didn't need a handrail back then and was astonished that his palm still moulded itself around the impression of the bark of an old tree. It was a miracle.'

When he knocked on the door, it was his mother-in-law who opened it. A crowd was there, but not his wife and daughter. He had to wait until his wife had finished work. His daughter recognized him immediately, but his lovely wife was silent, smothering hope.

Later, to be alone, he and Ksenia walked to a nearby garden square. Korolev noticed that the garden with its trees and shrubs, even the seats, was unchanged. Seven years had passed. Their lives had altered beyond recognition but the park was the same. Why had she never visited him in prison? Glushko's wife had visited him. Why did silence, full of things unsaid, hang between them now, when they had survived, when, unbelievably, they were together at last? They walked home silently offering each other no more than politeness.

The coolness of Ksenia's reaction may not just have been due to their prolonged separation and the uncertainties that had arisen during Korolev's imprisonment. There is evidence that Korolev had another lover at this stage. Although Yaroslav Golovanov has not published this information, his notebooks reveal that Korolev had a long-standing affair with Glushko's sister-in-law, Ketovan' Ivanovna Sarkisova: 'The affair lasted a long time, from 1933/4–1949/50. It was episodic and she did not have a hope because there was no future in it.' Intimate letters to 'Dear Ket' survive from this period. Korolev took care to let Ketovan' know when he would be in Moscow and urged her to write frequently: 'You know that your letters are precious to me.' He was anxious to catch up with what had happened to former colleagues and asked about the families of those that had been arrested, adding 'the thought of them does not let me sleep'.

Many more months were to elapse before Korolev was finally

permitted to leave Kazan in August 1945 and was sent to the People's Commissariat of Armaments in Moscow. He was informed that Soviet rocket technology was being developed and research on the German missiles was now a top priority. He was to fly out to Berlin at once. When Korolev protested at being obliged to leave his family yet again, he was immediately shouted down. 'You must understand,' said the Deputy People's Commissar, 'the Americans won't be resting. After the nuclear bombs of Hiroshima and Nagasaki, they will continue their work with nuclear arms. And now they have only one enemy – us.' Wernher von Braun, he explained, was in 'voluntary captivity' where he was 'enthusiastically helping them'. Korolev was assigned the rank of lieutenant-colonel in the Red Army, and his brief was to 'get to the bottom of the principles of construction of the V-2 and find out everything possible about how to arrange production'. As Korolev turned to go he added, 'it's the business of utmost governmental importance. It's under the personal control of Comrade Stalin.'

And so it was that on a cold September morning the former enemy of the state, now smartly attired in his new army uniform, took the flight from Moscow to Warsaw and then on to Berlin. Sergei Pavlovich Korolev had clear instructions. He had to catch up with and surpass the achievements of his Western counterpart, Wernher von Braun.

• • •

Within a month of von Braun's escape to the Western zone, the American War Department had approval for their plan for the German scientists. On 19 July, Operation Overcast was established by the Joint Chiefs of Staff with a view to 'exploiting German civilian scientists ... to assist in shortening the Japanese war'. Just over a week later, Colonel Holgar Toftoy arrived at Witzenhausen where many of the scientists were staying, determined to work with Staver to sign up the top rocket experts. Despite repeated requests to bring over five hundred scientists, as von Braun had recommended, Toftoy only secured approval for one hundred men. He had come up against resistance from politicians and certain army chiefs, who had argued strongly against bringing former Nazis to America. Compromises had been reached but he was conscious that he had little to offer the German team. The money was hardly an incentive, citizenship was far from assured and he only had six-month contracts with which to tempt them.

The Germans felt that anything less than a three-year contract gave a strong indication that the Americans intended to steal all their accumulated years of knowledge then discard them. Worse still, they bluntly refused to leave their wives and children behind in a country where there was no food and such chaotic law and order as only favoured the survival of the fittest. As negotiations dragged on, Toftoy knew that delay was dangerous. If he could not get the scientists to America soon under Operation Overcast, the Soviets could spirit them away. Only recently there had been an incident which confirmed his fears. Men wearing US army uniforms had turned up at the building where the Germans were housed. They had been very friendly and suggested a trip to the café in the village, but their US accents had a suspiciously Soviet sound and their offers had been declined.

With negotiations underway during the summer, Dornberger and von Braun were taken to Britain for a few days as the guests of British intelligence. The British were also hoping to set up a rocket research group and wanted to gather information at first hand. Although they permitted von Braun to return to the Americans, the British refused to part with General Dornberger, claiming he was wanted for war crimes for which hanging was the appropriate punishment. The British were hoping to track down General Kammler for supplying slave labour to Mittelwerk and Camp Dora, but Kammler could not be found. Dornberger would have to stand trial in his place.

Dornberger was incarcerated in a prison in deepest Wales and left to contemplate his future, which began to look exceedingly doubtful. He was certain that Kammler would never be found. While rumours were rife concerning Kammler's demise, Dornberger's own sources learned that the general had met his death in Prague. It was alleged that he had been sheltering in a bunker with about twenty SS officers when six hundred Czech guerrilla fighters had approached. Vastly outnumbered, Kammler was last seen smiling wildly and firing frantically as he led an assault. Major Starck, following him, as ever, gun in hand, decided that this was the hopeless situation for which he had been primed. Without waiting for orders, he shot Kammler in the back of the head.

Back at Witzenhausen, Toftoy at last had the formal approval from

Washington he needed to take 115 of the leading rocket scientists to America. The US army undertook responsibility for dependants left in Germany. Von Braun would be leaving his parents behind as well as his young cousin, Maria von Quistorp. She lived with her parents near Peenemünde and von Braun had often visited when he worked nearby. He went to see Maria again just before he left. Although nothing was said, they were aware of a mutual attraction but, as she was young and von Braun's future uncertain, they said their goodbyes, both feeling sure that this would not be final. Von Braun had a six-month contract in America. It was the chance of a lifetime. He was going to the New World where he hoped his ideas would be taken seriously. On 7 September 1945, he and the first members of his team departed secretly under escort for Boston harbour.

The day after Wernher von Braun left Germany, Sergei Korolev arrived in Berlin, charged with the responsibility of building von Braun's V-2 rocket. His private aim was to build a rocket that would make the V-2 obsolete and he immediately began to investigate the work of the Soviet team. One day he went to see Boris Chertok, who had no idea who Korolev was and how significant he was to become. 'I rose as a major should to greet a lieutenant-colonel,' recalled Chertok, who remembers this first meeting distinctly. 'He looked good in his uniform. It fitted him well ... Dark eyes with a jovial sparkle to them were looking at me attentively ... He looked tough and thick-set, like a fighter.' Korolev sank into one of the deep leather armchairs and relaxed as he enquired about the work at the Institute Rabe. He scrutinized a folder Chertok gave him 'irreverently and disapprovingly', making it clear he liked neither the German influence nor the German captions. Then he got right to the point. 'Who is in charge of rocket development and preparations for a launch?'

Chertok explained that Valentin Glushko at Lehesten was working on engine testing. Glushko's group were meticulously analysing components from the engines, including the combustion chambers, and making performance calculations. Korolev, however, had already been there. Only later did Chertok reflect that Korolev betrayed nothing of his complicated past, his research with Glushko in the 1930s, the nightmare of his denouncement by Glushko and others, the endless years of imprisonment. Instead he said simply: 'Yes – I've been to Lehesten already. The people there are doing a wonderful job – my old friends among them.'

It was soon apparent that there was little Chertok could tell Korolev. A few more polite questions followed, the pretty German secretary was noticed, and then Korolev was gone, driving at speed in his trophy car while Chertok was left wondering about the enigmatic stranger who appeared so well informed. At the time, Chertok did not know of Korolev's imprisonment in the Gulag. Later, when he learned of Korolev's past, he reflected on seeing Korolev driving off, taking the corner too sharply as he raced down the lane. He realized what a revelation life in Germany must have been for Korolev. 'He had freedom! How wonderful! He was not yet 40 – but there was so much to do! He had the right to take something from life for himself.'

And Korolev *was* in a hurry. He would master the German technology and go beyond it. He acquired a car – an Opel Olympia – and passes that would take him anywhere. The dark Gulag days were behind him; he could go where he liked, stay where he wanted. All of Europe lay in the haze of summer waiting to be discovered. The sense of freedom was exhilarating.

CHAPTER SEVEN

'Get dressed. You have one hour'

In no time Korolev familiarized himself with everything possible about the V-2. In October, the British invited all the Allies to test launches of the V-2 at Cuxhaven on the German North Sea coast. Only three Russian guests were permitted to attend; Glushko was included in the delegation but not Korolev. According to his biographer, Golovanov, Korolev was determined not to miss this opportunity. He offered to pose as the personal driver to one of the senior representatives of the mission, General Sokolov. 'Why do I need a driver when I'm taking a plane?' enquired the general incredulously. Korolev would not be stopped. 'But when you arrive, what then? What sort of general will you be if you don't have a personal driver? This will undermine your authority in the eyes of our allies.' It was soon agreed. Korolev could attend as driver, complete with the appropriate uniform.

In the event, none of them were able to act out their roles convincingly, probably because the general forgot that Korolev was acting as chauffeur. The British sent a car to meet them and the general immediately took the front seat, which was where Korolev, as aide, should have sat. He, the former enemy of the state, was relaxing imperiously on the back seat by himself. And when the car stopped General Sokolov opened the door without waiting for his 'aide' to rush around and assist him. If the British noticed any lapses in etiquette they did not reveal it, remaining 'moderately hospitable and business-like' throughout. However, they were intrigued at the level of learning among the Soviet ranks. The chauffeur, it appeared, had an air of authority, was well informed and asked more pertinent questions than the general. In fact, he made the English, who were running the show, look a little amateurish. The Americans, who were there in numbers and full of confidence, were disconcerted by the Soviets' first enquiry as to whether all the V-2s secretly shipped to America had arrived safely – and by the way they practically invited themselves to the US secret rocket testing ground at White Sands! (This may be an apocryphal story, however, since other sources, including one witness, have claimed that an irate Korolev was not permitted to enter and had to content himself with viewing the launch at a distance.)

Whatever the truth of the matter, as Korolev set about piecing together the V-2 he soon found himself puzzling over design flaws. He was intrigued as to why separate fuel tanks were contained within the rocket rather than using the skin of the fuel tank as the outer part of the rocket body. This would have reduced the weight of the missile. Much as he admired the German achievement, he felt there were many ways to solve a problem and the Germans had not necessarily found the best solution. Their cumbersome fuel tank design diminished the potential range of the V-2. 'He loved the V-2,' observed Boris Chertok, 'but at the same time it was highly irritating for him.' He was convinced that it was already obsolete technology and it would be possible to go beyond it. Yet he was ordered to study it in minute detail – and, ultimately, it was not his rocket, it was *von Braun's*.

Korolev inspired the young team of Soviet engineers and impressed military leaders with his management skills and vision for the work. He soon began to discuss more ambitious ideas with like-minded colleagues such as Vasily Pavlovich Mishin. He was particularly interested in ideas for a Soviet rocket that could travel much further than the V-2 – by perhaps as much as four hundred miles. More than ten years Korolev's junior, Mishin also came from a background troubled by state intervention. His father had been jailed when he was young, allegedly for not informing on someone who made detrimental remarks about Stalin. Like Korolev, he had been brought up by his grandparents. Mishin had soon excelled at aviation and became an acclaimed pilot at a young age. He and Korolev worked on all the myriad problems that arose in trying to produce and launch the V-2, constantly debating how the rocket could be improved: new fuels, larger rocket engines, advances in launch technology – no detail was too much.

With the shortage of components and materials, it was soon clear that they would only be able to gather materials for the creation of twelve V-2s. It was also becoming apparent to military leaders that a larger and more substantial organization was needed to enable the Soviets to exploit the missile technology fully. In March 1946, the Institute Rabe was subsumed into a much larger new organization headed by General Gaidukov: the Institute Nordhausen. Korolev had made such an impression that he was appointed as Gaidukov's deputy director and leading engineer. He had several major divisions to run, including the Lehesten plant for rocket engine testing under Glushko, who was rattled to have Korolev appointed above him, as well as divisions for launching, guidance, ballistics, technical design and pilot production. Korolev persuaded Vasily Mishin to act formally as his own deputy with the intention that they would fire a German V-2 successfully and then build a Soviet copy of it, the R-1.

To help them there was an ever-expanding workforce of German specialists recruited from the surrounding villages; staffing levels reached seven thousand during 1946. They set to work restoring the missing documentation and blueprints and building rockets for test firing. The relationship with the Germans was inevitably uneasy, the Soviets secretly fearful of acts of sabotage and the Germans 'worried for their future'. Nonetheless, Helmut Gröttrup was rapidly proving indispensable. Apart from running a division setting out the detailed technical specifications of the A-4, he was also asked to design a special train. The Red Army wanted a vehicle that would enable them to launch a V-2 rocket from any far-flung desert in the USSR. With more than twenty specially designed carriages carrying power equipment, a launch table with cranes and platforms, an armoured carriage for the launch control equipment – not to mention accommodation for top officials, including shower rooms and restaurants – it proved such a success that an order was immediately put in for another.

Surrounded by the bleak landscape of post-war Germany, the Gröttrups enjoyed a secure lifestyle. Irmgardt loved the twelve-room villa, the car and the chauffeur, the ponies in the stable, the cows in the meadow beyond the house, the generous extra rations and the maid. Helmut was inclined to rely on Irmgardt's judgement and she had been right: it seemed they had made a good decision. Soon Gröttrup was promoted to director of all missile development under German production. He expanded his team, employing still more German specialists, and started making improvements to the original V-2 rocket. In the hope of increasing the range of the V-2 from two hundred to four hundred miles, he wanted the propellant tanks to be placed under pressure and rehoused the control equipment behind the tanks. As his workforce increased, Gröttrup was soon employing about five hundred men. He enjoyed his work and his team came up with more than a hundred ideas for how to improve the design of the V-2. The Soviets were relieved that there had been no evidence of sabotage, and the Germans found the Soviets generous masters, with no hint of repressive behaviour.

Behind the scenes, however, the future for the Germans was less certain. With the unlikely wartime alliance of communist and capitalist regimes over, Stalin was in an isolated position. He was anxious to ensure that Soviet borders were not encroached upon again and endeavoured to strengthen his position in Eastern Europe, provoking hostility in the West. On 5 March 1946, Winston Churchill, speaking in Fulton, Missouri, made mention for the first time of an 'Iron Curtain' descending on Europe. 'From Stettin in the Baltic, to Trieste in the Adriatic, an iron curtain has descended across the continent,' he warned. 'Behind that line lie all the capitals of ... central and Eastern Europe: Warsaw, Berlin, Prague, Vienna, Budapest, Bucharest, and Sofia ... all subject ... to increasing control from Moscow.' In 1946, the Americans discovered that Soviet spies had passed the secrets of atomic power to Stalin and feared that the Soviet Union would soon be as powerful as America. Relations between the two power blocs deteriorated still further. The Americans were anxious to see Germany reinstated as an economic power to counteract the communist influence in Europe; Stalin was determined to see Germany weakened beyond recovery. Suspicion between the wartime allies soon grew into hostility.

In this atmosphere of mounting distrust, on 29 April 1946 the Allied authorities supervising occupied Germany issued a ban on rocket research for military purposes. This made it difficult to implement

Soviet plans for test launches of the V-2, especially so close to the American zone. In May, NKVD Colonel General Ivan Serov, Deputy Commissioner of Soviet Military Administration in Germany, wrote to Stalin highlighting his concerns. Stalin acted quickly. Secretly work was proceeding on a Soviet atomic weapon, but a mostly landlocked Soviet Union needed a missile that could swiftly deliver an atomic warhead thousands of miles into the enemy heartland. On 13 May, he issued a decree effectively creating a Soviet rocket industry. This was to be based in Podlipki, on the outskirts of Moscow, developing the site of an old artillery factory to form a new scientific research institute, to be called NII-88. Many key staff from the older organization, NII-1, such as Chertok and Mishin, were to be transferred to the new institute, which was to have high importance. Meanwhile Serov, who was an expert in deportations, was beginning to formulate a unique plan which would have considerable implications for the Soviet and German rocket team, who for the time being felt secure in the knowledge that they would never leave Germany.

The urgency with which he needed to implement his plan was brought home to Serov in early July that year when a British plane was seen flying over the Soviet facility where the V-2 rocket was being assembled. This was followed a day or two later by an American plane. Worse still, Serov accidentally came across plans that could rob the Soviets of their greatest prize. He hurriedly wrote to Moscow to inform senior officials at the MVD, the Soviet Ministry of Internal Affairs, which had replaced the NKVD.

TOP SECRET

From Berlin
Moscow MVD of the USSR, General Kruglov

… In recent days, the director of the Institute Rabe, engineer Gröttrup, has shown some passivity in work and has often been staying in bed ill. On 12 July, Gröttrup asked for leave from work due to family circumstances. We then conducted an interview with him and found there was no serious reason for his taking time off work. Gröttrup handed us a letter that he had received from an unknown German person who lives in Berlin. In that letter, Gröttrup was invited to take part in the work of rocket

> technology for the British ... From the above one can conclude on the basis of the letter Gröttrup has decided to take leave of absence and then change sides and go over to the English. Gröttrup is currently under observation ...
>
> COLONEL GENERAL SEROV

The author of the letter was arrested and questioned. Behind the scenes, as Serov worked on his master plan, his agents were compiling lists of those experts in the Soviet zone in Germany who would be 'co-operating' with the Russians in creating the Soviet rocket industry, willingly or otherwise.

• • •

While in Germany, Korolev had successfully re-established his career but the difficulties in his marriage had remained unresolved. Ideally he wanted to be reunited with his wife and daughter yet his relationship with his wife, Ksenia, was more and more defined by distance and silence. He was constantly busy in Germany; she was working long hours as a trauma surgeon at the Botkinskaya Hospital in Moscow. Cold indifference had taken the place of a once loving relationship. He was eager to see Ksenia and obtained clearance for his wife and daughter to stay with him in Germany. 'We will all be home together, and go to the sea for a couple of months,' he reassured her. 'I am awaiting your arrival with anxiety.'

Korolev was enchanted by his eleven-year-old daughter, Natasha. Trying to make up for lost time, he spoiled her and took her and Ksenia on holiday trips; they even explored Peenemünde and saw the gruesome Camp Dora together. Natasha was old enough to recognize that too much now separated her parents. 'What happened should not have happened,' she wrote years later. The eight-year separation had proved too much and 'this long awaited family life did not make them happy'. When school was due to begin in September 1946, Ksenia insisted on leaving and would not hear Korolev's protests. As he drove them to Berlin, 'we were all sad and my mother was crying'.

The following month, Serov put his plan into action. General Gaidukov called a conference for 21 October 1946 to discuss the proposed improvements to the V-2. It was a lively affair. Many new ideas were thrashed out and the meeting went on into the night. Afterwards, the general, who had been friendly throughout, provided a generous banquet for the two hundred German specialists. He insisted that everyone stay and celebrate their plans to test the first series of rockets.

The Germans, accustomed to frugal post-war rations, eyed the bounty displayed on the groaning tables with enthusiasm. The food was ambrosia, with mountains of unseasonal fruit and half a dozen glasses set by each plate. General Gaidukov toasted the clever Germans; in turn, the Germans toasted their excellent hosts. General Gaidukov was irrepressible, rising to toast his guests at every opportunity. Only the previous month Gröttrup had produced a design for a rocket based on von Braun's A-9 and A-10, which could potentially travel more than 1500 miles. The toasts were as abundant as the food, each demanding equally emotional responses from the guests. A great deal of vodka found its way down German throats and still the toasts continued. The vodka flowed copiously, but, strangely, only towards the Germans. If anything, the Russians were a 'little gloomy', according to Chertok, under strict instructions not to touch alcohol.

While the party was in progress and the senior staff occupied, the Red Army, under NKVD guidance, was struggling with logistical problems. Heavy lorries with armed soldiers were rumbling through the quiet villages. The crisp night air echoed with the sounds of banging on doors, shouted commands and cries and protests.

At three in the morning, the telephone by Irmgardt Gröttrup's bed rang. The caller was a woman friend, almost incoherent, screaming 'the Soviets are coming. They are taking us away.' More phone calls followed. Irmgardt tried to ring her husband but he could not be found. She remembers looking 'from every window: all I could see was the Soviet military. The house was surrounded by soldiers with machine guns. Outside were cars and lorries, nose to tail. Someone pressed the doorbell and kept his finger on it. Fists hammered on the door – the noise echoed through the house.' When she answered the door she was told: 'You are being taken to Moscow. You will be there for five years. Get dressed. You have one hour.'

Everyone was caught by surprise. One engineer, Hans Ulrich, was at home when the soldiers came. He told his wife that they were being moved to Moscow. Fearful that their three children would not survive the Russian winters, he and his wife pleaded to stay in Germany. His wife did not know what to do. 'For a moment,' she said, 'I tried to find an excuse for us. I looked at the three armed soldiers and the interpreter, who was trying to convince me that it was nothing scary, we were just moving house. I realized that there was no way out, we had to follow the orders.'

The exodus to Moscow was not confined just to rocket engineers. Up to seven thousand German experts from many branches of industry were transported to the Soviet Union within a couple of weeks. The wholesale transportation of German skill came under the terms of war reparation from which Stalin claimed he was owed twenty billion pounds' worth of goods. 'They may take all their belongings,' Serov is alleged to have said, 'furniture, cats, dogs, wives. If a wife is not available, any obliging female may be substituted.' Irmgardt Gröttrup was unwilling to comply with such Soviet unreasonableness, Chertok noted. She would not leave her comfortable home. Her furniture, however, was being piled on to the waiting lorries as she protested. She refused to endure the horrors of a Russian winter without her cows; her children would starve. Helmut Gröttrup refused to go without his family. Frantic calls were made to high command, who issued a guarantee that a freight car with hay was at the disposal of the Gröttrup cows. There really was no choice.

Throughout the night of 21 October, lorries delivered families to the waiting sixty-carriage train in the nearby town of Klein Bodungen. The journey, which lasted almost three weeks, ended with the train waiting silently overnight in a siding. Irmgardt remembers passengers being unnerved by the delay. There was much speculation. Were they going to the gas chambers or to a reception in the Kremlin? Some freshened up, putting on suitable clothes so as to be presentable for the gas chamber; others dressed for a fine reception. Irmgardt Gröttrup herself was intent on getting rid of a box in which were housed the bugs and lice she had collected on the train.

In fact it was neither the gas chambers nor the Kremlin to which the passengers were taken, but to an excellent meal with real coffee and sugar on trestle tables in a large hall decorated with frescoes depicting the five-year plan for industry. On the platform, the head engineer of

Scientific Research Institute NII-88, Colonel Yuri Pobedonostsev, was standing talking to Korolev. He greeted Irmgardt with a magnanimous gesture and enquired: 'What have you brought for me? A cheerful mood I hope,' whereupon she handed him the box of bugs.

The Gröttrups were treated well and given a six-room house in the leafy Moscow suburbs. Under Dimitri Ustinov, People's Commissar of the Armaments, who was running the newly created rocket sector, Gröttrup was assigned a German rocket 'collective' with the task of assisting in the creation of a production line and successfully firing the V-2. By Soviet standards, Gröttrup was paid well; he also had the usual extras such as a car with a chauffeur, and domestic help and extra food rations for his family. Housing, however, was in short supply in Moscow and many of Gröttrup's team found themselves much worse off, given one room to a family of three, sharing the faded glory with the ghosts of long-vanished aristocrats in the twenty-room mansions of pre-Revolution days. The less skilled Germans eventually found themselves on an island called Gorodomlya, almost two hundred miles from Moscow. It was a bleak place, not much more than a wasteland and the end of the world as far as human comfort was concerned. The housing was poor and services equally minimal since its few inhabitants held a fervent hatred for Germans, having survived particularly bitter fighting during the war.

Gröttrup soon found that he had come to a land where everything was in short supply. Despite the priority that Stalin had given to rocket research, the grandly named Scientific Research Institute NII-88, just a few miles north of Moscow, had a leaking roof and the heating was nonexistent. Invariably, temperatures were well below freezing. It lacked basic requirements such as tables, chairs and workbenches. After four years fighting Germany, life in the Soviet Union was grim. The winters were long. The summers were but brief forays into the sunshine to harvest crops. Finding food was a continual struggle and occupied most of the day. The freezing cold winds from Siberia brought temperatures to 30 degrees C below zero. Equally chilly fingers of austerity touched all life in Moscow. Everywhere dark shapes huddled, people waiting in the snow, forming long queues in the hope that the whispered rumour of a food delivery to an empty shop was correct and that there might be bread or sausage. Irmgardt Gröttrup, with her two cows producing milk, butter and cheese had, it seemed, found paradise.

Gröttrup soon found that the lack of materials and organization made his task of creating V-2s very difficult. There were few tools and essential plans were missing, lost somewhere in the Soviet system. Each day on his way to work, Gröttrup saw the V-2s and test stands brought all the way from Germany still waiting in the railway sidings – exquisite precision engineering, corroding and corrupting, slowly turning into rusting hulks, mournful reminders of life in Moscow. In these conditions, the Germans had to use whatever they had brought with them. 'One of the men even took the kitchen alarm clock to pieces because the clockwork had the very spring he need urgently,' Irmgardt Gröttrup recorded in her diary. 'When his wife asked why there were no springs of this strength in all of the Soviet Union, he replied "they were not in stock and it would take a year for the head of the buying department to get them".'

Apart from helping with the construction of a Soviet copy of the V-2, Gröttrup and his team were also starting to work on designs for the next generation of missiles. Gröttrup was developing a prototype that would go beyond the V-2 called the G-1. Gradually the Germans became aware of the logic behind Soviet working methods. Inexperienced Soviet engineers would work alongside the Germans and when the project they had been working on was completed and the Soviets had absorbed all the information possible, strangely they would then disappear and be replaced by another group of innocent Soviet recruits seeking information. It was almost as though a mastermind was stealing German brains.

When Korolev was permitted to return permanently from Germany early in 1947, he was promoted once again to Chief Designer for long-range ballistic rockets and given his own division to run within NII-88; this was known as SKB-3. It was with a sense of urgency that he began preparations for test launching the V-2 later that year and continued development on a Soviet version of the V-2, the R-1. He was not satisfied with making a straight copy of the V-2 as by this time he was critical of what he thought was an outdated vehicle. Nevertheless, he followed orders using the basic V-2 plan.

There were four main sections: the tail assembly and stabilizing fins, above which were the fuel tanks followed by the control equipment and finally the warhead. Korolev made modifications to the tail assembly to increase its range. Glushko's 25-ton thrust engines gave a longer range than the V-2. The lower fuel tank held liquid oxygen and the one above held ethyl alcohol and both were separated from the outer casing of the rocket and insulated with glass wool. In later rockets, to save space and weight the fuel tanks were incorporated into the shell of the missile. Above the fuel tanks the instrument section was redesigned for greater accuracy. It housed the controls and the gyroscopic instruments which guided the rocket. It stood an impressive 47 feet in height and had a range slightly further than the V-2 – just over two hundred miles.

Plans were soon underway to create detailed designs for the R-1 and even to develop prototypes later in the year. Even more important, Korolev was anxious to get the go-ahead to build a Soviet rocket of twice the range, now formally designated the R-2. He intended this to be a leap forward in design with improved fuels and fuel pumps, increased size and power of the engines and more advanced guidance technology to ensure accuracy. He was aware that he was in competition with Gröttrup's G-1, which had similar design features and might be favoured by the military in place of his own R-2.

As Korolev understood it, there was only room for one man to head the Soviet rocket programme – and that man would not be a German. Korolev's years in the Gulag system had taught him how to survive. The groups of inexperienced men who were sent to work with Gröttrup were absorbing all the information about German technology they could and systematically relaying it to Korolev's team. No details of Soviet plans and progress were ever made known to the Germans. All that was required of them was that they should transpose their knowledge – and Gröttrup had no idea what his fate would be when they had taken all he knew.

• • •

Meanwhile, Wernher von Braun and his group of scientists had been taken to the south-west of the United States, to an army base called Fort Bliss, in Texas. This was situated on the edge of the desert land known as the Tularisa, near the Mexican border. On a clear day he could see the little town of El Paso, high in the mountains to the south. Apart from this, the desert stretched away, a vast, featureless, sandy plain, to the San Andreas Mountains in the west. The flowers of summer had died, and mile after mile of sage and brushwood wearing the subdued colours of late summer carpeted the sandy plain to the Sacramento Mountains in the east. In every direction there was nothing to be seen but the shifting cloud patterns changing the shades of the desert floor. It was only a few months since von Braun and his team had been caught up in the horror of Germany's disintegration. Now, they were dwarfed by vast skies and an unpopulated landscape, the only noise the sound of wind in the brushwood.

The American army had brought them into the country unofficially and hidden them away at Fort Bliss. Home was a nucleus of shed-like buildings containing barracks, a mess hall, supply buildings and a popular clubhouse. The men had no passports or passes into the outside world and, except for a visit once a month to El Paso under guard, movement was restricted. No fences surrounded their new home – the desert came right up to the door – but it was understood that six acres was the limit of their freedom.

The hot days were without direction. There was a modest programme for rocket development in the Ordnance Department of the army but it was underfunded. America had just made a massive contribution to the war in Europe and nobody held the view that there would be another such demand in the near future. The war in Japan had been quickly settled with the atom bomb. The United States was the only country in the world with a weapon of such massive destructive power. Despite the urgency with which US Army Ordnance had wanted to bring von Braun and his men to America, in Washington developing guided weapons was not seen as a priority – not least because they already had bombers. The type of high-powered rocket programme the Germans had developed under the Nazis was not in existence – nor had it even been discussed. And there was little at Fort Bliss to encourage von Braun's group as they viewed the silent desert all around them from their wooden sheds. Refusing to be demoralized, however, they set about improving their lot.

They started to learn English and read books, as well as watching American films which brought curious idiomatic expressions and slang to their intellectual scientific vocabularies. Sport was popular when they were allowed to use the army swimming pool, bowling alley and tennis courts. An event anticipated with great pleasure was visiting El Paso. Once a month a group of four, under guard, would visit the town and enjoy the feeling of being part of the world again. They marvelled at the American idiosyncrasy of wearing ten-gallon hats and tall boots, assuming that this reflected the taste in men's fashion country wide.

Living quarters were in the former army junior officers' barracks and they were furnished with a very un-American lack of generosity. There was little comfort, no floor coverings or bedside tables and chairs. Some of the men discovered a new skill in joinery, making furniture from whatever scrap wood they could find. Confined to their six acres, so many brilliant minds found the restrictions irksome. Often the tension was relieved by lengthy pillow fights in the dormitories. Tempers were frayed as the wait in limbo continued. Being housed at such close quarters led to petty irritations, the personal habits of an individual often taking on huge importance. One of them, observed Huzel, 'had the habit of going to the washroom stark naked, save for a pair of wooden sandals, to the annoyance of his many co-tenants, both aesthetically and acoustically. One day, an unknown vigilante nailed his wooden sandals to the floor with six inch spikes, the protruding points of which they bent over in the ceiling below.'

At White Sands they had nothing more exciting to do than reassemble the rusting V-2s brought over from Germany. No American was remotely interested in von Braun's dream of a satellite orbiting the world as an observation platform to monitor hard-won peace. The Germans were merely guests of the US, expected to grow plump on good living, to advise and help, but, most of all, to be in America rather than the Soviet Union. Von Braun was in despair. He had brought his team to a 'lockup' in a desert with no future for the skills that had been so fêted in Peenemünde. To keep spirits up he kept his team busy on speculative space projects.

The firing of the first V-2 at White Sands before assorted top brass from the Pentagon would, von Braun hoped, be the springboard for an American interest in space, but after an impressive takeoff with thunderous noise and flames, it began an equally impressive disintegration three miles up and had to be aborted, not auguring well for the future. By June 1946, however, they successfully fired a V-2 to a height of sixty-seven miles, complete with equipment for taking measurements of the atmosphere. And in the following months they assisted the Americans in firing some twenty-five V-2s, but there was no money for new development, let alone a space programme.

In the absence of official backing, von Braun did find some local interest. On 16 January 1947, he spoke to the El Paso Rotary Club on the future of rocket development. He dazzled the elderly citizens of the town with his account of rockets 90 feet high, powerful enough to reach orbit around the earth. He described weightlessness in space, the theory of achieving earth orbit and even how to create space platforms that could serve as refuelling posts to reach the moon. 'The first man who puts his foot on the moon or another planet,' declared von Braun, 'will be in much the same position as Columbus when he discovered the New World. With mankind visiting and exploring other planets, the future history of our world is both unlimited and unpredictable.'

For the Rotarians of El Paso all this was a revelation and they showed their appreciation with a hearty round of applause, although a little unsure when it would all begin. The local papers were less appreciative. 'Builders of Nazi Secret Weapons Working for US' fumed the *El Paso Times.* The stunned readers of the *El Paso Herald Post* learned: 'Germans Scientists Plan Re-fuelling Station in Sky en route to the Moon.' The Germans, twiddling their thumbs in the desert, starved of resources, and in America on a short lease, could only wonder quite how their 'refuelling station in the sky' was to be achieved.

• • •

While von Braun had to be content with talking to the Rotarians of El Paso, that same spring, on 14 April 1947, according to his biographer, Golovanov, Sergei Korolev was summoned to meet Joseph Stalin himself to discuss the Soviet missile programme. In the dark and labyrinthine recesses of his mind, Stalin thoroughly understood that fear was a significant weapon. The Americans had shown their superior strength with the atomic bomb. The Soviet Union, too, would soon be in that position; but it would not be enough. Soviet weaponry must be feared. Methods of delivery of an atomic bomb were uppermost in his mind. He wanted an infallible means of delivering such a potent means of destruction. What was needed was a missile of deadly accuracy that could travel several thousand miles with its nuclear warhead and was immune to attack.

Korolev never forgot that meeting with Stalin. Everyone in the room was acutely aware of his power, so much so that Stalin could afford to be restrained in his manner. His voice was quiet and controlled as he cultivated a calmness full of threat, then would suddenly explode. In spite of the fact that this was the man who had authorized the purges which had led to his years of imprisonment, the man to whom he had written in despair from prison protesting his innocence, Korolev admired Stalin. Like millions of Russians, he did not hold Stalin personally accountable for his misfortune, convinced that NKVD henchmen and jealous colleagues were responsible. 'We all took Stalin as a "god" back then,' recalled Korolev's daughter, years later:

> Father was told not to ask questions and to speak as concisely as possible. He was surprised that he could take no papers with him; he had to know everything by heart. Stalin greeted father, but did not give him his hand. He slowly walked around the room smoking his famous pipe. He listened carefully, sometimes interrupting father to ask him questions in a reserved voice – and surprised Korolev with how competent he was ... all the questions were well thought over. Father did not know how much Stalin would agree with what he was saying, but he knew how much Stalin's word meant. Father counted on Stalin's support and he was not mistaken in his hopes.

The room was packed with military officials and senior ministers. Stalin proved to be extremely interested in liquid-fuelled rockets. Korolev could see that the ideas he had been harbouring for years might fit with Stalin's calculating vision. He put his case forward, while Stalin listened at first in silence, then interrupting with intelligent questions. Korolev felt 'terribly agitated', anxious to convince the premier that the V-2 was already obsolete – it was a matter of pride for Korolev to go beyond what von Braun had achieved. He wanted to gain approval for his own Soviet missile, the R-2. This, however, was a step too far for Stalin. He wanted the Soviet rocket team to show that they were masters of the

German technology. First they had to make a perfect Soviet copy of the German A-4.

For Korolev, who had still not been officially pardoned for his 'crimes', the meeting was a personal triumph. Despite the fact that he did not win approval for the development of the R-2, he told Golovanov years later, 'I was incredibly happy to be in the presence of comrade Stalin.'

CHAPTER EIGHT

'Did you understand about the warrant?'

With the Nuremberg trials in progress in Germany to try Nazi leaders for war crimes, von Braun and his men had not quite evaded the stigma of their Nazi past. By now, news of Operation Overcast, set up to exploit the knowledge of German scientists, had been leaked to the press and the operation had been discreetly renamed: 'Project Paperclip'. Under the expanded remit of Project Paperclip, approval was given for a thousand scientists and their families to reside in the United States.

For the American government the timing of Project Paperclip presented particular difficulties. At the time, in 1947, investigations were underway for the Dora–Nordhausen War Crimes Trial later that year. It was to be held at the former Dachau concentration camp where more than a thousand Germans, including senior SS officials, were imprisoned. Lawyers and investigators for the Military Court wanted to know who was responsible for the horror of the concentration camps that had produced von Braun's rockets. Since Hans Kammler had vanished and Albin Sawatzki, the technical director, had died, questions about who had been in charge and how this horror had come about could not be satisfactorily answered. Dornberger could shed no light on the matter and, from his castle in Wales, the charming folly of a Welsh coal 'king', he also successfully argued that he was no more responsible for V-2s landing on London than British and American pilots were for bombing raids on Germany.

There was a wall of silence from the Germans imprisoned at Dachau and sometimes there were flat denials – one even claimed, when confronted with evidence of atrocities, that 'the prisoners beat each other'. Gradually, the finger of suspicion began to point across the Atlantic to German scientists now working for the American government, notably the former general manager of the factory, Dr Georg Rickhey, now working for the American Army Air Force. Facing a potential death sentence, Rickhey was obliged to travel to Germany to be tried. For Germans working in America the fear was: who would be next?

Against this troubling background, unable to shake off his links with his Nazi past, von Braun, as well as the other German scientists, maintained at the army's insistence a low profile. The days at White Sands passed as unmemorably and as monotonously as life in the surrounding desert. Contracts were discreetly renewed. The scientists' families left behind in Germany came out to join them. In March 1947, von Braun married his beautiful eighteen-year-old cousin, Maria von Quistorp. Although marriage had not been mentioned when they had last seen each other in Germany, there had been an unspoken alliance and an understanding. Von Braun wrote to his father, Baron von Braun, asking him to find out how Maria felt. By chance, Maria was planning to visit his father, and the moment she arrived the baron ran out to see her in the drive. 'Wernher wants to marry you,' he announced elatedly. 'How about it? How do you feel? Will you marry him?' Maria smiled: 'I never considered anyone else,' she replied. They were married in Germany and just in case the Soviets attempted to get too close, they were chaperoned by an armed guard until they arrived back at White Sands.

In America, von Braun began married life modestly, on $675 a month, but with the definite feeling that the 'prisoner' status with which they had arrived in 1945 was less marked. He and his team were kept busy assembling and firing V-2s for the US military, but attempts to keep the German presence in Texas discreet failed when in May 1947 a spectacular mistake almost caused an international incident. A V-2 was fired, programmed to ascend north, but a fault in the gyroscope altered its direction south towards Mexico. The ground crew watched in horror. Within seconds the thud of impact was heard. It had landed on or very near the Mexican town of Juarez, crowded and spectacularly adorned for fiesta. A massive crater showed that it had missed the city by only a mile and a dynamite store by mere feet.

The following month, the trail from the war crimes investigation at Dachau led to White Sands. Military officials arrived from Germany wanting to know more about the role of the scientists and especially that of Georg Rickhey. Arthur Rudolph, the director of V-2 production, was also interviewed – in particular over exactly who had been responsible for the regular hangings in the tunnels. But getting information was extremely difficult; time had passed and witnesses were dispersed. Both Rickhey and Rudolph denied all knowledge of atrocities. Rudolph went one better, painting a picture of order and harmony, explaining that 'working conditions in the factory appeared good'. He claimed he had eaten the same food as the prisoners and had never witnessed a death in the tunnels.

Sixty thousand prisoners had worked at the Mittelwerk complex. More than twenty thousand had died. Towards the end the crematoria had consumed bodies four at a time, twenty-four hours a day. Yet the investigators could not find a reliable witness to explain how all this had happened. It was not long, however, before a cable arrived from Germany requesting von Braun's presence at the trial.

• • •

Sergei Korolev's meeting with Stalin had awakened in him a renewed sense of urgency. With the first launch of the A-4/V-2 planned for autumn 1947, he was always in a hurry, running, it seemed, to meet the next deadline. To his staff he appeared to have the inexplicable ability to be everywhere at once; with his restless, untiring energy they began to wonder if he ever went home to sleep. One of his designers remembered his first sight of Korolev: 'We were standing near the gates when they opened and a powerful American trophy car came in and passed us at high speed. Inside was a guy in a brown jacket who looked very intense. "Who is this crazy guy?" I asked. "That is the King", we were told. "He can't drive slower."'

Korolev's working day was long and arduous. Within the sprawling factory of NII-88, his department inhabited semi-derelict old buildings; packing cases were used for drawing boards and endless shortages slowed production. His three hundred dedicated workers were living in squalid conditions, but Korolev was generous with his time and tried to help in practical ways. 'People would come to him with all kinds of requests,' recalled his secretary, Antonina Zlotuckova. 'Every Thursday was reception day and he would begin seeing people in the morning and stayed until he had seen his last appointment.' Many times there were queues of more than thirty people needing urgent help to obtain housing or medicines for a sick member of the family.

Officially Korolev himself had still not been rehabilitated for his so-called crimes against the state, and was not a member of the Communist Party, a fact which gave those who opposed him ammunition for criticism. Valentin Glushko, who had also suffered a lengthy prison sentence, now, like Korolev, held a position as Chief Designer in his own right, presiding over a specialist rocket engine plant, OKB-456, near Moscow. The Lehesten engine testing plant that he had been using in Germany had been transferred virtually in its entirety. Rapidly establishing himself as the foremost Soviet rocket engine designer, he was now at work on improved designs.

Glushko had always felt when he first looked at the V-2 engines that it would be possible to achieve smoother and more efficient combustion. In particular, he studied the injector plate where the oxidant and fuel were injected into the combustion chamber. He knew that if he could improve the mixing, he would reduce the risks of creating pockets of unmixed fuel, which could explode, creating pressure waves which damaged the engine. To prevent this combustion instability he wanted to modify the injector plate. He also knew he could improve the fuel pumps and raise the combustion chamber pressure to increase engine thrust.

Glushko was prepared to collaborate with Korolev in his plans for a truly Soviet rocket, but their relationship, strained by the purges of the 1930s, was not an easy one. Korolev could hardly forget what he referred to as Glushko's 'despicable lie', but Glushko, too, had been told by the NKVD that Korolev had testified against him! Both were aware that they had been forced to name each other in their confessions. Whether Glushko also knew of Korolev's continued affair with his sister-in-law at this stage is not clear. Any lingering ill feelings, however, had to be cast aside as they were seen to cooperate fully to advance Soviet missile technology. The work continued under the watchful eye of the state security system. Ivan Serov, Beria's deputy, was a key member of Stalin's Special Committee for Reactive Technology supervising missile technology. He reported regularly to his master, Beria, on progress, or lack of it.

Korolev was well aware that some of his most secret ambitions could be considered anti-Soviet. Even though he was working for the military, he had not lost sight of his ideas, first inspired by Tsander, for space exploration. He was still in touch with another long-standing colleague from the 1930s, the brilliant engineer Mikhail Klavdiyevich Tikhonravov.

Tikhonravov was interested in the idea of launching a satellite. It had long been understood that in theory, if an object went high enough and fast enough, it should be possible to place it in an orbit around the earth like a second moon. To achieve orbit, Korolev and Tikhonravov knew that, according to Isaac Newton's law of gravity, the gravitational force that would pull the satellite back towards the earth had to be counteracted by a centripetal force, due the satellite's high speed, which tended to fling the satellite away from the earth and out into space. If these two forces were exactly balanced, the satellite would still be falling towards the earth, but its forward speed would ensure that its curving plunge would be equal to the curvature of the earth beneath it and it would therefore maintain a constant altitude in space. With the development of more powerful rockets, this goal now seemed within their grasp. Tikhonravov's team at NII-4 had tentatively begun studies on how to add stages to create a more powerful rocket that could reach orbital velocity – but this work was unaccountably stopped by the authorities. Much as Korolev loved his country, the Soviet Union in 1947 was not a place where dreams came to fruition.

Another thought preoccupied him. Half a world away, von Braun, he knew only too well from Gröttrup, had similar ambitions. America, the richest country in the world, could surely afford to fund a space programme. Von Braun could even now be working on plans for a satellite that would leave Russia unable to catch up. Korolev began to collect American scientific and technical articles hoping to search out clues from the Western press about possible developments in missile and space technology. He wanted to know what von Braun was doing. He needed a translator.

At NII-88, there was only one English translator: Nina Ivanovna Kotenkova. She had never heard of Korolev until she was summoned to his office. The articles he wanted translated were heavy with the difficult hieroglyphics of physics and engineering equations. She did her best, struggling with the confusing terminology. When she brought Korolev the translation, he asked her to read it out. She refused, conscious that she had done a poor job. Much of what she had written made no sense to her. He insisted and she read it out while he listened to the unfathomable explanation of Nina's interpretation on engineering. There were many articles to translate, she protested. She needed an engineer to help her explain the terminology.

An engineer was duly supplied and Nina's work improved. The journal articles came to life. Korolev began sending for her more frequently. It soon became clear that he was more intrigued by the interpreter than the journals. They would sit in his office and try to decipher the Western missile programme. One day, Nina recalled, Korolev took her hand. She withdrew it. Was she free on Sunday, he asked.

On Sunday his chauffeur drove them to a restaurant by the river. It was a day that would change their lives. Suddenly there was more in Korolev's life than a monkish existence in a lonely flat in Podlipki. On returning home, they found with surprise that they both lived in the same block of flats. In 1947, life in Russia was plagued with shortages. The penetrating cold of winter was an unrelenting enemy. Sometimes unaccountably there would be no water or the heating would fail. Most families lived in one room with few possessions and rations were frugal. Against this bleak background, they set up home together. Soon they wanted to get married.

Korolev's first marriage to Ksenia was over. He applied for a divorce but she refused. Ksenia wrote to his mother explaining her feelings: 'I am alone, forgotten and lost, not loved by the man for whom I had lived for the last ten years, who was the reason I could dream, think and live. I hate him for my broken life, for his lies, for his attachment to a younger woman … I still love him so much.'

• • •

'26.8.47 Helmut has gone, suddenly taken away,' Irmgardt Gröttup wrote furiously in her diary. 'Soviet pigheadedness has scored this time.' Managers congregated in the Gröttrups' house until 'half the ministry was there,' she complained and her beloved 'Helle' was whisked away yet again. Nothing was explained. Where was he going and for how long? She and the children watched as the car disappeared in a cloud of dust.

Helmut Gröttrup found himself bundled on to a special train bound for the barren steppes of southern Russia, in Astrakhan. The five-day journey south from Moscow ended in a bleak desert site fifty-five miles from the town of Volgograd, known as Kapustin Yar. This desolate region was to become the first long-range rocket-launching site in the Soviet Union and Gröttrup was to help preparations for the first series of launches of the V-2s in the autumn. The launch site was still in the process of construction. Over the summer, almost eight thousand army engineers were at work, battling against the heat of the day to create basic launch facilities. Nights were icy cold. On days when the rain fell horizontally across the wide featureless plain, the ground became a sea of mud.

After two months of 'endless prodding and negotiations', Irmgardt harried the right people until she was finally given permission to accompany her husband to the launch site. She was the only woman there and her first impressions were distinctly unfavourable. 'There seems to be no end to the vast plain,' she wrote. There was nothing: no roads, no sand or gravel for building, just dirt tracks in a world where 'the camel has not yet been replaced with the motor car'. Nor was there any housing; the village of tents for the army workmen stretched into the distance. Even water had to be transported to the site. Dilapidated test stands and machinery from Mittelwerk stood in the railway sidings looking incongruous, like bizarre pieces of sculpture. An old test stand from Germany quickly welded together by Soviet workmen stood nearby, its 80-foot ladder not quite vertical, at a drunken angle, inviting questions of safety.

Remembering life at Peenemünde, Irmgardt noticed that the Germans were 'nearly demented and I don't blame them – I wouldn't want to be involved with rocket launching with the Soviets':

> Their workmen are incredibly bovine and do exactly what they're ordered to … The other day, with high-ranking army people looking on, one of them fell 60 foot off scaffolding and cracked his skull. The VIPs didn't turn a hair, the workers went on as if nothing had happened and we were left standing underneath eyeing the scaffolding warily. Our engineers in Peenemünde would have been horrified at such makeshift carelessness … Helmut was talking to a couple of Soviet high ups when a loose girder fell within a yard of them, killing the leader of a brigade. I saw the three of them blanche but in a moment or two, they went on discussing aerodynamics and ballistics …

The importance of the first rocket of the V-2 series launched from the Soviet Union, built on Soviet soil, was such that it touched Germans and Soviets alike with an almost unbearable tension. The Minister of Armaments, Dimitri Ustinov, had surrounded himself with scientific experts and was making his presence felt. Serov, as Deputy Minister of State Security, also arrived with his entourage from the secret police, not to mention other leading military and state officials. The first launch was arranged for 18 October 1947 and every day seemed to bring insurmountable problems.

As the planned launch day approached, the Germans discovered a fault with the automatic control system. 'Our men have been out of their minds all day,' wrote Irmgardt. Their endless tests yielded nothing. The gyro platform was sensitive to vibration and out of twenty thousand connections they wondered if it would be possible to discover which was faulty with a day to go before the launch. The tense atmosphere grew to boiling point as the Soviets glowered at the Germans and muttered about sabotage. Korolev, never a man of equable temper, was driven to exasperation.

With the problem finally solved, the day of the launch arrived with bright, clear skies. Irmgardt Gröttrup recorded the momentous event:

> We are off! Zero minus 10 … Zero minus 9 … Zero minus 8 … Zero minus 7 … Zero minus 6 … Zero minus 5 … Suddenly the launching platform collapses sideways and with it the fully loaded rocket. One leg of the platform has given way – a rivet is broken

… Zero hour – stop!

We make a dash for the bunker while the workmen run towards the platform and, with absolutely no sign of fear, winch the whole thing back into position, platform, rocket and all and prop it up with girders. There's Russia for you.

All clear – Zero minus 4 … Zero minus 3 … Zero minus 2 … Zero minus 1 … FIRE.

With a tremendous roar and a burst of fire, the long, silver shape trembled then slowly rose straight and true into the immense blue. There was sudden, lunatic madness, as the Soviets in particular understood what had been achieved. Everyone was hugging someone, regardless of rank. Ustinov embraced Korolev, Korolev embraced Gröttrup and anyone else within range. Hats were thrown into the air. Tears of joy, dancing, manic laughter and screams of jubilation resounded over the primitive test site in the middle of the empty desert. Readings confirmed the rocket had landed within twenty miles of the intended target.

Two days after the initial success, however, the next rocket launch was a failure, veering wildly to the left and heading straight for Saratov, a heavily populated area. 'Everyone looked stricken in horror,' observed Irmgardt. 'The Soviets shot sidelong glances of suspicion at Helmut as he wiped the perspiration off his forehead … The Soviets believe the German "devils" capable of anything.' In the event, although no one was killed, the rocket was over a hundred miles off target. This was followed by yet another failure. Each failure was reported by Serov to his master, Beria. Evenings were devoted to dissecting the carcass of the day's mistakes. Korolev understood that each error, however large or small, was an opportunity to learn. For Serov it was unacceptable. 'Why is there such inaccuracy?' he would demand. 'Whose mistake is this?'

Isolated and under pressure, Korolev missed Nina. He would write to her from his carriage on the stationary train imagining her in the Moscow flat as he looked out at the wind whipping up dust clouds on the wide plain. 'This morning before we left I sent you a telegram. It is Sunday today. How we used to enjoy our free days together. Do you remember our swimming trip? I was thinking about that today and I so wanted to see you just for a moment and to embrace you tight … How much I miss you … How dear and close you are to me … I think about you a lot.' Korolev wanted to marry Nina, but Ksenia still refused a divorce. He planned to try again when he returned to Moscow.

• • •

While Korolev was building his reputation in the Soviet missile programme with each successful firing of von Braun's rocket, in America the illustrious name of von Braun was in danger of becoming tainted as investigations continued into Nazi war crimes. He was asked to give evidence to the Dora–Nordhausen War Crimes Trial which was underway in Dachau. Colonel Toftoy refused permission for any of his scientists at White Sands to appear in the Military Court in Germany. Von Braun, however, made a full deposition statement in Rickhey's defence.

He denied ever working at Mittelwerk, but conceded that he had visited '15–20 times', to discuss technical matters. He did acknowledge that in the beginning working conditions were 'extremely primitive since the tunnels … were not fit to absorb many thousands of workers'. Nonetheless, he claimed that conditions 'were continuously improved during the entire period from the last months of 1943 up to my last visit at the plant'. He claimed that the late Albin Sawatzki had 'exclusive authority' on management issues at Mittelwerk, and was 'personally responsible' for the feasibility of the programmes. Kammler had seized power and taken charge of production in a 'dictatorial manner'. Sawatzki had carried out Kammler's orders 'expertly but with the utmost ruthlessness'.

In making these claims, von Braun hoped to spare his former colleague Georg Rickhey, who, if found guilty, could face the death penalty. However, in his deposition he failed to mention the role of one of his own colleagues at Fort Bliss, Arthur Rudolph, director of V-2 production at Mittelwerk. In 1943, Rudolph had arrived with sixty thousand men from Buchenwald concentration camp to build the mine at Nordhausen into a vast underground factory. Slave labour was used to dig out the tunnels that formed the production line and, when the factory was finished, slave labour produced the V-2 wonder weapon. In his deposition, von Braun also overlooked the role of others in his rocket team at White Sands. This included his own brother, Magnus von Braun, whom he had sent to work at Mittelwerk managing gyroscope production, and Dr Kurt Debus, who had been in charge of test launches of the V-2 at Peenemünde. Debus was identified in US government reports as an 'ardent Nazi' since it had come to light that he had denounced one of his own colleagues to the Gestapo.

At the trial, Georg Rickhey was acquitted, fourteen others received prison sentences and one received the death penalty. As for von Braun and his colleagues, a full public investigation as to whether or not they were guilty of war crimes never took place. It was in the interests of the US army under Project Paperclip to draw a veil over the past. In September 1947, the Office of Military Government submitted its report on von Braun:

> Based on available records subject is not a war criminal. He was an SS officer, but no information is available to suggest he was an ardent Nazi. Subject is regarded as a potential security threat by the Military Governor, Office of Military Government for the US. A complete background investigation could not be obtained because the subject was evacuated from the Soviet zone in Germany.

However, a few months later, on 4 December 1947, a letter from the Director of the Joint Intelligence Directives Agency shows that he was not happy with this conclusion and wanted von Braun and other senior scientists' cases completely cleared. Pointing out that in the above report Wernher von Braun and other scientists had been identified 'as potential security threats' to the US, he requested that their cases 'be reviewed and new reports submitted'.

On 26 February 1948, the US Office of Military Government obligingly resubmitted their investigation. This time they concluded that 'no derogatory information is available on the subject' and it was likely that von Braun was 'a mere opportunist'. If his conduct had been exemplary while in the US for two years, 'it is the opinion of the Military Governor, that he may *not* constitute a security threat to the United States'.

With the investigation complete, all evidence that could expose the extent of the White Sands rocket team's involvement at Mittelwerk was discreetly classified. Cynics would later claim that Project Paperclip was so named because a paperclip was attached to each file which the American authorities sought to whitewash. For forty years, the records that could shed light on von Braun's role at the V-2 factory and his professed innocence of any knowledge of atrocities were locked away in a mountain of dust-gathering archives. For the time being, the matter was well and truly buried – almost.

• • •

Despite some setbacks, the test launching of a series of V-2s in the autumn of 1947 was a major leap forward for the Soviets. Even with the lack of V-2s, their blueprints and the German experts who had originally drawn them up, Korolev's team had used its ingenuity to work out solutions to these problems – and all this in a country devastated by war. Buoyed up by this success, in April 1948 Korolev at last received formal approval to start production on the R-1 for launch testing in the autumn. Development work on the R-2, the Soviet missile that he had first begun discussing with Mishin in Germany, could also begin.

The R-2 represented a considerable advance over von Braun's V-2 and Korolev felt his future was bound up with it. It would be bigger than the V-2, standing 57 feet. Designed to fly twice the distance of the V-2, it would be much lighter and more accurate. A major innovation gave the rocket more space to carry fuel, increasing range but decreasing weight. This was achieved by designing the upper fuel tank, containing ethyl alcohol, as part of the structure of the rocket. The walls of the tank became the walls of the rocket. Achieving this posed considerable problems as this feature changed the external temperature of the rocket, thereby altering pressures in the tank. Another intriguing problem for Korolev was separating the warhead from the main body of the rocket, but keeping the warhead stable. If this could be achieved, another decrease in weight would be a huge benefit. Glushko was able to raise combustion pressure in his new RD-101 engine, achieving 35 tons of thrust and increased power. Finally, the guidance system was much more accurate than that of the R-1. Korolev was confident in the R-2 and keen to test design features but he was constantly thwarted by the impoverished state of Soviet industry. Raw materials were in short supply and supporting industries almost nonexistent.

It was becoming increasingly clear that the Soviets had learned all they usefully could from the Germans and the time had come to dispense with their services. 'It's final: our German working community is to be transferred to the island, lock, stock and barrel,' Irmgardt wrote despairingly in February 1948. Their destination? The comfortless Gorodomlya Island where the less senior German technicians had first been housed. The train journey took several days through marsh and forest and ended with a boat trip across the Seliger Lake. 'We were so curious about the island that we hardly noticed the dreaded barbed wire,' she added. Irmgardt soon found she hated the place. Her husband's status was diminished, his salary reduced, the car and chauffeurs a distant memory. Although Gröttrup could continue with plans to refine the G-1, over the following months it was looking doubtful that this rocket would get the go-ahead for production. The Gröttrups were housed in something little better than a hut and although water did actually flow from the tap – Irmgardt was not reduced to carrying it in buckets from the lake – living standards were basic and fear about the future now they were of no use to the Soviets was all-pervasive.

The Germans were not invited in the autumn of 1948 when the launch testing of the R-1 began. At first the test series did not go well for Korolev. Equipment failure prevented the first two planned launches in September. The next rocket did fly, but veered more than 50 degrees from the planned flight path to land only six miles from where it was launched. This was followed by yet another failed attempt before, finally, on 10 October 1948, there was a successful launch and the R-1 travelled almost 190 miles. This success was followed by others – but there was a recurring problem. The missiles were several miles wide of the target. As ever, Serov was on hand to scrutinize any failure and report back.

When he returned to Moscow, Korolev received an unexpected summons to 'the Big House' – the Kremlin. He was instructed to go to 'Lavrenti Pavlovich, office 13'. According to his Russian biographer Mikhail Rebrov, Beria himself demanded to see Korolev. At the time,

Beria was primarily preoccupied with controlling security issues surrounding the Soviet atomic bomb, which was being developed by the leading Soviet atomic physicist, Igor Kurchatov. Although security in the Soviet missile programme had been delegated to Serov, there is evidence that Beria found time to intervene personally.

In Rebrov's colourful account, as Korolev approached Beria's room he passed his assistants, their normally impassive faces heavy with foreboding. He entered office 13. Beria did not look up. Everything about him conspired to give an image of obstruction. Korolev saw a big man behind a massive desk, his large, balding head bent over papers. When he finally looked up, Korolev noticed 'his eyes were almost completely obscured by the gleaming glass circles of his pince-nez'. It was difficult to gauge a man whose eyes could not be read. Beria threatened Korolev by creating the impression that there were 'doubles' – another team of designers working on exactly the same problem and whose work was much more satisfactory. 'Why were Korolev and his colleagues working so badly?' he demanded angrily. For Korolev it brought back sharply the fear of his early years, the feeling of 'conveyer belt-supplied suspects' whose lives could just be wiped away. He suddenly felt struck by 'his complete helplessness and impotence and the lawlessness of the people in power'.

The threats did not end with the meeting. Korolev began to receive unexpected telephone calls in the middle of the night. At first this was to demand explanations. 'Have you had an explosion again? Why? Whose fault is it?' On one occasion, claims Rebrov, Korolev was terrified:

'Comrade Korolev?' The voice was faint but nonetheless recognizably Beria's.

'Speaking.'

'I've been sent the protocol of the latest tests. Another failure. And again no one is to blame? Some people are soliciting an award for you – but I think you deserve a warrant!'

Beria laughed. Korolev fell silent. He knew further failures were unavoidable during development testing.

'We are doing our work honestly ...' he tried to explain.

'It has to be proved. Did you understand about the warrant?"

The line went dead. Korolev, alone in the darkness, found 'his hands were shaking and his mind was in chaos'. Fear and the experience of repression he had endured for seven long years were never far from his mind. He was very aware 'any dissent with official policy was fraught with annihilation, if not physical then moral. The powers hugged with one hand and maimed with the other.'

Korolev had another reason to be anxious. Despite the pressures on him to prioritize military goals, he had not been able to dismiss his all-consuming interest in space. He had been encouraging his old friend Tikhonravov to present his ideas for launching a satellite at the prestigious Academy of Artillery Sciences in Moscow. They hoped to create an open forum for discussion. In the aftermath of the Second World War, however, the notion of allocating scarce resources for something as apparently inconsequential as a satellite was considered so heretical that at first Tikhonravov had not even been granted permission to speak about it publicly. The very idea of it, says Golovanov, was seen as 'strange if not wild'. After all, the Soviets were in urgent need of a military programme – not a space programme. What use was a satellite? Yet after some campaigning, that autumn, while Korolev's R-1 test series was underway, Tikhonravov was finally permitted to give his presentation.

Tikhonravov had about him the air of a man who had perhaps already sampled the mysteries of another planet. Although his manner was usually mild and retiring, when he talked about space he came alive with a conviction that could not be doubted. He explained that it had long been predicted that if an object could achieve sufficient speed, it should theoretically be possible to create a balance between the outward force generated by its velocity and the earth's gravitational pull. This would enable a rocket to stay in orbit around the earth, held by exactly the same forces that keep the moon in orbit. These inspiring ideas came from the Russian space pioneer Konstantin Tsiolkovsky, the mathematics teacher whose formulae working out the relationship between a rocket's weight, fuel and speed had shown that in principle it would be possible to launch a satellite. Now, at last, these ideas could be put to the test.

Tikhonravov believed that it was possible to create a rocket that might have enough power to place this intriguing prospect within their grasp. There were countless unknowns: they did not know what the upper atmosphere was like, how cold it was, or whether a craft's delicate instrumentation could be destroyed by radiation. However, from temperature and pressure measurements of the lower atmosphere they could estimate the thickness of the atmosphere at forty to fifty miles. To orbit the earth, they needed to get above this to reduce the friction and drag that the atmosphere would have on any vehicle. Tikhonravov had a solution. He outlined what he called his 'packet theory', showing how three or more rockets strapped together could provide enough power to achieve the necessary speed – estimated at 18,000 mph.

Many in the audience dismissed these ideas outright as 'the realm of fantasy' and certainly anti-Soviet. Tikhonravov was accused by his superiors of not focusing on 'real' work and was quietly demoted. The usually unassuming President of the Academy, Anatoli Blagonravov, who had given permission for the presentation to go ahead, was fearful that they were at risk of being 'accused of getting involved in things that they did not need to get involved in'. But Korolev was fascinated. Tikhonravov was setting out ideas that might enable them to make a rocket powerful enough to blast into orbit. The prospect was tantalizing, and Korolev, although aware of the risks, hoped to raise it at the highest level. He would choose his moment carefully.

In July 1949, he thought he had his opportunity when he was summoned to meet Stalin once more. It was an important occasion, less than a month before the testing of the first Soviet atomic bomb. Stalin wanted a full report on the status of both his atomic and missile programmes. Waiting in reception with Ustinov, Korolev found himself terribly nervous at the thought of mentioning his ideas to Stalin. Eventually he was shown in. The atmosphere was tense. Stalin was pacing around the room, dressed in military style: high, stiff collar, the epaulettes of a general, trousers with wide, red stripes tucked into leather boots. Korolev noted that Ustinov 'did not take his eyes off Stalin for a single moment'. Stalin happened to question Ustinov on the best means of transporting rockets, and 'Ustinov didn't just stand up, he almost bore himself into the ceiling'.

One of the senior military officials present was critical of allocating scarce resources to rocket development since they could be as much as three miles off target. What use was such an inaccurate missile to the military? When he had finished, Stalin looked around the table. 'Does anyone else want to say something? Please, Comrade Korolev …'

Korolev stood up 'without taking his gaze off the chestnut eyes of Stalin' and launched into a passionate retaliation, accusing the official of being 'short-sighted, technically backward and lacking a sense of innovation'. He explained that they would soon be able to test launch the R-2 and felt sure the answer to go beyond even this new design was on his drawing board already. His latest design, the R-3, would have a range of 1800 miles and the new engines, in development by Valentin Glushko, would be capable of at least 120 tons of thrust. This rocket would have the ability to strike at military bases in the heart of England. When he had quite finished, Korolev went suddenly 'white as chalk', realizing that he might have gone too far.

There was a pause as Stalin digested the conversation. He seemed to savour the discomfiture of the people in the room, almost prolonging his deliberations. At first he appeared to agree with the military but then, after pacing the room once more, he announced: 'I think that there is a great future in rocket technology. We need rockets as part of our armament. Let our comrade soldiers gain experience in using rockets. Let us ask Comrade Korolev to make the next rocket more *precise* …'

For another long moment he remained silent, aware of his ability to create fear, then he launched into an attack against Churchill and Truman. They were warmongers, he said, who would like to use their atomic weapon on the Soviet Union. It was essential that men like Kurchatov, Ustinov and Korolev speeded up their efforts.

The moment passed. There was no chance for Korolev to raise his vision of a space satellite. The meeting was entirely focused on critical military goals. As he left in silence, Korolev impulsively decided to sound out a senior member of the military, Colonel General Mitrofan Nedelin, head of the Chief Artillery Directorate, whom he hoped would be sympathetic. But Nedelin, too, was dismissive. He warned Korolev against too close an association with Tikhonravov, who was widely seen as 'a dangerous dreamer', leaving Korolev in no doubt that talk of space flight or a satellite was foolhardy. The Soviet Union was full of ears listening for such subversive talk. If Korolev persisted, he was likely to lose more than his job. Bringing the conversation to an abrupt end, he warned that higher generals had 'called for the dismissal of

Blagonravov' at the Academy of Artillery Sciences for providing some support for the satellite notion. 'Your name has also not been left out.' Korolev listened to the warning, but was undeterred. With plans in development not just for the R-2, but a whole new family of Soviet rockets – the R-3 and the R-5 – he would be more careful, but nothing would stand in his way.

Two months later, on 1 September 1949, Korolev's ambition to marry Nina was at last fulfilled. It had been hard for his first wife, Ksenia, to acknowledge the marriage was over and the divorce came at the high cost of estrangement from his daughter, Natasha. 'Everything inside me turned to stone,' Natasha recalled, years later. When her father remarried, her mother made her agree never to see her father's new wife. 'I adored my mother and agreed to the promise,' Natasha says.

Almost immediately after his marriage to Nina, Korolev was off to Kapustin Yar. He had less than a year to prepare for a series of launches on his first Soviet rocket to go beyond the German work – the R-2. Facilities there were still basic. In the autumn and winter, the nights were so bitterly cold that people slept fully clothed, while the wind, moaning and spitting, whipped the sand into a weapon of torture. Communal high spirits and alcohol-induced warmth could be found in the canteen – but not by Korolev. He told Nina, 'I can't have a good time any longer if you are not with me.'

Despite the inhospitable surroundings, he worked with complete dedication and in April 1950 was promoted within NII-88 to run the Special Design Bureau – now renamed OKB-1. By the autumn he was ready to embark on test launches of the R-2. The first launch was a failure but on 26 October 1950 a feeling of triumph raised spirits when the R-2 travelled, as planned, 370 miles from the launch site. Glushko's new powerful RD-101 engines with their 35 tons of thrust combined with the innovations in the design of the missile at last enabled a greater range.

Korolev's R-2 set a new Soviet record and established his position as the leading Soviet missile designer. In just five years, the former 'enemy of the state' had unknowingly won the race for supremacy and stolen a march on his rivals in America. His team had created the foundation of a significant Soviet rocket industry, having developed a rocket of twice the range of the V-2, with many more in the planning stages. And

Korolev secretly hoped that this was just the beginning.

As for von Braun, Korolev could have spared himself any concerns that he was pioneering an American space programme. The nearest von Braun had got to the stars was when he wrote a science fiction book in 1948. Entitled *The Mars Project*, it described the fantastic adventures of a journey to the planet. Seventeen publishers turned it down.

PART THREE

The Race to Space

'Dreams, dreams, without dreams man is a bird without wings. And now I'm very close to the greatest dream of mankind. In every century men were looking at the dark blue sky and dreaming.'

Sergei Pavlovich Korolev, to his wife, Nina, 1957

'Sweeping around the earth in a fixed orbit, like a second moon, this man-made island in the heavens … could be the greatest force for peace ever devised or one of the most terrible weapons of war.'

Wernher von Braun, *Collier's*, March 1952

CHAPTER NINE

'A second moon'

In August 1949, the bright-ringed flash and the mushroom cloud billowing high in the sky over the wilderness of Semipalatinsk in Kazakhstan told of the Soviet success in testing their atomic bomb. American complacency was halted. For so long apparently impregnable, separated from Europe by a vast ocean, Americans now began to fear a Soviet nuclear strike. Cold War divisions between the capitalist West and the communist East hardened as the two superpowers sought to strengthen their positions. Western nations formed a military alliance, NATO, and in due course the Soviet-controlled Eastern Bloc united under the Warsaw Pact. As the rhetoric of the Cold War intensified, fears grew on both sides that the opposing superpower might attempt a pre-emptive strike.

In the West there was a fear that communism was spreading across the globe and would in time undermine the capitalist way of life. China became communist under Mao Tse-Tung and other countries in South-east Asia had communist uprisings. Then, in June 1950, communist North Korean forces launched a devastating attack into US-backed South Korea. Amid fears that Stalin had masterminded the attack, the American-led UN forces beat back the communist troops into North Korea. Soon, Chinese troops joined the North Koreans and, as the crisis escalated, the American General Douglas MacArthur called for the use of atomic weapons against China. In this explosive political climate, as the world apparently edged its way towards World War Three, defence spending rose 350 per cent in America. And at last von Braun was presented with an opportunity. He was ordered to develop a missile which could propel a nuclear warhead two hundred miles. The waiting was over.

The US army had transferred von Braun and his team to live among the community in Huntsville, Alabama, working at the Redstone Arsenal. They began designing a rocket combining V-2 technology with an atomic warhead, which became known as the Redstone. Like the V-2, it was a short-range tactical battlefield rocket that could carry its warhead over two hundred miles, but it could deliver a much heavier payload with greater accuracy. New features were incorporated, many based on ideas originally devised at Peenemünde and similar to those developed by Korolev's team for the R-2. The large fuel tanks became an integral part of the body of the missile, and were not contained separately within it. The nose cone and warhead could separate from the main body of the rocket and the guidance system was greatly improved with more lightweight modern electronics replacing the old valves. However, although this provided von Braun with new research in rocket technology, funding was low and the modest aims set for the Redstone were far removed from his vision of space exploration.

While research was underway, von Braun fell under suspicion again, not for Nazi links this time but communist ones. The US Senator Joseph McCarthy initiated witch-hunts against 'card-carrying members of the Communist Party' in senior government positions. 'McCarthyism' was rife. German atomic scientists working in America had been exposed as Soviet spies. Klaus Fuchs, a physicist, was tried at the Old Bailey in London. His trial caused a sensation as he admitted passing on the secrets of the American nuclear programme to the Soviets, enabling them to catch up fast in atomic weaponry. Also in the spring of 1951, the Soviets had begun to release many of the Germans who had been working for them. Back in the West, some had made contact with their old Peenemünde colleagues. The finger of suspicion began to be pointed at the Germans at Huntsville.

In June 1951, the Director of the FBI received intelligence suggesting that the Soviets were interested in the German scientists in America and feared that 'attempts might be made to develop them as espionage agents'. File notes show that von Braun was quizzed in detail about his

Wernher von Braun's A-4: a test launch at Peenemünde.

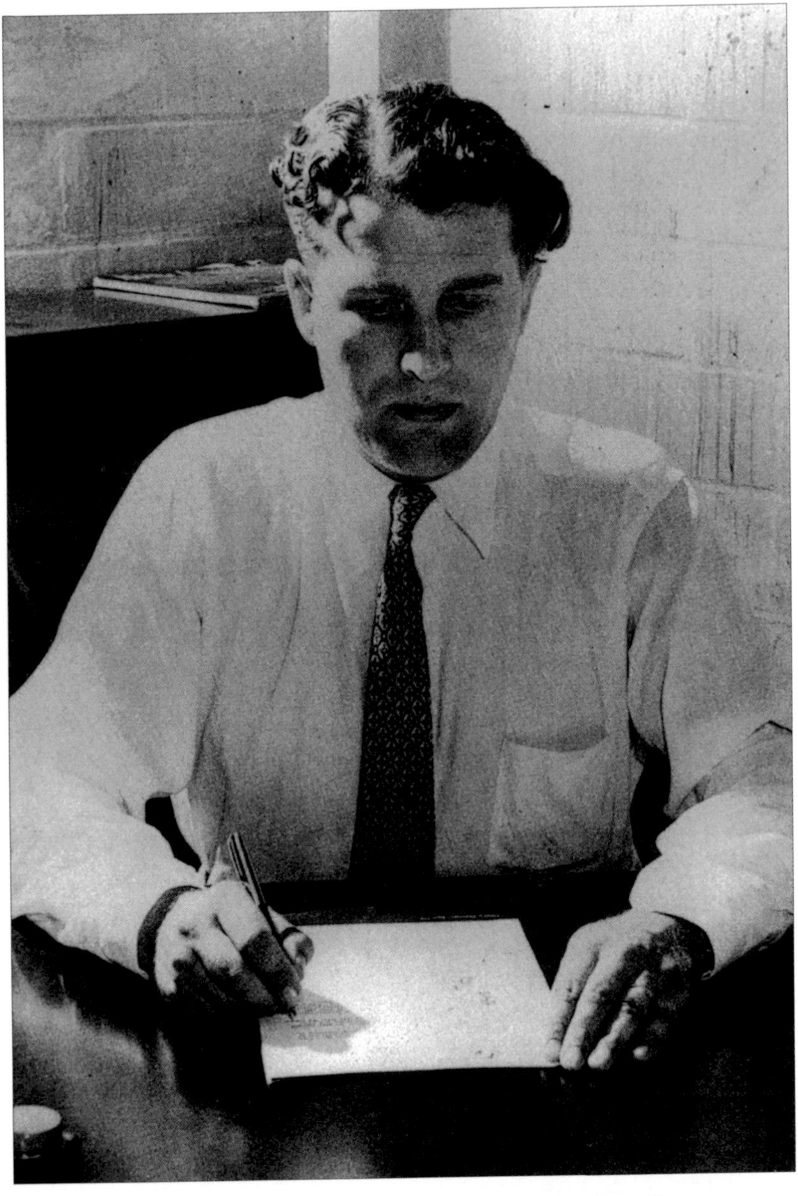

The young Wernher von Braun, chief designer of the V-2 rocket.

ABOVE: Production of the V-2 in the tunnels of Mittelwerk, deep underground in the Harz mountains.

BELOW: Over 60,000 concentration camp prisoners were used as slave labour to build the V-2.

More than 20,000 concentration camp prisoners died building Hitler's V-2.

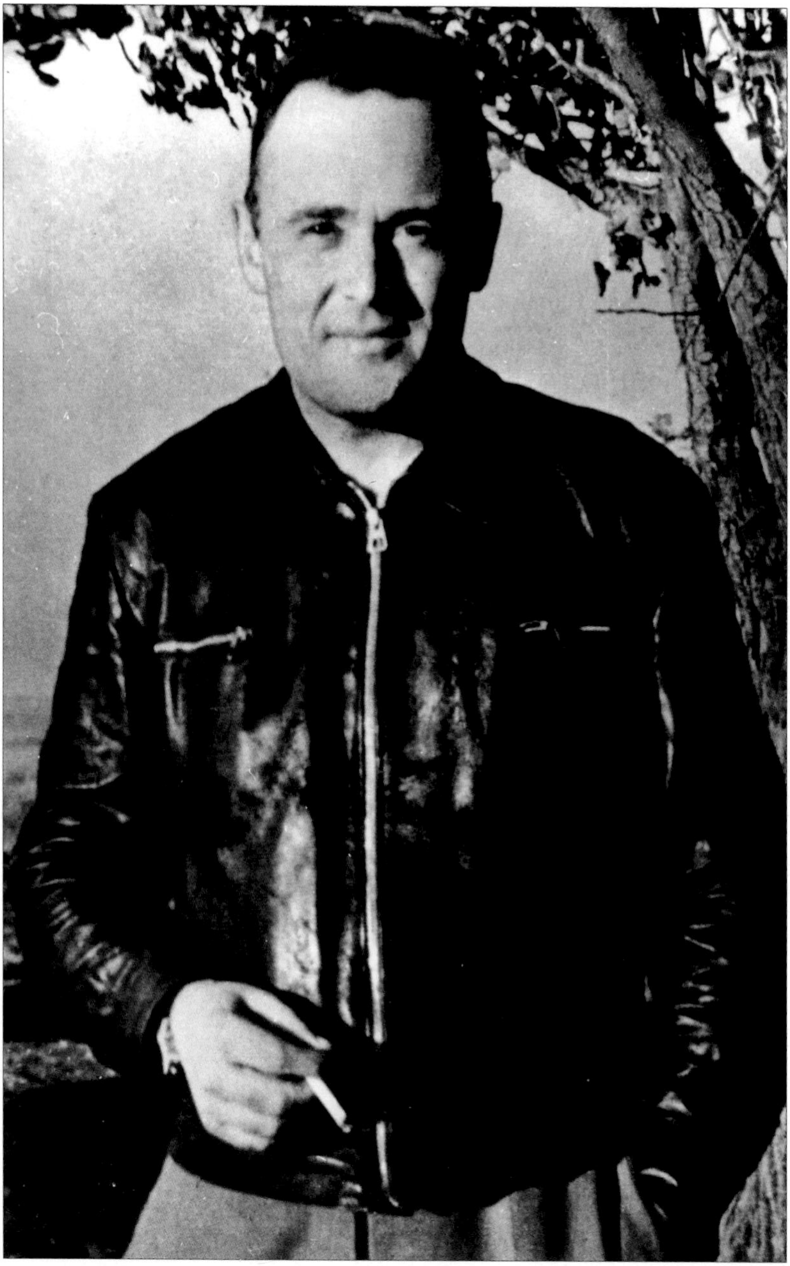

Sergei Pavlovich Korolev survived Stalin's Gulag to become the anonymous Chief Designer of the Soviet Rocket Programme.

Korolev's revolutionary R-7 rocket launched the first satellite and the first man into space.

'Kaputnik!': The Americans' first efforts to launch a satellite on 6 December 1957 ended in humiliation.

Sergei Korolev was the first to launch a live dog into space.

Ham, the space chimp, who could do an astronaut's job 'even with his feet on fire'.

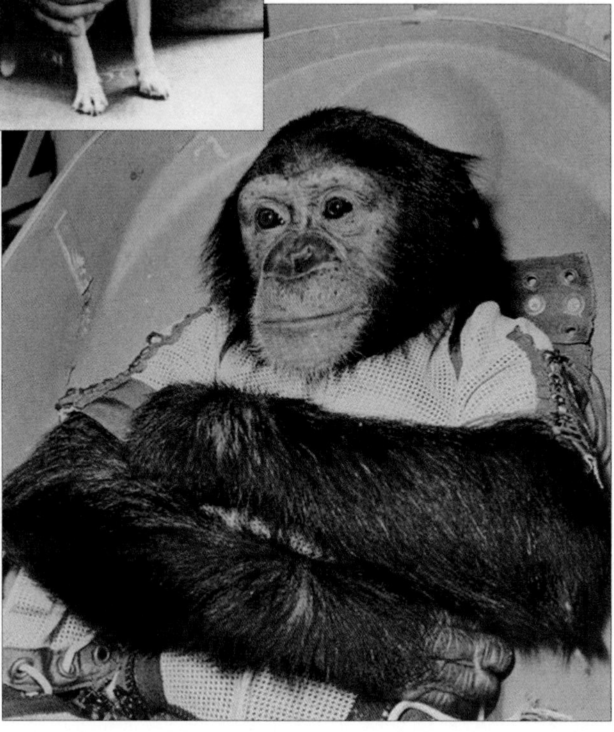

relatives, especially those residing in the Soviet zone such as his father-in-law, Alexander von Quistorp, formerly president of a Berlin bank. Von Quistorp had been invited to the Soviet zone in Berlin after the war, ostensibly to attend a high-level meeting with leading bankers to discuss currency stabilization in Germany. They had all been arrested and deported to an unknown destination in the Soviet Union. After a two-year wait for news, he was finally traced to the Waldheim camp in East Germany. Such was the fear that von Braun might be recruited as a double agent that within a few months he was interrogated again. He was even quizzed as to whether he 'received unsolicited copies of the *Daily Worker*' or whether he 'had been the recipient of any threatening letters'. This time records show that von Braun was satisfactorily 'reoriented' as to his duties and responsibilities.

The continued FBI investigations were dispiriting and the prospect of backing for a space programme seemed more remote than ever. 'We can dream about rockets and the Moon until Hell freezes over,' one close friend confided to von Braun. 'Unless the people understand it, nothing will happen.' Von Braun seized the initiative and decided to set out his ideas in the popular press. Eight articles were published in *Collier's* magazine. The series, 'Man Will Conquer Space Soon', started in the spring of 1952 and showed his fascination with mastering that remote and infinite frontier. The idea of space travel was made real as he described space stations, shuttles, the effects of gravity, astronauts and their special suits. It was hugely successful – more than three million copies were sold – and at last America woke up to the presence of Wernher von Braun.

Eight thousand miles away at Podlipki, near Moscow, Sergei Korolev and Nina pored over the pages of *Collier's* like misers fingering gold. They marvelled at the detailed picture it presented of life in space. 'What you will read here is not science fiction,' pronounced *Collier's*. 'It is now possible to establish an artificial satellite or "space station" in which man can live or work far beyond the earth's atmosphere … the first nation to do this will control the earth. And it is too much to assume that Moscow's military planners have overlooked the military potentialities of such an instrument.' Dr Tikhonravov was mentioned by name, pointing out his claim that rocket ships could be built and the creation of a satellite was now possible. 'A ruthless foe established on a space station could actually subjugate the peoples of the world,' said von Braun. 'Sweeping around the earth in a fixed orbit, like a second moon, this man-made island in the heavens could be used as a platform from which to launch guided missiles. Armed with atomic warheads, radar-controlled projectiles could be aimed at any target on the earth's surface with devastating accuracy.' This 'fast moving star' sweeping around the earth 'could be the greatest force for peace ever devised or one of the most terrible weapons of war'.

Von Braun described the theory behind the launching of a satellite. 'Nature will provide the motive power,' he wrote, 'a neat balance between its speed and the earth's gravitational pull will keep it on course … circling according to the same laws that govern the moon's path around the earth.' He wanted to achieve a stable orbit, where the effects of earth's atmosphere would be negligible: ideally 1075 miles above the earth. At this magic altitude, moving around the earth at more than 15,000 mph – twenty times the speed of sound in air – the satellite would complete a tour around the globe once every two hours and could see any spot on earth within twenty-four hours: the perfect spy. To launch a satellite into this two-hour orbit it would be necessary for a rocket to achieve a staggering speed of 18,000 mph – almost five times as fast as the V-2 – to attain the right balance between speed and gravitational pull. He set out how a three-stage rocket would achieve this, each stage designed to project it forward at ever-greater speed before falling away. And he was clearly entranced by the romance of it. 'The earth, more than a 1000 miles below, will appear as a gigantic glowing globe … The continents will stand out in shades of grey and brown bordering the brilliant blue of the seas. One polar ice cap will show as blinding white, too brilliant to look at with the naked eye … while the whole earth will be framed by the absolute black of space.'

Korolev and Nina followed the series of articles in which von Braun went on to show how a lunar voyage could be achieved. He described the rockets that were needed for the five-day, 239,000-mile trip and the return journey, predicting that man would 'set foot on the ancient dust of the moon' by 1977. The intricacy of the space suit, designed to resist radiation and micrometeorites, with its layers of insulation to provide a living temperature, was convincing. The concept of man living in the empty territory between the earth and the stars as they would in any city was described as though the problems of living in space were already solved. Each month as the articles elaborated on the theme, Korolev became convinced that von Braun was involved in a space programme. It was obvious to him that von Braun shared his dream – but, unlike Korolev, he had found the backing to make it real.

Worried now that von Braun was in the lead after all, he continued discussions with Tikhonravov on how to launch a satellite. Tikhonravov's team began to research the predicted trajectories of rockets and their satellite payloads launched at various speeds using complicated mathematical models that took into account the variation of fuel expenditure, rocket mass and engine thrust over time, as well as rocket attitude or direction. They predicted how to get the satellite into orbit using different missiles, how to re-enter the earth's atmosphere and even how to recover the satellite. However, Korolev knew he could not get any satellite into orbit until the R-3 left the drawing board and they had run into seemingly insolvable problems.

The engine for the R-3 was intended to have 120 tons of thrust – more than triple that used in the R-2 with its 35 tons of thrust. As the leading engine designer, Glushko faced the difficulty that in trying to scale up, the larger the combustion chamber the greater the chance that the fuels would not mix properly. This would cause combustion instability with pockets of unmixed fuel suddenly detonating and creating pressure waves that could bounce around the combustion chamber and even rip it to pieces. And quite apart from these difficulties with the R-3, Korolev faced another, very personal setback in May 1952. To his astonishment, one of his subordinates, Mikhail Kuzmich Yangel, was promoted above him to director of NII-88. Whether this was because of Korolev's passion for non-military projects, or because he still carried the taint of being an ex-prisoner and was not a member of the Communist Party, was never made clear.

However, events on the world stage were to overtake Korolev's struggles at OKB-1. In November 1952, America detonated the world's first hydrogen bomb. It was many times more powerful than the atomic bomb, which was hard to contemplate given the scale of the destruction and annihilation at Hiroshima, which had wiped so much of humanity away without a trace. The new H-bomb, tested on Eniwetok Atoll in the Pacific, erased the island from the face of the earth. In January 1953, the new President, Dwight D. Eisenhower, threatened to use an atomic bomb against China. The Soviet Union began to prepare for nuclear war.

In the freezing cold of March that same year, Stalin died, as if bowing out of a situation that could no longer be controlled. He left a power vacuum almost as terrifying as Stalin himself. The man who had ruthlessly killed an estimated twenty million of his countrymen and sent eighteen million more to the Gulag, had been left to die slowly, unattended. In the early hours of Sunday 1 March, he had said goodbye to his colleagues and drinking friends, Malenkov, Beria, Bulganin and Khrushchev. They had eaten a late meal and watched old American movies, as they did most nights at Stalin's dacha. Stalin slept all the next day and had still not emerged by evening. This was most unusual, but no one dared enquire whether all was well for fear of punishment. At 10 p.m. a brave official found Stalin lying on the floor in a puddle of urine, unable to move or speak coherently. His drinking friends from the previous evening arrived but, for reasons of their own, did not call a doctor and berated the guards for making a fuss when Stalin was clearly 'sleeping'. By the second day, with Stalin still unable to move and soaked in urine, the guards feared for their lives should he die. Eventually doctors were found to attend him, trembling with fear as they did so. Stalin was suspicious of doctors; his own was currently under arrest and subjected to torture.

Stalin had had a stroke. He took several days finally to relinquish life, surrounded by doctors, guards and what was left of his family and his fearful, scheming colleagues. When he was unconscious, Beria screamed abuse at him. But when Stalin opened his eyes, looking as though he might recover, Beria fawned in obeisance, shouting orders at the doctors to save him. Nothing could be done. Stalin understood this and a terrible fear was in his eyes, especially when he looked at Beria. The days passed interminably slowly until finally death arrived, bringing agony. His face was contorted. He could not breathe. His lungs filled with fluid. At the very last moment, he opened his eyes. It was an awful look, his daughter, Svetlana, recalled; either mad or angry and full of terror. Then he fell back, a surprising serenity of expression that he had never had in life stealing over his face.

Stalin's death brought a power struggle within the Kremlin, with Beria, Khrushchev and Malenkov vying to take charge. The outcome would be critical for Korolev. Stalin had been in complete control of matters of defence and he had often backed Korolev; now the future looked more uncertain.

There was another area in Korolev's life that troubled him greatly. Still estranged from his daughter Natasha, he was rebuffed every time he attempted to contact her. With the approach of her eighteenth birthday, he decided to write to her:

> I don't think your behaviour towards me is right, dear Natasha ... My love for you is genuine and strong. I often think about you and so wish that you would see me again and that the distance which has developed between us would cease to be ... I am very far away from you, but on 10 April I will be thinking about you, here in the desert. Don't forget your father who loves you very much, who is always thinking about you and will never forget you ...

She did not reply. Korolev confided to Nina, 'it hurts me so much that she does not want to know me, my own daughter'. Nina herself had been unable to have children. On Natasha's birthday, Mishin overheard Korolev try to speak to her by telephone from the launch site. When Natasha realized it was her father, she put the receiver down 'and he just sat there crying'.

Korolev's persistent alliance with Tikhonravov and his unwelcome ambitions for space flight had been noted by senior party officials. The R-3 was running into such difficulties that Korolev wanted to scrap the project entirely and move swiftly ahead with a bold new missile design incorporating many of Tikhonravov's ideas. When he had made this proposal at a meeting earlier in the year, he had run into serious opposition from one of the most senior figures in the Soviet defence industry, Vyacheslav Malyshev. Whatever the difficulties with the R-3, Malyshev was appalled that Korolev would even contemplate scrapping a programme which was so crucial to Soviet military planning in favour of something more speculative. He turned on Korolev, severely criticizing him in public, accusing him of seeking to pursue the more ambitious project because he wished to get into space. Korolev was angry and unguarded, refusing to comply. Malyshev now made darker threats. 'People are not irreplaceable,' he said. 'Others can be found.' The warning was clear and delivered in such a forceful manner that everyone in the room fell silent.

An unexpected twist, however, was to hand Korolev the lead in the space race. On 12 August 1953 – less than a year after the Americans – the Soviet nuclear scientist Andrei Sakharov succeeded in creating a hydrogen bomb which was exploded in the wastes of the Semipalatinsk test site in Kazakhstan. In just four years the Soviets had created a weapon twenty times more powerful than their first atomic bomb – but Vyacheslav Malyshev was already looking beyond this. He wanted a bomb with even more power: a second-generation thermonuclear bomb. According to Sakharov's memoirs, he received a visit from Malyshev, who had now been promoted to Deputy Chairman of the Council of Ministers and charged with leading the hydrogen bomb and intercontinental ballistic missile (ICBM) programme under the intriguing title of 'Minister of Medium Machine Building' – a deliberately ambiguous label designed to confuse the West. Malyshev urged Sakharov to write a report immediately on the specifications for the next generation of thermonuclear device. In particular, he was required to estimate the weight of such a weapon. Sakharov was hesitant, but evidently did not feel he could refuse such a senior minister. 'I had no one with whom to consult,' he later admitted. 'But nevertheless, wrote a report on the spot and gave it to Malyshev.' Sakharov's spur-of-the-moment estimate – 5 tons – was to have a lasting effect on the Soviet missile programme.

Armed with this information, the minister paid Korolev a visit. In October 1953, Malyshev came alone to Korolev's division, OKB-1 within NII-88. His face, usually clouded with mistrust, was wreathed in unfamiliar smiles. He put himself out to be pleasant, which immediately put Korolev on his guard. The usual group of engineers, including Mishin, were in attendance, wondering what motive lay behind Malyshev's good humour. He enquired about the work in progress and asked what the lifting potential was of the new rocket. He did not like what he heard: a lifting potential of 3 tons. That would not do. *Five* tons was required. Korolev was up in arms at such an impossible demand, but not for long. After all, Malyshev was the one making the request, and he seemed all the more sinister when he was smiling.

Korolev and his team were amazed: at the time, rocket technology could barely haul 1 ton. Three tons would be a breakthrough, let alone five. The military order, however, was soon confirmed. Korolev was to design a new rocket capable of carrying a 5-ton warhead and travelling five thousand miles. He could not be completely sure that such an order could ever be fulfilled. This could not be a pale imitation of the V-2 – it would require rethinking almost every aspect of design. There was one small hope. The revolutionary ideas of Tikhonravov, which carried no weight with the orthodox, might give him the impetus for a completely new approach.

Korolev had already been contemplating a missile incorporating many of Tikhonravov's ideas. Tikhonravov wanted to cluster entire rockets or boosters together – either in stages one on top of the other or side by side – to make one massive rocket. He had put forward several different configurations for how to achieve a missile that could carry a 3-ton warhead some four thousand miles. These plans had to be hastily scaled up into an even bolder scheme.

Korolev's team developed a revolutionary design for a massive rocket made up of a central core with four booster rockets strapped around it – known as the R-7. The five main engines – one on the central core and each of the strap-on boosters – would themselves be built to a bold new plan: they would be *multichambered* to provide even more power. The aim was to create a rocket with a total thrust of 390 tons, nine times more powerful than any other Soviet rocket.

The idea behind the multichambered engines was simple. If they tried to scale a conventional combustion chamber up to the required volume, they would greatly increase the risk of combustion instability – creating damaging pressure waves inside the chamber that might destroy the engine. The aim instead was to combine several combustion chambers together, fuelled by one pump, to create a much more powerful engine. In fact, Korolev envisaged a new engine comprising *four* combustion chambers, each one of which had as much thrust as a V-2 engine. With five main engines in the R-7, it was like combining the thrust of *twenty* V-2 engines.

Steering on the R-7 would not be achieved by simple rudders, as it was on early missiles, but with the steering or vernier engines, which could change the direction of the thrust. The boldness of the vision and the sheer scale of engineering involved were breathtaking. It was a quantum leap forward in design. Nothing like this had ever been created before – and some believed it was impossible.

Korolev was reliant on the leading Soviet engine designer Valentin Glushko, and their relationship, always awkward since the purges, was now tinged with jealousy as Glushko adjusted to Korolev's success. They had already clashed when Glushko had initially refused to design the next generation of liquid-oxygen engines, the RD-105 and RD-106, which had aimed to provide a thrust of 50–60 tons with a much larger combustion chamber. This carried the risk of vibration and combustion instability and Glushko had opposed Korolev, arguing 'that he was violating the bounds of his own professional competence'. Yet even these would not be powerful enough for the new R-7, and Glushko's team were to work on the multichambered engines, the RD-107 and RD-108, which would combine four combustion chambers together to produce a thrust of more than 75 tons.

However, Glushko refused to build the small steering engines that Mishin had put forward, which were to control the direction of the missile once the strap-ons had fallen away. In these vernier engines, nozzles for the exhaust could swivel, so, in theory, it should be possible to control the direction of the thrust. Tsiolkovsky had long predicted this would be possible and Korolev was keen to try it – not least because the old methods of using stabilizing fins and exhaust vanes would create extra drag and reduce the speed of the rocket. Glushko did not want to be responsible for a failure. He insisted that it was risky to introduce so many innovations at once and declared 'it would be impossible to control the rocket with these thrusters'. But Korolev had complete faith in the designs proposed by his staff. It was what the military needed and there was, of course, another benefit of developing such a powerful rocket of which Korolev was well aware: it gave him the chance to realize his aspirations for a satellite. Mishin brought in a group of young engineers from another institute to help solve the problem of these small steering engines.

A meeting was held at the Kremlin at which Korolev outlined his bold new ideas for the R-7 to the military grandees. To his delight, the design excited much interest and there was general approval for him to proceed. For reasons of security, the rocket was officially designated the codename 'Product 8K71'. For Korolev, the most exciting aspect of

'Product 8K71', or the R-7, was that it would have the power to put a satellite into orbit. Fired up by the support for the new rocket, he wrote to the Central Committee of the Communist Party putting forward proposals to launch a satellite, but his request was unaccountably removed from the draft. Undeterred, Korolev continued to lobby party leaders, government officials and heads of various ministries. With the approval of the R-7, despite continued opposition, he hoped he was on the threshold of at last realizing his ambitions for conquering space.

But his hopes for official approval of a space project proved premature. Stalin's death signified the end of an era. The political situation was unstable. The would-be power brokers were stalking each other with treacherous intent. That spring and summer, in the scramble for power after Stalin's death, it looked as though Beria would seize control. With Malenkov's support, he had taken over all security services. Beria, however, had underestimated Khrushchev, who had so often played the part of the buffoon at the late-night drinking sessions at Stalin's dacha. Khrushchev had a plan for Beria's destruction. With the military on his side, he persuaded Malenkov and others to his view.

Unaware of the danger, Beria felt secure and was looking forward to a meeting of the Presidium in June. His wife, Nina, was uneasy and with prophetic vision warned him that he could be in danger. Beria had no time for his wife's nervous anxiety, but at the meeting Khrushchev denounced him and called for his arrest. In the following months in prison, he was to learn about fear, but was still unprepared for the end. On 23 December 1953, he was stripped, handcuffed and shackled to a hook on the wall, his screams stopping only when a rag was shoved into his mouth. Before he received a bullet in the head, he shared something of what he had inflicted on his numberless victims.

• • •

Pentagon officials had chosen Cape Canaveral in Florida as the best site for developing a new missile launch facility. The Cape was a landscape of desolation: flat, arid, sandy soil and scrubland as far as the eye could see. No one would want to live there or build a farmstead among the alligators and particularly voracious mosquitoes. This vast expanse of barren land near the Atlantic Ocean, glittering here and there with water, seemed the ideal place for the US government to launch its rockets into space. In August 1953, von Braun and his team drove down the long, straight roads barely raised above the surrounding salt marshes towards the sea for the first test flight of their Redstone. The new launch towers stood, like intricate modern sculptures, among the hangars, concrete bunkers and sheds.

The Redstone was launched but the flight was a spectacular failure, veering wildly off course. Problems had arisen in the guidance system. The engineers watched its drunken course for five miles, then aborted it, and the rocket crashed ignominiously into the ocean. Although later in the year they did launch the rocket successfully, many of von Braun's team were becoming disenchanted. After the ceasefire between North and South Korea a month earlier, their budget for missile development had been slashed once again.

The lure of America for many had been the possibility of a space programme. Yet despite Colonel Toftoy's almost constant pleading, senior military leaders were only interested in the use of rockets as weapons. At the 1952 Annual Symposium on Space Travel, von Braun tried again to point out the 'tremendous potential' of space technology as a 'deterrent of war'. A space station, he argued, could serve as a bombing platform, allowing first-strike capability with minimal response time. And circling miles above the earth, equipped with powerful telescopic cameras, satellites could take detailed pictures of 'any suspect area on the face of the globe', giving the US the ultimate spy, silent and undetectable. 'Thus they can pull up any Iron Curtain no matter where it is lowered.'

Yet he met serious criticism from others in the field. Milton Rosen, head of rocket development for the Naval Research Laboratory and in charge of creating the Viking rocket to study the upper atmosphere, was blunt about the problems. Von Braun's much-publicized plans for space travel were based on 'a meagre store of scientific information and a large amount of speculation!' He told *Time* magazine he was 'frankly aghast at the difficulties that von Braun lightly brushes aside ... von Braun's 7000-ton shuttle rockets, to say nothing of his space station – would be a reckless leap into the blind future', and this could only lead to a 'gigantic fiasco'. Any sort of space travel was such a leap that it would take 'some basic novelty equal to Faraday's discovery of electromagnetism'. Furthermore, there was a real danger that space projects could only be developed at the expense of the missile programme. Von Braun had always wanted to be the 'Columbus of Space' warned Rosen in *Time* magazine, 'but the feeling of many practical missile men is that von Braun's satellite proposal would fail and leave the US without the new weapons it needs'.

With the lack of interest in space projects, the German rocket team from Peenemünde, which had been held together for so many years by assurances from von Braun that America would one day fund space exploration, began to break up. Von Braun's ability to charm people and keep them in his orbit had held the group together for nine years, but patience was now wearing thin. His budget for missile development had all but dried up. Commercial industry had more to offer with larger salaries and greater chances of promotion. One by one members of the team drifted away. 'We must remain strong,' von Braun would say, urging them to stay. 'Some day Congress has to ease up and grant funds for research.' But even his own brother Magnus had had enough; he left to join the Chrysler Corporation. Von Braun himself secretly tried to resign, but his requests were simply dismissed. 'He wrote me notes threatening to leave,' admitted Major Hamill, who had been in command of the German group. 'I always ignored what he said and what he wrote I threw into the waste paper basket.'

Having all but given up hope of a serious space project, von Braun was astonished in June 1954 to receive a telephone call asking if he wanted to put a satellite into space. Two days later, he found himself ensconced in Washington discussing satellites with Commander George Hoover of the Office of Naval Research and other scientific research teams. Despite the obstacles that Milton Rosen had outlined, Hoover himself was enthusiastic. 'Gentlemen, the time has come to stop talking and start doing,' he began. 'We will now go ahead and build a satellite.' Hoover wanted to coordinate efforts to create an unmanned satellite carrying scientific instruments. It looked as though the empty years of waiting might be over.

America would be taking part in the coming International Geophysical Year, planned for 1957, he explained. This was timed to coincide with a period when the eleven-year sun spot cycle would be most active, with high radiation. Western scientists were planning a range of experiments to study the upper atmosphere. A satellite equipped with scientific instruments would be the centrepiece of a whole array of dazzling events. Was von Braun interested and what would he suggest? With the ease of a man telling a familiar story, von Braun explained how the Redstone rocket could be easily adapted to carry a satellite of about 5–10 pounds. Stages could be added to boost power with a cluster of solid propellant rockets.

The scientific enthusiasts meeting that day were keen to work collaboratively, with the army and navy working together to launch a satellite. Professor James Van Allen from the University of Iowa was nominated to head the scientific research for the satellite, and was to begin on designs at Huntsville. Von Braun asked for $100,000 for development, arguing in his report for the Department of Defense that 'a satellite vehicle circling the Earth would be of enormous value to science, especially to upper atmosphere meteorological and radiological research … Since it is a project that we could realize … it is only logical to assume that other countries could do the same. *It would be a great blow to US prestige if we did not do it first.*'

But President Eisenhower soon introduced an element of competition. The CIA had received reports that the Soviets were also working on a satellite, and suddenly this project assumed a new importance. The President set up a top-level Technologies Capabilities Panel to advise on how to meet a potential Soviet missile threat. Among their recommendations in February 1955, the panel proposed a programme to develop a reconnaissance satellite to spy on the Soviet Union, but they cautioned that America should launch a civilian satellite first, to establish the 'freedom of space' precedent – the right of a satellite to fly over a foreign country. The International Geophysical Year provided perfect cover for this civilian satellite project.

Eisenhower adopted this recommendation and the Department of Defense was requested to set up a committee to investigate how to launch this first satellite, led by Dr Homer Stewart, a physicist at the Institute of California. Homer's Committee on Special Capabilities decided that, instead of the army and navy working in collaboration, all three services – army, navy and air force – should pitch individual satellite proposals. Only one would be chosen.

The Huntsville team was confident that theirs was the best project and their army proposal would be selected over the navy's, which would be led by Milton Rosen, and the air force. However, there had been some criticism that the Germans were too closely involved with the space programme that would represent America. In view of this, the Germans decided that now was the time to appeal for citizenship. Since arriving in the small town of Huntsville in 1951, they had made a big impact on their surroundings. With limited resources, they had bought a field and proceeded to build themselves houses. They formed clubs and musical societies and set up a town orchestra which proved immensely popular. The locals, unsure at first about having so many people in their midst from a country with which they had so recently been at war, had been completely won over.

Once again, there were the inevitable FBI investigations. This time von Braun had to take a lie-detector test. A steady number of his former colleagues had returned to the West. Rumours abounded about Helmut Gröttrup: that he had committed suicide, or divorced Irmgardt – who had married a Soviet official – or was in a concentration camp. In fact, he and Irmgardt and their two children had finally been released from Gorodomlya Island in November 1953 and were now living quietly in the West. The CIA had interrogated him and gained some insights into the early Soviet missile programme, but the FBI wanted to find out what else von Braun might know. In December 1954, he was given a polygraph test and was once more interrogated about any possible communist links or affiliations, contact with foreign governments, or any other violations of security regulations. There was much concern about his failure to have sent a single classified letter through the official Classified Mail Section at Redstone. The screening also included vetting his neighbours and friends to check on his potential communist or Nazi affiliations. He has 'no bad habits ... is a Church going man, and fears God,' said one.

Von Braun's team swore the oath of allegiance en masse in a ceremony in Huntsville High School on 14 April 1955. Huntsville, showing its appreciation, saluted the German scientists and their families with a special dinner. Later in a speech, von Braun thanked the Huntsville community, acknowledging his debt 'for the understanding and encouragement which has greeted us everywhere in the US ... We feel genuine regret that our missile born of idealism had joined in the business of killing. We had designed it to blaze a trail to other planets, not to destroy our own.' Becoming an American citizen, he said, was the happiest day of his life.

That spring von Braun began hosting a series on the Disney Channel. He set out to inform America about the wonders of space and how to get there. With Disney's drawing expertise, the programmes soon captured the imagination of Americans. 'I believe a practical passenger rocket can be built and tested within ten years,' announced von Braun as he proceeded to show them a four-stage orbital rocket ship travelling to a space station, the shape of a wheel 250 feet across, powered by an atomic reactor. The pilots in their special suits, facing the probabilities of weightlessness and travelling to the moon or even Mars, captivated audiences of fifty million. Almost overnight, von Braun became the face of space travel for Americans and space became the next frontier to be explored.

With this success, combined with the complete certainty that his proposal was better than his rivals', von Braun was hopeful that his team would finally win its opportunity to launch a satellite. In early July, each of the armed services had its chance to pitch its vision.

The air force plans were running into trouble. They wanted to launch a large satellite containing an ambitious programme of scientific research using their Atlas intercontinental missile, which had been under development for several years. The Atlas was the most powerful American rocket yet built, with a range of five thousand miles. Although not as bold a design as Korolev's R-7, the Atlas had an elegant engineering solution to the problem of reducing the weight of a missile. The outer skin of the central core was so thin that, with its payload in place, it could only stand up and maintain its rigidity if it was pressurized: it was like a great metal balloon. Although the pressurized shell was a quantum leap in missile technology, in the short term the engineers had so many teething problems that it looked like a non-runner. Von Braun thought it was clever, perhaps too clever.

The navy proposal, led by Milton Rosen, was also facing difficulties. Rosen aimed to put a satellite into orbit using the Vanguard rocket – as yet unbuilt. The first stage would be adapted from the navy's Viking rocket, which had been used in high-atmosphere research to reach an altitude of over 150 miles and had successfully taken the first photographs of the earth from space. Two more stages would be required to provide enough power, and although the third stage had not yet been designed, Rosen claimed it would be ready by 1956. Von Braun thought this was stretching credibility.

With the 'old reliable', as the Redstone was nicknamed, von Braun was convinced he had the most plausible bid. With his customary assurance, he spoke for twice as long as the others, providing convincing details that the Redstone could have a 15-pound satellite carrying scientific instruments ready to launch within the year simply by adapting the existing rocket. The booster and the guidance systems had all been successfully tested. There was nothing to stop them. Furthermore, the matter was urgent: 'We've got mighty little time to lose, for we know that the Soviets are thinking along the same lines. If we do not wish to see the control of space wrested from us, it's time, and high time, we acted,' he argued passionately. Everyone had complete confidence that the army design must be the winner.

A few weeks later, President Eisenhower formally announced that America would launch an earth-circling satellite: a 'second moon'. 'For the first time in history,' declared his press secretary, James Hagerty, 'this will enable scientists throughout the world to make sustained observations … beyond the earth's atmosphere.' At last, it seemed, von Braun's chance had come.

• • •

While Wernher von Braun and his team enjoyed the comfortable living conditions of 1950s America, a very different lifestyle in a hostile environment awaited Sergei Korolev. During the spring of 1955, he began to spend several months of the year at a new launch site being built in a remote desert region on the equator in Kazakhstan, around a hundred miles from the Aral Sea. Kapustin Yar did not have the facilities to cope with the giant R-7 and was too close to Western radar stations in Turkey and Iran. A site far from possible Western listening posts was chosen: this was 'Research Range 5' of the Ministry of Defence. The nearest settlement was Tyura-Tam, a bleak place, little more than a small railway station, a pump house and a few houses and cabins for railway workers. The town of Baikonur lay 220 miles to the north-east

– although this name was used for the launch site in an attempt to confuse Western intelligence of its precise location. Desolation and emptiness stretched for thousands of miles east and north across Siberia, marked only by the railway line crossing from one empty horizon to another. Few had ever chosen to live in this barren waste. It was like a dry, dead planet; a place of parching heat or freezing cold, salt marshes, thorn bushes and an uneasy wind that whipped the sand and dust into gritty, blistering weapons of attack.

In 1955, the desert wasteland around Tyura-Tam was transformed when more than five hundred soldiers and a small army of construction workers poured into the area. The first workers were housed in railway carriages or tents, sleeping in their clothes as a protection against the cold at night. Later would come the luxury of barracks shared by two families. During the day they could labour in temperatures as low as 40 degrees C below zero in the winter and as high as 50 degrees C in the summer. 'Our conditions were very poor,' recorded one worker. 'Fear was all-pervasive of getting things wrong ... Water was brought in cisterns. It smelt of kerosene because the same hose was used to distribute water as for kerosene.' In the early days the work was often manual. Using shovels and hoes the men and their wives built flood protection on the banks of the Syr Darya River. Sickness was prevalent, alcohol their only comfort. A much-appreciated refinement was the arrival of the sauna and laundry train, strangely, it seemed, always in need of repair when it was the women's turn to use it.

The site took shape steadily, the monolithic buildings rising like some incomprehensible new Aztec city. At 'site 1', preparations were underway to build the largest launch complex yet created. More than 30 million cubic feet of soil had to be removed for the exhaust gases that would be generated at liftoff and endless tons of concrete poured. The massive launch platform itself would require a million cubic feet of concrete. At 'site 2' workers were constructing a vast hangar, more than 330 feet long and 160 wide, where the R-7 would be assembled. Engineers tested each structure at a full-scale mock-up of the launch facility that was created at the Leningrad Metallurgical Plant to ensure that all equipment would function correctly. The aim was to build a site that would dominate the future, with a well-equipped laboratory and test site for the larger military rockets. Slowly, huge steel towers rose up with solid concrete bunkers hunched behind them, their basements going down five storeys into the desert floor. A new road and a rail network crisscrossed the site from the railway station, which was always busy.

At first, when Korolev visited the site, he lived in a railway carriage until accommodation could be arranged for him. When he wrote to Nina, he would sometimes sit by the window viewing the characterless landscape, smoking here and there with swirling dust storms, but seeing only the future: the large rockets, the cosmodrome and the journey into space: 'Here in these sands and winds, I think about everything and about you,' he wrote. 'No matter how hard our separations, it has to be in the name of a great work for our Motherland ... Once I had a dream that there will be a moment when a white machine – our dream – will launch from the Earth.'

When Korolev was back with Nina in Moscow, he continued with his ritualistic scanning of information from the West on the American space programme. The Americans, it seemed, were as open about their space projects as the Soviets were secretive and von Braun had become the natural figurehead. Over the years, he had become familiar to Korolev, almost like an acquaintance, but there was no chance of any exchange of views between them, working as they did on opposite sides of the Cold War divide. Von Braun was often in the limelight and was able to speak his mind. Necessity had made Korolev the invisible man. They would never once meet nor speak to discuss the subject that consumed them both. Von Braun would never know the sense of urgency that ruled Korolev's life.

In late July 1955, Korolev learned of Eisenhower's announcement that the US would be launching a satellite, which would orbit the earth during the International Geophysical Year. It was as he had feared: America was in the lead. Eisenhower's 'moon' would shine on the Soviet Union. Now that Stalin and Beria were dead, Korolev could not resist stating his beliefs more openly, warning that an American satellite would soon be trespassing above Soviet soil.

• • •

By August 1955, the Committee on Special Capabilities in America was ready to vote on which of the armed services would win the opportunity to put the first US satellite into space during the coming International Geophysical Year. Publicly the emphasis worldwide was on peaceful scientific study; more than seventy countries were involved. A US satellite in orbit furnished with scientific instruments for measuring radiation in the upper atmosphere could both contribute to scientific study and underline American prestige. The upper echelons of power in the US navy and air force were concerned at the dominant role the US army was seeking to take in space. Records show that some of the members of the committee expressed concerns about launching the first *American* satellite made by *German* engineers – with a modified Nazi vengeance missile at that. Consequently, the navy, under Milton Rosen, with its Vanguard rocket was selected instead.

Von Braun managed a face-saving smile when he informed his team of the committee's decision. 'They stopped us in our tracks with the satellite,' he said. Behind the scenes he pleaded with the committee, warning that the Soviets would be bound to get into space first if they went ahead with the Vanguard proposal. 'They did not realise the tremendous risks they were taking,' he said later. 'The Vanguard project had to start virtually from scratch ... Any such thing as developing a three-stage missile with three brand new and unproven stages on a time schedule of two years was absolutely unheard of.' He begged that he should be allowed to continue his work as a backup. This only served to arouse the suspicions of the officials in the Department of Defense. One of them became so convinced that von Braun secretly planned to launch a satellite ahead of the navy that he invited him to his office. 'Let me make it very clear, Dr von Braun,' he warned. 'You have absolutely no authority to do any work on the satellite!'

Despite the setback, the team at Huntsville was given a consolation prize: designing a larger rocket known as the Jupiter. It would have a much greater range than the V-2 or the Redstone – 1500 miles and travel at speeds of 16,000 mph. Soon they were also at work modifying the Redstone to create the Jupiter C, which was designed to solve the military's problem of delivering warheads to distant targets and re-entering the earth's atmosphere. On re-entry, the nose cone carrying the warhead at great speed met such friction that it would overheat and melt.

Von Braun was aware that tests in this area could conveniently serve a dual purpose. A space ship would also need to re-enter the earth's atmosphere, and information gleaned for the military programme could be useful if ever he got the go-ahead. His team had been exploring the problem of the nose cone overheating for several years. Countless materials had been subjected to the flame from rocket motors and they had found that the best solution was to use an ablative shield: multiple layers of a slow-burning ceramic-based material that melted equally as slowly as each other on re-entry, effectively insulating the nose cone and preventing it from burning up. The design of the nose cone was also critical. A blunt nose cone forms a layer of compressed air ahead of it, which also serves as insulation and, since it has higher drag, speed is reduced.

The Redstone was used as the basis for the new Jupiter C. It had three stages; the first one used liquid oxygen and kerosene in powerful new engines. The two upper stages were boosted with fourteen small, solid-propellant rockets, and, topping it, the nose cone. As the team worked on refining the Jupiter C, it was becoming hard to hide a sense of excitement. Their brief had been centred on solving the problems of nose cone re-entry, but in the process they had also built a rocket capable of putting a satellite into orbit. If on the trial launch the nose cone were connected on top of an extra rocket stage and loaded with a small satellite, the Jupiter C could put the world's first satellite in space. The thrilling possibility of taking the world by surprise had not gone unnoticed by von Braun.

Quite apart from the US navy, von Braun now had a Soviet rival. On 30 January 1956, there was a dramatic U-turn for Korolev. Comments about Eisenhower's 'moon' appeared to have worked like magic on the Soviet bureaucracy. Korolev and Tikhonravov's campaign for a Soviet satellite now found a more receptive audience. Eventually, the Academy of Artillery Sciences was able to give its blessing to the project and, with this support, other senior figures were gradually won round. On 30 January 1956, at a session of the Presidium of the CPSU Central Committee, the Soviet leadership made its U-turn. An ambitious plan was approved to create a large satellite, weighing 2650 pounds – bearing the hopes of numerous experts involved in scientific research institutes and design bureaus. That same day, the CPSU Central Committee and the Soviet Council of Ministers issued a regulation, 'On the Operations to Create an Artificial Earth Satellite'. The Soviet satellite was assigned the name 'Object D'.

To ensure that the decision was given priority, Korolev seized his opportunity to discuss the satellite with Nikita Khrushchev. The next month Krushchev came to OKB-1, surrounded by army chiefs, to see more of the Soviet ballistic missile programme at first hand. Korolev and the director of NII-88 showed him around the institute, saving the most exciting development until last: a full-scale replica of the R-7. Suddenly, Korolev was the impresario with all the wonders of space on display. Silence descended as the Soviet leaders tried to take in the overwhelming size of the rocket. Just as Korolev had hoped, this towering spectacle had a great impact. Khrushchev's own memoirs – although describing another occasion – highlight how amazed he was by Korolev's work:

> I don't want to exaggerate, but I'd say we gawked at what he showed us as if we were a bunch of sheep seeing a new gate for the first time. When he showed us one of his rockets, we thought it looked like nothing but a huge cigar-shaped tube, and we didn't believe it could fly. Korolev took us on a tour of the launching pad and tried to explain to us how the rocket worked. We were like peasants in a market place. We walked around and around the rocket, touching it, tapping it to see if it was sturdy enough.

Korolev's succinct presentation on the R-7 was followed by Glushko, who gave such a detailed, technical account of the engines that Korolev began to feel anxious the Soviet leadership might lose interest before he had had a chance to mention the satellite. He stepped in swiftly to wrap things up and move the party on. He assured them that the biggest wonder was yet to come. Waiting in a dark corner of the factory was a model of the satellite, which, after the grandiose possibilities of the rocket, looked singularly unimpressive. But it had unlimited potential, Korolev explained. Flying around the world nothing could escape its photographic eye. It was the perfect Soviet spy. America would hate it, especially if the Soviets beat them to it and launched first. Khrushchev, delighted by the idea of the perfect, silent Russian spy hated by the Americans, had only one question – would putting a satellite into space detract from the important work of missile development? Korolev assured him that it would not, and Khrushchev, flexing his muscles of absolute power, declared: 'If the main task does not suffer then do it.'

At last, Korolev had the full approval he needed. No one would be likely to dispute Khrushchev's decision. He was able to move his long-standing friend Mikhail Tikhonravov into his design bureau to help him. To beat the Americans into space before the International Geophysical Year, everything would now depend on the performance of his revolutionary new rocket, the R-7.

CHAPTER TEN

'Close to the greatest dream of mankind'

By 20 September 1956, von Braun was ready for the first test launch of the Jupiter C. He and his team knew that if the final, or upper, stage could be live, it would be powerful enough to reach into orbit. Kurt Debus, who was in charge of launch procedure, later admitted that he 'jokingly discussed several times … the prospect of surreptitiously or secretly arming the upper stages'. They could take the world by surprise and shoot the rocket straight into history as they launched the world's first satellite.

However, this possibility had not escaped officials at the Pentagon. Von Braun was summoned to Washington and duly informed that there were to be no 'accidental' satellites. The upper stage must be filled not with fuel – but with sand ballast. Pentagon inspectors attended the launch just to make sure.

When the rocket was fired it did indeed set a new record, reaching a height of 682 miles and travelling 3335 miles with a speed of 1600 mph. It was a perfect launch and a great achievement. The team was jubilant. Von Braun was ready to approach the Pentagon again for permission to launch a satellite. 'We flew higher, faster and farther than any rocket had flown,' he claimed. 'If we had just one more rocket on top we could have placed a satellite in orbit around the earth.' Once again his request was met with a blank refusal.

Von Braun's team was still bathing in the glory of the launch and still hoping for eventual permission to launch the world's first satellite, especially since the navy's Vanguard was behind schedule, when an unforeseen problem arose which killed off all hope. The Secretary of Defense, Charles Wilson, restricted the army to developing weapons with a range of no more than *two hundred miles*. At a stroke, von Braun's team was out of the running.

Ironically, when Korolev studied the Western press he arrived at exactly the opposite conclusion, unaware that von Braun's plans had been put into jeopardy. Korolev came across a news report on von Braun's Jupiter C launch, saw the altitude record and was convinced that he had tried, but failed, to launch a satellite. He had no way of knowing that the upper stage had been deliberately filled with ballast. The only logical conclusion was that the Americans were very close to a successful satellite launch.

No one was more aware of the need for speed than Korolev but he was still a long way from launching the R-7. The design was such a bold leap forward that there were many innovations that still needed to be tested. New materials were being developed to protect the nose cone on re-entry – based on carbon and silicon. Work was also in progress to create a guidance control system for longer-range flight. This was achieved by combining improved inertial guidance systems as used on the A-4, with a radio-controlled system to correct the trajectory, all of which guided the small vernier or steering engines. Arguably the greatest complications arose with Glushko's ambitious, multi-chambered engines.

Initial static tests on the efficiency of the engine showed it fell far short of requirements. Measurements revealed that the acceleration achieved in relation to the fuel consumed was low, which meant the rocket would not have enough energy to reach orbit. Further difficulties arose when these engines were clustered together. In a static test of the first stage, where all four main engines around the central core were fired together, a problem with the oxygen supply line created an explosion. Apart from the damage to the engines, even the test rig was put out of action. Test launches of the R-7 planned for early in 1957 were put back to March.

As for 'Object D' – this, too, was causing major delays. Numerous scientific research institutes were taking part in the satellite experiment and nothing was running smoothly. They were all falling short of their specifications and asking for more time. 'Object D', it seemed, had an overload of scientific data to digest. Three-quarters of its weight was dedicated to equipment for analysing cosmic and ultraviolet rays, atmospheric density and the earth's magnetic field, but it was hard to design the equipment to be light and compact enough to fit into the small space. None of the different faculties could synchronize their efforts. Korolev's temper was rising as delays and excuses became the norm. Basic mistakes brought him to boiling point. He simply could not understand what he considered to be 'sloppy workmanship'.

During the cheerless autumn and winter months at the launch site in Kazakhstan, life was as dismal as extreme cold and long hours could make it. After each shift everyone would meet in the canteen. Korolev arranged for the limited diet to be enhanced by fresh fruit and vegetables to prevent scurvy but the excessive consumption of alcohol was a cause for concern. It was the only easy way to forget the atmosphere of tension, the heavy workload and comfortless conditions. Those desperate enough to forget the world for a few hours could resort to a flask of stolen industrial alcohol. Korolev was inclined to turn a blind eye to it as long as work did not suffer, but his punishment was severe for sloppy work or for anyone displaying a hangover. 'We have no right to make mistakes and no one to turn to for help,' he wrote to Nina, 'yet I myself am expected to answer everything and help others all the time. My mood is good but I cannot hide the fact that it is very hard for me to take failures … I do believe in what we do and in our "lucky star"'.

As time passed, Korolev was becoming increasingly concerned about the delays in the assembly of 'Object D'. The factory at Podlipki outside Moscow offered continual excuses. Eventually Korolev lost patience and told them to send it as it was; it could be assembled at the launch site. The factory, glad to be rid of it, packed it up and sent it to the airport in a lorry. Boris Chertok recalls that the driver, with a hearty breakfast of industrial alcohol inside him, drove like a maniac, unaware of the delicacy of the precision instruments in the back, until he eventually crashed into a tree. This episode not only severely tested the shockproof quality of the instruments but also 'the nerves and health' of the Chief Designer.

Korolev was furious. Anyone who had the remotest connection with 'Object D' felt the lash of his tongue. However, now that 'Object D' was at last under his control, Korolev could see it was never going to work; it was too big and complicated. The many agencies involved with its experimental flight were uncoordinated and unaware that time was running out. Then Tikhonravov came up with the brilliantly simple answer. Abandoning the idea of taking all the complex scientific research equipment, he proposed that they build a much smaller, lighter satellite – just a simple silver sphere containing a radio transmitter, 150 pounds or less. The earth would still have its very first artificial satellite, and, equally importantly, the entire world would be able to hear it. Korolev decided that in the circumstances this was the right decision and immediately asked his engineers to work on a design that would not leave him dependent on many different subcontractors.

As they began work on the designs for the simple satellite, preparations were also underway for the arrival of the R-7 by rail. Originally the intention had been to build the rocket vertically on site. This had not proved possible. It would be assembled horizontally at the vast hangar and then manoeuvred into an upright position at the launch pad. The launch pad itself was being built to a revolutionary new design proposed by Mishin. His ingenious idea solved many problems. The rocket would be effectively suspended above the ground and held in place by four great 'petals' that would swing open on launch to release the rocket. This petal-shaped framework of girders contained all the systems for fuelling the rocket and gantries for providing access to the massive structure. It became known as 'Tyulpan' – tulip – because of the way the 'petals' peeled open simultaneously at liftoff to release the missile.

For Korolev, the demanding workload at the new launch site at Baikonur was eased by the delight of knowing that he had finally achieved reconciliation with his daughter. He had taken a chance and surprised her with a visit at Hotkova, where she was working as a doctor, after four years' training at university. To please her mother she had promised not to see him. At first, she had been restrained and uncertain in her manner, but after a while the sheer thrill of seeing her father could no longer be reined in by arbitrary rules. 'We walked through the forest for more than three hours and talked,' Natasha said. 'We had never been one to one like this before and we could not stop talking. He was interested in my studying and soon my father was close to me just like in my childhood.' He told her what he could about his work and she could see how it obsessed him.

From her practical world of medicine it was difficult to take his stories of satellites and trips to the moon seriously. 'I can see you don't quite believe me,' he said. 'But it will happen.' What she saw very clearly was how tired he looked.

Korolev still had to win approval to launch the simplified satellite. 'Soviet Union *must* be first,' he appealed to the State Commission, explaining the need for swift action. He wanted to supervise the project personally. By February 1957, his request to build a *prosteishy* – simple – Sputnik reached the USSR Council of Ministers. Korolev secured permission to launch the Soviet's first satellite later in the spring, assuming that all went well with the launch of his new rocket. Although it had taken Korolev less than three years to get the massive R-7 from the design phase to this point, he was frustrated to have to delay the test schedule by another couple of months. The first launch was put back to May 1957.

• • •

It was becoming clear to von Braun that Soviet capability was growing. Reports from US bases in Iran showed Soviet missiles reaching speeds of 5000 mph, which almost certainly meant that they could launch a satellite. It looked as though the Soviets had caught up with the Americans. Von Braun confided to a friend that he 'was convinced the man behind the Soviet programme just had to press a button and he's supplied with all he wants'. By contrast, his team faced further budget cuts – which he saw as punishment for his leaks to the press that his team was better placed to launch a satellite than the navy. With the support of Brigadier General John Medaris, the newly appointed Commander of the US Army Ballistic Missile Agency at Redstone, army officials appealed once again for permission to launch a satellite on a Jupiter C. They could be ready by September – early in International Geophysical Year – and hopefully before the Soviets. Yet again, von Braun's team was refused. America, it seemed, was throwing its chance away.

In the Soviet Union, during the final preparations for the first test launch of the R-7, Korolev received some good news. In April he was notified that he was now rehabilitated and that he had been wrongly and unjustly imprisoned. He was also now allocated a modest cabin at the launch site comprising three small rooms, a bedroom, kitchen and office. It was bleak and sparsely equipped. 'The walls were decorated with dark wallpaper,' his daughter recalled. 'There was no stove in the kitchen. No way to warm up food or make tea or coffee in the morning.' Wild pigeons nested in the porch. But Korolev liked simplicity – it left him free to devote himself exclusively to his work. There was nothing to disturb his concentration. Pre-flight checks for the first R-7 launch were due to start on 5 May and he supervised every last detail.

Many of the men, including Korolev, had been away from home for five months, often working for fifteen hours a day. Yet Korolev's demands for high standards were relentless. He was totally consumed by his vision; he pushed himself to the limits of his tough physique to achieve his goal. His temper and harsh words for the lesser vision of a colleague or the betrayal of his dream by shoddy workmanship were well known. 'His diatribes were the stuff of legend,' recalled one engineer, Anatoly Abramov. 'He was a master at it. His eyes would flash, his words would destroy yours, he would threaten to send you home walking between the railway tracks ...' But he was so respected that 'no one took offence'. Behind his back he was called the 'Iron King'.

Once installed at the launch site, suspended by Mishin's 'petals', the R-7 looked like something from science fiction. Towering 108 feet, the central core was surrounded by the four strap-ons, each 62 feet long. On 15 May, liftoff began with the usual fearsome noise, the ground and buildings reverberating to a volcanic eruption of white-hot flames. Everyone sheltered in the bunkers. Mishin's 'petals' opened perfectly and the rocket rose slowly with a faultless liftoff – total thrust nearly 400 tons. It gained height steadily. A full minute passed. Korolev could see it in the periscope moving fast through a luminous blue sky, trailing flame. He dared to hope. Then a sheet of fire tore down its length, encasing it as he watched. It disappeared, leaving a long trail of white cloud, hit the ground and exploded in flames. One of the four strap-ons had torn away from the core less than two minutes after liftoff. The many months of work and preparation had ended so quickly in a defeat that it was hard to bear. Nedelin, standing near him, was clearly shaken. When Khrushchev heard the news, he was disappointed but held fire on criticism. Korolev refused to show his frustration. 'This rocket will fly,' he said.

Over the following weeks, the strain of the long hours and the responsibility for the success of the R-7 was beginning to take a toll on Korolev's health. He had a sore throat and, unable to shake off the infection in spite of the summer heat, was taking penicillin. It was a stifling 55 degrees C. He worked at night to avoid the worst of the heat. Tempers were short. There seemed to be no end to the grinding workload. Nedelin's staff were always looking over his shoulder. Glushko was angry about the destruction of his engines. 'We are working under a great strain both physical and emotional,' Korolev told Nina. 'The temperature is 35 degrees in the shade … everyone feels a bit sick … I am getting ready with some slight misgivings to have a cholera shot … we work till the middle of the night … If only I could be with you. I wouldn't leave you for a minute. I want to hug you and forget all about this stress.'

There were three more failures on 9, 10 and 11 June as launches had to be aborted. This reduced Korolev to a new low. Nedelin was in an ugly mood and endless autopsies apportioned blame. Tempers were heated. The causes of the failure were often due to small, careless mistakes. On one occasion a valve was installed the wrong way round; on another a different valve was stuck in a closed position. There was plenty of criticism for Korolev. Feelings were painfully raw. 'Glushko arrived today, and to everyone's amazement – including mine – began to tell us all that our work was utterly worthless, using the dirtiest language and the crudest phrases,' Korolev confided to Nina. 'This created a terrible impression on everyone … His tirade unfortunately could not be considered criticism but simply mindless malice …' Doubts about the next launch were expressed.

Korolev tried to conceal his own worries and keep up everyone's spirits. He called an informal meeting but despite his efforts to keep things calm it was not long before tempers erupted. Forcing a change of mood on the gloomy scene, Korolev produced a big parcel and declared: 'I've another problem for you here.' He rummaged in the box and pulled out a large chocolate rabbit. It was a present from Nina that was much appreciated. It disappeared very quickly as sweet things were a luxury rarely seen at Tyura-Tam.

Every day the sun rose in a sky drained of colour. The heat was like an oven, the ground too hot to walk on with bare feet. Sandstorms covered everything with dust. Through it all everyone worked for the next launch, scheduled for 11 July. Korolev wrote to Nina on 15 June: 'It probably won't be a success ... the truth is that our goal has never been reached before in all the history of technology.' Preparations for the launch were triple checked. The rocket lifted off perfectly into a clear sky. It looked good, but then it started moving erratically to the left as though it was knocking against an unseen barrier. The failure was spectacular as it exploded like a giant firework leaving a cloudy trail of vaporized fuel.

The launch site was steeped in gloom. It was discovered that someone had made yet another small mistake. A worker had connected a battery wrongly to the control system with the disastrous result that all four strap-on boosters broke away from the core just thirty-three seconds into the flight. No one had the heart or the energy to be angry with the man. At a meeting to discuss results, it was clear that Nedelin had finally had enough. He wasn't interested in who was to blame. He wanted immediate results. 'The army needs just one thing,' he declared, 'a rocket that will work!' He threatened to stop the tests at once, suggesting Korolev go back and rework the missile in OKB-1. This would almost certainly cost him his chance to launch a satellite before the Americans. And Korolev's relationship with Glushko seemed irreparable. The latter agreed with Nedelin that there was no point in carrying on. 'Forty of my wonderfully designed engines have been broken during the tests,' he complained. 'If things keep going the way they are, the production line will collapse.' Korolev confided in Nina: 'Things are very, very bad.'

As the summer wore on and the desert shimmered in the unrelenting heat, an infection of hopelessness spread insidiously through the workforce. There had been five attempted launches, each one a calamity. Korolev refused to admit defeat; there was one more rocket available to launch. He persuaded the military to give him one more chance. The launch was set for 21 August. If it ended in the usual fireworks display, he knew he would have to give up hope of launching a satellite before the Americans. He was well aware of talk behind his back which questioned the design of the R-7, sceptical that such a massive complex of engines would be able to synchronize as one. He wrote to his wife: 'We have such a short period of our life for creating something and every step to something new and unexplored is achieved by a narrow margin with a high price to pay.'

The launch on 21 August began with the usual roar and thunder and terrifying flames. The rocket seemed to balance undecidedly on the fire and fury beneath it, then slowly lifted, straining to take its great weight improbably up into the heavens. Those on the ground looked, waiting in silence, wondering if it would soar without faltering into the distant blue. They watched until it was out of sight. It flew towards its target at Kamchatka: an almost perfect flight. Disbelief turned into euphoria with much celebration. Korolev was the man of the moment, so happy there was no chance of sleep. He talked and partied all night, drunk on dreams of the future.

On the 27 August 1957, the Soviet news agency TASS reported the successful flight. To von Braun, this could only mean one thing: the Soviet Union was on the brink of launching their satellite for the International Geophysical Year and America was just sitting back watching them do it. General John Medaris ordered that two Jupiter C rockets be maintained and kept in a state of readiness. In desperation, von Braun and several members of his team actually hid 'Missile 29' – Jupiter C components and a specially modified upper-stage rocket – in a shed at Cape Canaveral. The team would be ready as and when a change of policy came.

That same month, von Braun's team was the first successfully to recover an object from space. In heat-shield testing with the Jupiter C, for the first time they recovered a nose cone that had returned through the atmosphere having reached three hundred miles into space and travelled more than a thousand miles from where it was launched. 'Everything went exactly as planned,' said John Medaris. 'Each step of the complex operation in the nose cone was precisely on time … It was a little over an hour until the signal was flashed: CONE ON BOARD. APPARENTLY UNDAMAGED. We were jubilant.' It was for von Braun a crucial first step on the way towards launching and returning a craft from space.

Korolev thought the KGB far too cavalier when they informed him that there was no hurry to get a satellite into orbit. He did not believe it and began to call in every day at the workshop where the new satellite was taking shape, urging everyone involved to hurry, insisting that the 'Americans are ahead, we could lose face'. When the simple Sputnik was completed he was invited to inspect it. It was a perfect silver ball, 23 inches in diameter, weighing only 184 pounds and sprouting four elegant radio antennae. But Korolev was not satisfied. The join of the two half-spheres 'looked rough'. He demanded that the satellite itself take pride of place on a special stand and that the working conditions around the satellite be improved. 'Coats, gloves, it's a must! It must be extremely clean everywhere.' For Korolev the metal sphere was the physical embodiment of his dream. He loved it; and everyone else connected with it was expected to revere it. 'Don't forget, this is the Earth's satellite,' he told the engineers, 'and it is the first one, *the very first one*. It must also be beautiful.' As he left, he turned with one more instruction: 'Place a velvet spread under the satellite ...'

In the weeks before the launch, Korolev was uncharacteristically reserved, 'rarely smiling'. There were so many unknowns. He wasn't quite sure where the atmosphere ended and space began: would micrometeorites damage the surface of his satellite, transforming the shining silver ball into a pockmarked mess? And it seemed almost impossible to expect the radio transmitter to send signals when in the ionosphere – the part of the upper atmosphere that has been ionized by solar radiation. Would the seal be affected by the blistering array of effects to which it would be subjected, the extremes of temperature and speed? The satellite was such a small, fragile object; soon it would be on top of the massive R-7 rocket. If it exploded on the launch pad, as the majority of them had, the rocket, the satellite, his hopes, everything would be smashed. The race would be lost. On one occasion as testing continued on a mock-up of the satellite, a problem arose when joining the two halves together. They would not fit and the satellite was not airtight. It was taken apart and a piece of thread was found, wedged under the rubber padding. Korolev was incandescent. 'Do you realize what you are creating, what you have been entrusted with? If you can't do better, then turn in your pass,' he exploded.

The complex calculations needed to work out the trajectory of the rocket and satellite were virtually complete. Each calculation was a balance between the rocket's speed and its weight, which rapidly reduced due to consumption of its fuel, the thrust of the engine, which depended on the pressure in the combustion chamber, and the outside atmosphere, which in turn depended on altitude, and so on. The aim was to devise the correct trajectory that would get the final stage and its satellite payload coasting perpendicular to the earth's surface at the desired altitude and a speed of 18,000 mph. To accomplish this feat, Korolev's engineers used one of the first computers in the Soviet Union, at Moscow State University.

'It was based on lamps,' recalled engineer Georgi Grechko, 'and took up the space of an entire room. The programme was written on a paper roll. Holes were made in the paper according to the programme and I had a hole puncher in one pocket – to punch the necessary holes which were missing in the programme – and glue in the other pocket to get rid of any extra holes in the paper.' Nonetheless, this 'computer' could perform ten thousand calculations per second.

When Korolev could not sleep, he sometimes visited the laboratory in the middle of the night where the engineer in charge of the radio transmitter, Vyacheslav Lappo, was still at work. Huddled over the equipment, they would listen to the *beep, beep* of the satellite as it circled the world. They heard the dull, insistent note echo in the darkness. The sound would be slightly different just before it died, Lappo explained. This dull, repetitive note was magic for Korolev. He confided to Nina: 'Dreams, dreams, without dreams man is a bird without wings. And now I'm very close to the greatest dream of mankind. In every century men were looking at the dark blue sky and dreaming.'

One day while Korolev was working on the launch site, one of the technicians heard a radio report which confirmed the Americans were planning to give a talk entitled 'Satellite over the Planet' on 6 October at the National Academy of Sciences in Washington at a conference held as part of International Geophysical Year. 'We ran to Korolev saying that maybe the Americans were planning to launch their satellite on 4th October and tell everyone about their success on the 5th,' recalled Georgi Grechko. 'Could we be one day too late?'

Korolev turned to the KGB to find out what they knew. But their answer was as mysterious as they were: 'We have no information that the Americans plan to launch on the 4th October but we have no information that the launch is not planned for that dated either.'

This was no time for hesitation. Korolev decided to act. After urgent enquiries, he found that his next R-7 could be rushed to the site and launched by 4 October.

A few days before the launch, Korolev was 'almost flying' between the different operations at the launch site. 'You could see him everywhere; he checked, dispensed advice, got anxious and swore.' He allowed himself no rest. He followed an absolute rule: 'It is better to check ten times than to forget once!' It had become an established tradition for senior members of the team to escort the rocket on foot to the launch site. With great solemnity, early on 3 October, the massive rocket was slowly taken to the pad. Korolev led the procession, his hat in his hand.

An hour before the launch, a final meeting was called. It was a distinguished gathering; the chief designers had arrived from Moscow as well as members of the State Commission, including the head of NII-1, Mstislav Keldysh, the influential mathematician who had provided critical support for Korolev's plans. All eyes were on Korolev as the company waited for his precise and detailed deliberations on the imminent launch. Slowly he rose from his chair and looked around. For once, he could not speak. So many eyes were staring up at him, offering encouragement – even Glushko's. At last he began, sparingly, as though his words were rationed. The rocket and satellite had passed their pre-launch tests, he said simply. 'We will launch today at 22 hours and 28 minutes.'

The countdown began. All those involved were concentrating on their part in the drama. On the brilliantly lit launch pad, the massive rocket was wreathed in misty shafts of liquid oxygen. The familiar thunderous roar of its power was even more terrifying in the darkness. On the observation deck, people watched it lift imperceptibly from its bed of flames. Smoke, dust and vapour swirled at its base as it slowly rose, bathing the launch pad in white brilliance, robbing the world it left behind of colour. It looked like a perfect liftoff evolving into the perfect flight until the missile suddenly seemed to fall. For a second, people cried out that the rocket was falling, but then realized that the rocket, because of its satellite, was taking a different trajectory to the earlier R-7 test launches. The rocket was aiming for the earth's orbit.

When the satellite was due to pass overhead – assuming it had successfully achieved its first orbit – everyone crowded into the radio station situated in a van, some way from the pad, waiting to hear the little voice from space. Slava Lappo sat attentively by the receivers, straining for sound. Finally he heard it – a faint sound growing steadily louder and clearer until it could not be mistaken; an insistent '*beep, beep, beep* …'

Korolev turned round and saw everyone – Tikhonravov, Keldysh, Glushko and all the team – hugging and clapping and cheering. One senior member of the State Commission, Nikolai Pilyugin, was scrambling around, trying in vain to find something. 'Give up!' Korolev yelled, 'and listen to the music!'

'Music! Music!' Pilyugin replied. 'You are a romantic, Sergei. An incurable romantic.'

'This is music no one has heard before,' came the reply.

Success was confirmed. For several glorious hours, everyone at the site gave in to wild celebration. The disappointment of the previous launch failures that had characterized the summer months was unleashed in a frenzy of exhilaration.

Now Korolev could find the words he needed as he rose to thank everyone on an improvised platform. Some of the men had worked without sleep for several days, but the ingrained tiredness that everyone felt seemed to fall away. Korolev was transformed, almost lit from within, unable to contain his elation at the momentous step forward made in the name of Russia.

'Today, something has happened that the best sons of mankind and our wonderful scientist Konstantin Tsiolkovsky had dreamed about,' said Korolev. 'A genius, he had predicted that mankind won't be eternally confined to earth. The sputnik is the first confirmation of that prophecy. The conquest of space has begun – a big Russian thank you to you all!'

The next day, before he left by plane for Moscow, Korolev gave orders that all the scientists who had been at the firing range be given a teapot of alcohol to celebrate. He used the flight to catch up on sleep but was woken by the pilot who had exciting news. He went to the cockpit to hear more details and returned elated: 'Well, comrades, you can't imagine. The whole world is talking about our satellite. It seems we have caused quite a stir.'

CHAPTER ELEVEN

'A Race for Survival'

On the night of 4 October, as Sputnik was blithely crossing the American continent, von Braun happened to be hosting a cocktail party at Huntsville to welcome the new Secretary of Defense, Neil McElroy. The press officer at Huntsville ran in with unwelcome news. The Russians were coming: a Soviet satellite had the run of the American skies.

Silence descended on the party. The guests were stunned, the top brass visibly diminished. Von Braun was furious that the Soviet Union had been allowed to take the lead: 'I was in a position to have done this a year ago with Jupiter C.' He turned on McElroy and made it very clear what he thought of the way decisions on space had been handled. 'Everyone is counting on Vanguard. I'm telling you right now, Vanguard will never make it!' When asked by McElroy if the US could respond, von Braun was unhesitating. 'For God's sakes turn us loose … I can launch a satellite in 60 days,' he replied. General Medaris modified this; he thought it might take ninety.

Having the Secretary of Defense as a captive audience, von Braun urged the need for an immediate US space programme. The balance in the Cold War was changing. The Soviet Union appeared to be visibly dominating the skies, and, by implication, the earth. 'The launch of sputnik has revealed Soviet capability to fly to the moon, or orbit the sun or even go for a manned flight,' he declared. 'It could well be over five years before we could catch up.'

The following day, America awoke to the electrifying news that a Soviet 'moon' was now watching over them. On 5 October 1957, this was the front-page headline in the *New York Times*:

SOVIET FIRE EARTH SATELLITE INTO SPACE;

IT IS CIRCLING THE GLOBE AT 18,000 MPH.

SPHERE TRACKED IN 4 CROSSINGS OVER THE US

Journalists led with an almost irrational howl of horror. If the Soviet Union could put a satellite in space, then it could deliver a nuclear warhead: American cities were in their sights. The US was now in a race for survival, continued the *New York Times*. 'Whoever controls space will control the world,' warned Senator Lyndon Johnson. In the face of mass hysteria from the public, the Pentagon stayed calm while von Braun continued to state the urgency of the situation: 'Failure to be the first in orbit is a national tragedy that has damaged American prestige around the globe.' Elsewhere in the West there was praise for the Soviet triumph. In France, *Le Figaro* declared that gravity was now conquered: 'Myth has become reality.' In England, the *Manchester Guardian* announced: 'the achievement is immense. It demands a psychological adjustment on our part towards Soviet military capabilities ... The Soviets can now build ballistic missiles capable of hitting any chosen target anywhere in the world.' The underlying message behind all the headlines was that communism seemed superior to capitalism. The Soviets had triumphed where the West had failed.

Presiding over the fairy-tale minarets and golden domes of the Kremlin, as Sputnik ringed the earth, the solid, practical Nikita Khrushchev suddenly found himself metamorphosed into a leader of considerable stature. The world appeared to be at his feet, acknowledging the potent, almost mythical power of the Soviet Union, which without fanfare, in a throwaway gesture, had apparently conquered space. The end of the war had seen the Soviet Union broken and impoverished. Now, not much more than a decade later, their totalitarian system had, it seemed, produced a legendary flag to wave at the West.

Enjoying both his success on the world stage and the discomfort of

America, Khrushchev consulted Korolev to find out whether another such landmark event could be produced for the fortieth anniversary of the October Revolution, which would occur on 7 November. To Khrushchev's amazement, Korolev suggested putting a living creature, a dog, into orbit around the earth. For Khrushchev, this was a welcome publicity stunt that would affirm his position as world leader of standing, but for Korolev there was a fundamental issue at stake. Could a living creature survive in space? Animals had been launched into the upper atmosphere but never before experienced a prolonged state of weightlessness in orbit around the earth. There could be unknown problems that might prevent a living creature surviving in space.

With less than a month to create a satellite, all those on holiday were summoned back to work. Korolev held an impromptu meeting with senior members of his team to discuss how they could design the second satellite to support life. With great delight, they savoured the headlines. Mishin began by reading from the *New York Times*: the achievement of Sputnik was the 'greatest deed of Soviet science', which 'could only be achieved by a country with first-rate conditions in a vast area of science and engineering!' In other papers, it was nothing less than 'a turning point in civilisation' in which for the first time 'man is no longer confined to his planet'. The comments of Hermann Oberth – who had joined the Germans in Huntsville – were particularly appreciated. 'Only a country with a large scientific and engineering potential could solve the most intricate problem as launching a satellite,' he said. 'I am impressed by the talent of the Soviet scientists.'

Korolev explained that their next challenge was to modify the satellite to support a living creature for several days in space. There was no time to design a completely new piece of equipment; instead they would adapt the nose cone of the R-7. The dog would be held in a cylindrical container, fitted with life-support systems and monitoring equipment. The life-support system would provide oxygen and absorb carbon dioxide and water vapour from the small cylinder. Drawing on expertise from previous high-altitude testing on animals earlier in the 1950s, a special pressure suit would be adapted for the dog and a system provided to keep it cool; there would also be a tailor-made automatic feeding trough. Scientific equipment was to be placed on board to study radiation and cosmic rays. Although this satellite would weigh 1120 pounds, six times more than the first, room was found for a television system to show pictures of the dog in space.

Several dogs had been in training as part of the high-altitude testing programme and one of these was chosen for this extraordinary journey. Although her name was Laika ('barker'), she had a particularly sweet-tempered nature. One of the air force doctors on site even took her home to play with his children. Korolev visited the institute where her flight was being prepared and with his usual relentless thoroughness insisted on scrutinizing every last detail of the life-support system and the measuring devices.

On 31 October, once sensors had been put in place to monitor Laika's heart rate, blood pressure and respiration, she was strapped into her container. Over the next two days, as the R-7 was prepared, she waited patiently 70 feet up in the nose of the quiescent rocket, oblivious of her sacrificial role; and during the long night prior to takeoff, the tiny scrap of life, of bone and fur and trusting eyes sitting astride the rocket, was mercifully unaware of the brilliant trail she would be blazing.

Finally, on 3 November, Laika was launched from Tyura-Tam. Her heart raced to 260 beats a minute during the launch, but no abnormalities were detected and she made it successfully into orbit. In weightlessness, too, there were no signs of adverse effects. Laika seemed calm and she was breathing normally. Her main difficulty was coping with the heat. Despite the cooling system, however, the temperatures in the cylinder rose steadily and Laika died of overheating after about six hours. For years Soviet propaganda concealed her true fate, claiming that she had survived until the fourth day in orbit.

Pictures of a Soviet mongrel dog travelling in space looking as comfortable as if she were on a car journey were beamed down to an incredulous American public. Khrushchev boasted insufferably and American humiliation was complete. Twice in a matter of weeks the Soviets had beaten the US. Now they had another first: a live creature in space. Von Braun could not be silenced. He gave lengthy interviews to the press explaining the need for America to act quickly. 'Our own work has been supported on a shoestring,' he said angrily, 'while the Soviet Union has emerged more powerful than ever before. The thing that worries me most is their rate of acceleration.'

Medaris received furious calls from Washington, urging him to stop von Braun talking to the press. Worse still, von Braun found himself actually blamed for the American failure and an alleged 'missile gap' – with the Soviets suspected of having more powerful missiles, and many more of them. A Senate Armed Services Preparedness Committee was hastily convened to investigate how the Americans had apparently fallen so far behind the Soviets. Von Braun was principal witness. As chief of missile development at Redstone, von Braun 'was in charge of the whole programme so if there is any responsibility it rests with him,' declared the democratic National Chairman, Paul Butler, to the press. 'Von Braun himself should have told Congress and the President that we were not making satisfactory progress!' Von Braun in turn was vociferous in his criticisms of the Defense Department and the Pentagon, pointing out the repeated frustrations he had experienced and how officials turned a 'deaf ear to pleas' that 'we build a really big rocket engine'.

As if to rub salt into their wounds, on 7 November 1957, President Eisenhower went on television to reassure Americans that there was less to fear from Soviet missile and satellite developments than was thought. In an attempt to highlight American successes in this field, Eisenhower even showed the Jupiter missile nose cone that had been successfully recovered from space – but he failed to credit von Braun or the army team for this work. Von Braun and Medaris watched this televised display in disbelief. It was the ultimate insult. They had been repeatedly denied the opportunity to launch a satellite and were not even credited for the successful work they had done. They both drafted letters of resignation.

The very next day, there was suddenly a change of heart. General Medaris and von Braun were informed by McElroy that approval had come from the Pentagon for the army to launch a satellite, with the proviso that it must contain equipment for scientific experiments in accordance with the spirit of International Geophysical Year. The press had a field day reporting on the major policy switch. 'Army told: Join Race. Get Satellite Up There' blazed the *San Francisco Chronicle*, reporting on the odds in favour of the army. The 'Jupiter C is considerably bigger and substantially more powerful than the Navy's Vanguard device', declared the *Chronicle*. It weighed significantly more, and while Vanguard could 'only manage to send a satellite up 200–300 miles', the Jupiter C could potentially reach more than six hundred. Nonetheless, the navy had the key advantage: they were given the go-ahead to launch at Cape Canaveral first. Von Braun's team would go second.

Preparations for the firing of the Jupiter C went ahead smoothly. It was modified into a four-stage rocket, Juno 1, which had the power to reach an orbital velocity of 18,000 mph. There were a few delays while scientific equipment was installed in the modest 30-pound satellite, 'Explorer 1'. Scientists at the Jet Propulsion Laboratory in Pasadena, California, led by Dr William Pickering, designed the 34-inch satellite to carry sensors that would measure temperature inside and outside the satellite and record micrometeorite impacts. A special cosmic ray counter, devised by the physicist Dr James Van Allen, was installed to measure radiation around the earth. Chemical batteries were fitted to provide power to operate the transmitters.

Meanwhile, in less than a month, on 6 December, Milton Rosen and his navy team were ready to launch their satellite, to much flag-waving and high expectations. The event was to be televised and the country waited for the wonder that would show the Soviets what the Americans could do. The countdown began. The whole world was watching as liftoff started with a fury of flames, the huge 70-foot rocket balancing for an infinitesimal second or two, lifting a drunken couple of feet before collapsing into its own inferno. In a final irony, the modest satellite – not much bigger than a grapefruit – had tumbled clear of the nose section, and, not far from the blaze, its transmitters could be distinctly heard broadcasting – not from space but from the surrounding scrubland: *beep, beep, beep*.

LIGHTING-UP TIME?

The disaster, witnessed not only nationwide but worldwide, was humiliation on a grand scale. The press hummed into life with endless inquests on the fiasco: 'Oh what a flopnik!' There was no shortage of names for disaster: 'Dudnik … Puffnik … Oopsnik … Stallnik … Goofnik … Kaputnik!' Even the director of the Vanguard project conceded it was the worst 'humiliation since Custer's last stand'. This was not good enough for the politicians. 'How long, how long, Oh God,' cried Senator Lyndon Johnson, 'will it take us to catch up with Soviet Union's two satellites?' Khrushchev was thrilled, revelling in the 'Vanguard' shame. Had the Americans thought they would be first in space, he wondered out loud to the listening world.

Kurt Debus and others from von Braun's team had been at Cape Canaveral preparing their own test launches during the navy's Vanguard launch. They had heard the countdown through the loudspeaker and could not stop themselves smiling when it all failed. At last they would get their chance. Von Braun had a three-day launch window at end of January. If the army failed to launch in this time, the navy would get a second chance in early February. On the 29th, von Braun's enormous Jupiter C, carrying the satellite, was sitting on the launch pad, a thin column of steel rising 70 feet out of the flat landscape waiting for the drama to begin. The countdown started. To his great frustration, von Braun had been ordered to Washington, leaving General Medaris and Kurt Debus in control at the Cape. Since it was anticipated that the event would cause a great stir, von Braun had been requested to join senior army officials at the Pentagon's communications room to deal with the press.

Warnings started to come in showing adverse weather conditions. At high altitudes winds were a menacing 170 mph. This would affect the rocket's course, and the flight had to be cancelled. If von Braun's rocket could not be fired within three days, his flight would be deferred and the navy would get their second chance. Next day there were long faces as the weather deteriorated even further. At 40,000 feet the winds were an unrelenting 230 mph. Another cancellation was inevitable. Time was running out and the high winds were forecast to continue. Then on the last day, with just a few hours left for the launch window, General Medaris was faced with a choice. He was informed that one of the younger meteorologists had predicted that the high winds might possibly drop for a few hours towards the end of the day. The long wait had been gruelling. Medaris made his choice.

At 8.30 p.m. on 31 January 1958, countdown started once again. At 9.45 p.m., with barely two hours left of their launch window, a potentially serious problem was found – a possible fuel leak directly beneath the rocket. Someone would have to go out to the pad and check underneath the fully fuelled 'live' rocket. The call went out for a volunteer 'without any dependants'. Quite apart from the dangers of an explosion, there were uncertainties about breathing in the fuel, dimethyl hydrazine. One young engineer did not hesitate. He ran out towards the rocket, a solitary figure in the darkness, and disappeared among the clouds of venting gases. When he finally reappeared, he was able to confirm there was nothing wrong – it was just a spillage. The countdown continued. Then, with just an hour and half to go, a question mark hung over a jet vane. Other data did not confirm a fault and Debus continued the countdown. At 10.45, with just over an hour remaining, it was time to press the switch for ignition.

Amid the noise and hubbub surrounding him in the communications room in Washington, von Braun became increasingly anxious. An hour and a half had passed since the launch and there was no confirmation that the satellite was in orbit from the Pasadena tracking station, which would be the first to come into range if Explorer was in orbit. They should have heard by now. It was beginning to look as though Khrushchev would be dancing with joy once again and saying something painfully witty at America's expense. Minutes passed. Explorer was overdue. They rang the tracking station in Pasadena – still no signal. Eight minutes had elapsed since the signal had been due. Failure looked to be on the cards.

Suddenly all the telephones in the room started ringing at once. Four tracking stations had a signal loud and clear. Explorer was in a higher orbit due to a slight excess of speed, which accounted for the delay. The sound of Explorer's signal was the sound of success; its unmelodious messages sweet music to everyone concerned.

At last the US was in space and America was not shy to celebrate it. Huntsville erupted into spontaneous jubilation; their man, their team, had won. The whole town turned out and went wild, behaving like children at a party. The public appetite for the coming space age was voracious. Every small-town newspaper carried the story. *Time* magazine featured von Braun on its cover. Americans could hold their heads high again. They had even made a scientific discovery that the Soviets had failed to announce: James Van Allen's Geiger counters had revealed that a belt of radiation enveloped the earth. Honours were showered like confetti on the team. Von Braun was invited to the White House where he hoped to talk to the President about the US space programme.

Dressing at his hotel, he could not find his white tie – obligatory for dinner at the White House. He would be meeting the President without a tie! When informed of the problem, Press Secretary Jim Hagerty told von Braun not to worry: a tie would be found. Later, in the exquisitely pale and resplendent reception room, the President entered, striking an offbeat note, wearing the only black tie in the room and muttering that he couldn't find his white one …

While the glamorous image of von Braun shone from the front pages of the Western press, Korolev remained completely anonymous, referred to only as the mysterious Chief Designer. In America, von Braun became *the* space expert, the details of his apparently glittering life familiar to many. For Korolev, the triumph of the Sputniks altered nothing. His name was still unknown to the Soviet people, his success never publicly acknowledged. He was never quoted in the papers, nor did his photograph ever appear. Such was the Soviet fear that there might be assassination attempts from the West, they could not afford to put their top men at risk; better if they did not appear to exist.

Korolev was permitted to publish the odd article under a pseudonym – Professor K. Sergeev – but at no stage did he reveal his true role as the leader of the Soviet successes. After the second Sputnik launch, Korolev was close to exhaustion. His heart was giving him trouble. He spent almost a month in a sanatorium where 'arrhythmia and overfatigue' were diagnosed.

While resting near the town of Kislovodsk, he decided to visit the grave of his old friend and mentor Fridrikh Tsander, who had been such an inspiration to him. Tsander had died of typhus at Kislovodsk and been buried in the city cemetery. Korolev fretted when he could not find his friend's grave. He insisted on an extensive search. Experts were even summoned from Moscow. Eventually Tsander's grave was identified, overgrown and neglected, as though the man had never lived to inspire with his ideas. Korolev was troubled by the knowledge that Tsander, who had done so much for Soviet science, was simply not acknowledged, as though, just like himself, his very existence had been wiped out. It hurt him to see the indifference that death had brought to a man he remembered so vividly. He therefore arranged for a petition for 60,000 roubles to pay for a tombstone for Tsander's grave.

PART FOUR

The Race to Orbit

'It seems that nature jealously keeps its secrets and even here, where the mind of human beings is able to open it up – every step to something new and unexplored is achieved by a narrow margin with a high price to pay ... We have such a short period of our life for creating something.'

SERGEI PAVLOVICH KOROLEV to his wife, NINA,
28 September 1960

'If we do not match the ambitious Communist programme to visit the Moon and the planets ... we may in the not too distant future, be surrounded by several planets flying the Hammer and Sickle flag ...'

WERNHER VON BRAUN interviewed by VICTOR RIESEL
for the Hall Syndicate, 21 January 1959

CHAPTER TWELVE

'America sleeps under a Soviet moon'

For Sergei Korolev the success with the Sputniks was just the beginning. The images of rockets travelling with ease through dark, velvet skies had been the original impetus for his career and now, almost thirty years later, the urgency of transforming that youthful vision inspired by Tsiolkovsky into reality was undiminished. 'The road to the stars has been opened,' he wrote under the pseudonym Professor K. Sergeev in *Pravda* in December 1957. 'An important bridge to space has been built.' A universe of possibilities was revealed. Encouraged by the response to his success, he requested permission to send out a series of unmanned lunar probes and won approval. 'Comrades,' he told his team with some satisfaction in March 1958, 'we've just received an order to deliver the Soviet coat of arms to the Moon.' Not content with this, that spring Korolev and Tikhonravov began to set out their grand vision for a unified Soviet space programme.

It was ambitious, embodying Korolev's long-held dreams. In their report 'On the Prospects of Mastering Outer Space' they set out an agenda that went well beyond artificial satellites and the exploration of the moon. They argued for settlements in the earth's orbit populated by scientists busy with investigations into colonizing space. They envisaged an entire transport system around the earth. Reaching even further in preparation for interplanetary travel, Korolev could see a thriving colony on the moon, as if he were a latter-day Columbus setting out for America. Preliminary research for this could start as early as 1960. Preparations for unmanned return flights to Mars and Venus from 1963 onwards would pave the way for man to make the journey to the planets. Each stage in achieving these goals was scheduled, and included research on energy sources, space suits, docking of craft, assembly of rocket components and their integration into a space station, not to mention consideration of the different types of engine to use in space. Their comprehensive plan, which Korolev recommended should start immediately, was sent to the Military-Industrial Commission for consideration.

Korolev did not wait for a response. Pictures of Explorer 1 were a reminder that the Soviet Union did not have sole ownership of the heavens. The next achievement would be to put a man into space, someone who could take that reckless step between earth and infinity. This was uncharted territory. Nothing was known for certain about the effects of space travel. Would a man survive the conditions in space? What kind of craft would be needed to take someone out of earth's orbit and bring him back to earth safety? The R-7 – once adapted with an extra upper stage – had the power to put a man into orbit but as yet there was no way of bringing him back.

For all of Korolev's confidence, he was aware that there were many problems to solve. First, they needed a particular kind of individual who would willingly face the rigours of space travel: obviously fit and dedicated, brave without being rash and calm in an emergency. There would be some hard lessons to be learned before space travel became as easy as boarding a train. The delicate human frame was not designed for cutting through the atmosphere at thousands of miles per hour inside a vast rocket filled with fuel. The rocket's acceleration, acting like the force of gravity, put enormous stress on bone and muscle and could crush vital organs. Extremes of heat and cold could kill. Space, essentially a vacuum with no atmosphere, could cause the astronaut's blood and bodily tissues literally to boil. The space vehicle would need its own reliable, pressurized atmosphere. No one had given any serious thought as yet to the design of the vessel and how to return it to earth.

Tikhonravov was assigned the task of overseeing the designs of the first spacecraft. The most critical time for the returning craft would be during re-entry to the earth's atmosphere. Temperatures generated by friction on re-entry through the earth's atmosphere could reach as high as 5000 degrees F, nearly half as hot as the surface of the sun (10–11,000 degrees F). Tikhonravov's team had to resolve many issues such as the best shape for the craft, the navigation system and the insulation that would keep its occupants from burning to death on re-entering the earth's atmosphere. New breaking rockets, known as 'retro rockets', had to be designed to guide the craft through the exacting narrow re-entry corridor. If the angle of re-entry was too steep, the craft would burn up as the friction of the atmosphere would be too high for any material to withstand. If the angle was too wide, the craft would bounce off the atmosphere and ricochet into a higher earth orbit, never to return to earth.

Tikhonravov promoted one of his talented young engineers, Konstantin Feoktistov, as head of the group that would tackle the challenging problem of designing the space ship for safe return. Feoktistov's team began by working on the optimum shape of the capsule to minimize heat. Months were spent considering the relative merits of cones and spheres. They concluded that a sphere was the best shape that could be designed in the time. The blunt nose of a sphere creates a shock wave just ahead of the surface of the craft, deflecting the hottest area of air flow safely away from its surface. They also developed a simplified attitude control system – for orientation in space – and chose an asbestos-based material to build the craft, making it thicker than was strictly necessary for extra protection. After investigation of a number of different landing methods, they selected a straightforward parachute for the passenger who would be ejected in a cosmonaut's chair through a hatch. Research was also underway into heat protection, orientation systems and tracking and communications with the craft.

During the spring, competition with the Americans intensified. Although a second US Vanguard had failed in February, by March Vanguard 1 and Explorer 3 had reached orbit and transmitted a wealth of data. However, on 15 May 1958, yet more brilliance shone on Korolev when Sputnik 3 was successfully launched. This was a backup of the original 'Object D' that had been delayed in delivery. With a massive weight of 1.3 tons, it eclipsed the US efforts – Vanguard 1 weighed a modest 3 pounds. Khrushchev once again revelled in the impact of Soviet rocket wizardry on the Americans, who were wondering how Soviet rockets had lifted such a weight in space. To the chagrin of the

Americans, Khrushchev, putting on his master of ceremonies' hat, took centre stage and gleefully informed the world that 'America sleeps under a Soviet moon'. Korolev's 'star' was at its zenith, and it was at this point, when it looked as though he could do no wrong, that unseen forces of destruction began insidiously to undermine his position.

While Khrushchev appreciated the brilliance of his Chief Designer, who had so often given him an opportunity to crow over the Americans, he was beginning to question the success of the R-7 as a missile. Quite apart from the teething troubles in the test schedules, it was proving costly, requiring specially built launch pads. More important still, it did not fit military requirements well enough. The army needed a missile that could be fuelled quickly, and, once fuelled, could be stored for long periods in a state of readiness. The R-7 took nearly a day to fuel, and if a launch was cancelled, the liquid oxygen and kerosene propellant had to be removed. At a meeting in May 1958, Khrushchev drew Korolev's attention to the R-7's drawbacks. He wondered if Korolev could suggest a more ingenious alternative to the needs of the army.

Korolev defended the R-7, pointing out that the Americans had nothing as powerful, certainly nothing that could put up anything as heavy as the recent Sputnik 3. Khrushchev argued that it must be possible to design a rocket that could take advantage of the new, more efficient fuels becoming available which appeared to have all the properties that the military needed: greater economy, higher energy and a quicker fuelling time. But Korolev was adamant; new fuels such as unsymmetrical dimethyl hydrazine and nitric acid were 'the devil's venom', a toxic and highly explosive brew, which quickly corroded fuel tanks and was sensitive to too many variables. Furthermore, there was no time to research and test new engines with an unknown fuel. He was in a hurry to have a man in space. He wanted a fuel he knew and the low-temperature cryogenic fuels he used, such as liquid oxygen, were safer and had greater lifting power, essential for a space launch. He was not prepared to compromise.

Khrushchev found Korolev uncooperative; he did not like his attitude. The next day he decided to sound out Glushko on the subject. He soon found that Glushko actually preferred the new fuels, and gave him the very answer he wanted to hear. The new fuels such as unsymmetrical dimethyl hydrazine were much more efficient, he argued, especially for ICBMs. He explained that Mikhail Kuzmich Yangel, who had been in charge of his own design bureau at Dnepropetrovsk since 1954, had successfully designed medium-range missiles using the new fuels. A brilliant designer, Yangel had ambitions to build a longer-range ICBM. Khrushchev was determined to see this through, and made a point of consulting Yangel, who acknowledged that, while there were difficulties, they could certainly be overcome.

At his next meeting with Korolev, Khrushchev returned to the question of the new propellants – but found that Korolev had not changed his opinions. Khrushchev mentioned Comrade Yangel's work and pointed out that now was the time to build a new generation of ICBMs using Glushko's engines. Khrushchev was effectively threatening Korolev with Yangel. Suddenly the goodwill he had enjoyed from Khrushchev was paper-thin. Khrushchev later recalled Korolev's desperate reaction as he abruptly changed his tune: 'I propose that you give this acid-fuelled missile project to me,' he persisted. 'Besides that, I will *also* make an oxygen-fuelled missile that will be capable of nearly instantaneous action.'

Khrushchev was unimpressed. His former easy attitude suddenly disappeared as he cut Korolev short and reminded him sharply where he was and to whom he was speaking. 'We made a decision, Comrade Korolev,' he announced curtly. 'You are assigned the oxygen version and Yangel is entrusted with the acid missile. That decision will not be cancelled ... We will see who wins.' Yangel was to build the R-16 with the new fuels and, if he succeeded in meeting military needs, the R-7 could become out of date – possibly even obsolete. With that Khrushchev abruptly ended both the interview and any possible feeling of friendship with Korolev. When he later related the story to his son, Sergei, he added that he thought Korolev was about to hit him.

The fact that Khrushchev was now favouring Yangel was a double blow to Korolev. Mikhail Yangel, brilliant engineer and party man, had come to the attention of Ustinov in the early 1950s for 'fast-track' promotion, heading a guidance systems department under Korolev. He was the one who was promoted over Korolev in 1952 when the post of director of NII-88 had fallen vacant, a position most people assumed would fall to Korolev. Now here he was in favour once more.

Korolev found that he could no longer pick up the telephone and expect easy access to Khrushchev, bypassing red tape. He was frozen out. Glushko had been instrumental in pushing Yangel forward, and was now designing engines for the new R-16 that would function with the new fuels. To Korolev it seemed as if Glushko had betrayed him. In a meeting between Glushko and Korolev, words flew like knives. Glushko was as vitriolic as Korolev, no longer prepared to stand in the shadow of the great Chief Designer. The breach between them, which had existed for so long, had now become a chasm. Nothing could be salvaged. Although Glushko had virtual monopoly in his field of engine design, Korolev was so angry he swore he would never use Glushko's engines again.

And Korolev's loss of patronage continued. In March 1958, Khrushchev's own son, Sergei, interested in a career in aviation, chose not to work for Korolev, but for an ambitious chief designer who was gaining influence: Vladimir Chelomei. Chelomei had enjoyed an astonishing rise to power during the 1950s, winning support from Khrushchev himself to develop a new type of naval cruise missile. His design, which could be launched from a warship, did away with costly launching sites. Chelomei was a master at diplomacy. 'He was very stylish,' recalled Sergei. 'He looked like an artist, dressed well, wore natty ties … he was very cultured and had everything in his head. He could talk for two days about his ideas.' The calculating Chelomei decided to challenge Korolev in his own area, and began to develop bold ideas for a space programme of his own. He used his polished charm in senior circles, and, by cultivating Sergei, found that he had the ear of Khrushchev himself. Winged, piloted spacecraft, winged rockets and voyages into deep space: nothing was too ambitious for Chelomei.

No longer having direct access to Khrushchev, Korolev was now forced to go through official channels to get approval for his plans. He was subject to the wishes of his army superiors such as Ustinov and Nedelin whose paramount interest was defence. Rather than sending a rocket to the moon, their greatest concern was to launch a reconnaissance satellite. Korolev faced a dilemma. Tikhonravov had designed the space ship and was now seeking approval to proceed but Korolev knew he would not get the funds to build both a spy satellite and the manned space capsule. 'Spy capsules won't work yet,' he had tried to persuade the military leaders. 'We have to develop a manned capsule first.' He was determined to find a way to press ahead with a manned craft, but, if he did so, he could be accused of neglecting defence needs, which might be construed as a treasonable offence. Party politics and the devious scheming of unknown enemies could not be discounted or dismissed as he had found out before, to his cost.

On top of all this, he faced a serious challenge from America. Public fears of Soviet dominance in space continued to be a hotly debated issue. 'There is something more important than the ultimate weapon,' Lyndon Johnson, Senate majority leader, had declared. 'That is the ultimate *position* – the position of total control over Earth that lies in outer space.' To answer concerns that the Americans were losing the lead, Eisenhower decided to create a civil establishment dedicated to space research. On 29 July 1958, the creation of an American space agency, the National Aeronautics and Space Administration – or NASA – was approved. 'Space exploration holds the promise of adding importantly to our knowledge of the Earth, the solar system and the universe,' claimed the President grandly. More importantly, NASA would preserve America's position as a leader in space science.

Project Mercury was soon approved. It aimed to put a man into orbit around the earth and recover him safely. NASA's Chief Administrator, Keith Glennan, was under no illusions as to the enormity of the task. He identified the key problems to a White House committee: dealing with re-entry and safe landing to return the astronaut to earth; increasing the reliability of the launching rocket; protecting the astronaut from high-energy radiation; and ensuring his psychological and physical ability to cope with the strange new environment. In 1958, $89 million was approved to help NASA tackle these problems, a budget that was to increase exponentially, reaching $740 million in four years. Quite apart from the rivalry he faced within the Soviet Union, Korolev now found himself a diminutive and untried David facing the legendary might of Goliath as he was ranged against America, the world's industrial giant.

Undeterred by the scale of the challenge he had set himself, Korolev, the 'Iron King', pushed himself tirelessly, frequently flying between Moscow and Baikonur at night to avoid losing any work time. That summer he was preparing to send his lunar probes to the moon. Reaching that remote lunar world was the vision that drove him on relentlessly, giving him no peace. Yet launching a rocket on a trajectory to the moon was a monumental task, requiring complex mathematical models to gain pinpoint accuracy, not to mention rockets in which nothing malfunctioned. Simply to crash a rocket into the moon – itself a rapidly moving target – required aiming the rocket at a point ahead of the moon such that the moon would cross that point at exactly the same time as the rocket. To achieve lunar orbit was even more complicated. As the rocket approached just ahead of the moon, a small rocket engine would have to fire at precisely the right second – no sooner, no later – to slow the rocket down so that the moon's gravitational force would just balance the rocket's momentum, ensuring that it entered lunar orbit and was not flung out into deep space.

In September, Korolev's attempt to send his first probe failed. To rub salt into the wound, that same month, at an exhibition of military hardware at Kapustin Yar, Khrushchev spent an hour inspecting Chelomei's work and an insulting few minutes walking around Korolev's rockets. Korolev tried again for a lunar launch in October and once again faced disaster. After two minutes the rocket exploded spectacularly. It was shattered into a thousand fiery pieces, raining down on the desert like a shower of meteorites. A team was sent to recover it, obliged to live like nomads for several months in freezing conditions as late autumn turned into winter. And Korolev still had no response to the document that he and Tikhonravov had set out for the Soviet leadership, proposing a coordinated Soviet space programme like NASA.

Korolev confided in Nina: 'It seems that nature jealously keeps its secrets and even here, where the mind of human beings is able to open it up – every step to something new and unexplored is achieved by a narrow margin with a high price to pay … We have such a short period of our life for creating something.' She would write back with encouragement, asking him to take care of himself and reminding him to take the pills for his heart condition.

• • •

In October 1958, NASA came into being and a core team preparing for manned space flight was soon established. Robert Gilruth, a brilliant engineer from the former National Advisory Committee for Aeronautics at the Langley Research Center, Virginia, was put in charge of the Space Task Group to run the Mercury programme. He had been working on new experimental craft with test pilots and his role was pivotal in design improvement. Gilruth was passionate about engineering design and spent his spare time creating his own hydrofoil hulls which he sailed at the weekends, the perfect form of relaxation for life in the hectic space industry.

The fledgling new space agency rapidly grew to incorporate numerous other research teams and laboratories across America such as the navy's Vanguard programme and the Jet Propulsion Laboratory in Pasadena, California. Yet, unaccountably, the man who had done most to promote the reality of space travel in America was not part of it. Gilruth appointed his colleague Maxime Faget as Chief Engineer to run the Flight Systems Division, which would design the Mercury spacecraft. Other appointments soon followed. Despite his high profile, fears were growing in von Braun's team that they would find themselves left to rot in some political wasteland, still considered more German than American.

As usual, von Braun was at pains to reassure his team. He was in 'no doubt' that there would eventually be a 'Moon landing project' and that 'our team would play a major role in that project'. In public, too, he continued to advance the significance of moving into space. 'Man must establish the principles of freedom of space as he has done with freedom of the seas,' he urged in *Time* magazine. 'And like everything else this can only be done from a position of relative strength.' In the world of long-range ballistic missiles and thermonuclear warheads, there might be no warning. 'Either we will be ready at a moment's notice, or historians may conclude over the ruins of our cities that we were "weighed and found wanting".' Yet the NASA project to launch and build a spacecraft was marching on swiftly without von Braun. For almost fifteen years, his team had lived on promises. This was no longer enough. His assurances that their role in the space industry would be important began to sound less convincing. Inevitably the team shrank as members left for the irresistible private sector.

Von Braun, however, was assigned to improving the Jupiter still further. His remit was to produce a more powerful rocket capable of lifting heavier payloads. It would have a central core formed from the

Jupiter's fuel tank, with eight Redstone tanks clustered around it and eight Jupiter engines. Eventually it was renamed *Saturn* – von Braun's suggestion, as Saturn followed Jupiter in the planetary system. Its upper stages would house powerful engines designed to use the new high-energy fuels. These new fuels, which combined liquid hydrogen and liquid oxygen, could sometimes prove problematic; liquid hydrogen in particular needed to be kept at extremely cold temperatures.

Meanwhile, despite public denials that the US was racing the Soviet Union, the Americans were trying to beat them with an unmanned probe to the moon. The first attempt to reach a lunar orbit, in August 1958, proved an embarrassment when the air force's Thor-Able rocket failed even to leave the pad. That autumn saw three further attempts to reach the moon. Although they also failed, Pioneer 1 set a new record, travelling more than 70,000 miles into space, and Pioneer 3 reached almost as far and identified a second belt of radiation. But 18 December was the moment for a particularly sweet US victory as Eisenhower's disembodied voice, orbiting in space in an Atlas rocket, wished the world a 'Happy Christmas' – not a message that Khrushchev believed or necessarily wanted to hear. America seemed to be catching up. Anyone with a radio could hear the American President conveying 'America's wish for peace on Earth and goodwill towards men everywhere'.

So sure were NASA officials that they would eventually achieve a manned space flight, that, in January 1959, the call went out for men who would be capable of handling a space vehicle. The search began for men with jet pilot training and a university degree between the ages of twenty-five and forty. Size and weight were important: no one could be taller that 5 foot 11 inches or weigh more than 180 pounds. There was a considerable response to NASA's request but by February this was narrowed down to thirty-two men. Since it was not known exactly what the effects of travelling in space might be on the human body, NASA wanted men at the peak of health, with quick reactions and the ability to deal with and endure any hazardous situation. Experts could only speculate what lack of gravity, or increased gravity, might do to the human body. How would blood circulation be affected? Would the brain be suddenly drained of blood? Would the pilot, alone in space, looking down on the world, so completely divorced from human kind and human scale, lose his reason?

The hopeful sky riders found themselves subjected to a battery of tests seemingly quite unrelated to flying a plane and applied with excruciating thoroughness. For these good-looking and fit young gladiators, dignity was abandoned. There were blood tests, urine tests, fertility tests, body fat tests, stool tests: they were wired up to machines capable of measuring almost anything measurable in the human body. And still the measuring and monitoring continued, with tests to observe the clutching of hands, or exactly how much the subject's eyes fluttered and rolled when cold water was poured into ears. It was never explained why the white-coated technicians were so absorbed in the men's stools and so persistent in administering enemas, or exactly what relevant information was gathered by the doctor's rubber-gloved hand feeling through the rectum with a maddening interest in the bowel and the prostate. Worse still, did he really have to insert a balloon there and inflate it?

To test their state of mind, they had to face hours in a dark and silent room, completely lacking in sensory stimulation. At other times they were subjected to volleys of quick-fire questions and expected to respond instantly, while, simultaneously, a soft and sultry voice insinuated the wrong answers, and control switches and buttons from some infernal machines demanded attention. They had to endure extremes of heat and cold, deathly silence or a cacophony of noise; the thrill of the treadmill until legs were worn down and the heart gave up; or the threat of being dropped in the desert with minimal survival gear and expected to return to the centre like a homing pigeon. At least the 'one to one' interviews with the psychologist provided the safety valve of ridiculous answers to stupid questions until the cost of such flippancy in the face of authority was registered. Then it was back to the comparative simplicity of being spun like a top at the end of the long arm of the centrifuge and various other well-thought-out exercises that made the individual horribly aware of human frailty and desperately determined not to let that show.

At the end of it all, when mind and body were sufficiently pummelled, intrusive and embarrassing photographs were taken. No odd angles were spared – an exhaustive record of the candidate's physical condition was to be used as a comparison with whatever state the traveller might be in on return from space. As the unendurable tests gradually began to reduce their numbers, seven men were chosen, emerging from the grisly process like Greek gods, ready to accept the unknown challenge without question and represent America. Their next ordeal was to be thrown to the press, which they decided was infinitely worse than the mission impossible from which they had just returned.

While prospective candidates were being selected, the Mercury capsule was under development. Unlike the spherical Soviet design, Max Faget's team in the Space Task Group at Langley judged that a cone with a blunt face would be the safest vehicle to cope with the great heat of re-entry – which was enough to melt metals. 'You would think that that is a horrible way to enter, you really should have a nice point,' admitted senior design engineer Caldwell Johnson. Their research, however, revealed the exact opposite. If the capsule was pointed, then it would have little drag and would accelerate into denser parts of the atmosphere creating ever more heat until it burned up. Counterintuitively, the blunt face of a bell-shaped Mercury, slamming into the atmosphere broad end first, created a tremendously compacted shock wave of air ahead of it. That 'cushion' helped insulate the actual skin of the craft from the heat of re-entry and also slowed the capsule down much faster than if it had a more streamlined needle shape. 'It would get so hot that a shield of gas at the front part of the cone essentially insulated it,' explained Caldwell Johnson, 're-radiating its heat back out.' The heat shield was designed so that the outer layers would ablate or burn away under the intense heat of re-entry, dissipating much of the heat.

Research was underway into how to fit retro rockets to the capsule for attitude control – to orientate the craft in space. It was vital that the craft be correctly positioned on its approach back through the earth's atmosphere. The vehicle had to be aerodynamically stable, enabling it to stay balanced and not flip over as it sped through the atmosphere at 15,000 mph. At the apex of the cone, a cylinder housed the parachutes, which would slow the craft still further in the lower atmosphere in preparation for landing gently in the sea. To protect the astronaut in case of a serious malfunction of the rocket, Faget and his team also designed an escape tower – a girdered, lightweight tower strapped to the Mercury spacecraft at the top of which was a cluster of solid rockets. At the first sign of the imminent destruction of the main booster, the escape rockets would fire, which would pull the spacecraft off the stricken rocket to safety. Tests were also underway to determine the best means of regulating temperature, pressure and oxygen supply within the capsule. The decision was made that, rather than air, the astronaut would have a pure oxygen atmosphere. This was a decision that would prove to be fatal.

• • •

Despite the developments in the American space programme, to his great frustration Korolev still could not win the backing from the Soviet leadership for a manned space flight programme. In November 1958, he did persuade the Council of Chief Designers to approve his designs both for the Vostok capsule and a reconnaissance satellite. But just to make sure that military needs took precedence, some wit coined the phrase 'Reconnaissance satellite is more important for the Motherland', which infiltrated the society at Baikonur like an unwelcome echo.

New Year 1959 saw Korolev preoccupied with a fourth attempt to land an object on the moon. He was desperate for success, knowing that the Americans were trying to get there first. 'I am nervous,' Korolev wrote home, 'but what can I do if all our work is a search for something new, and because of this we may have failures and misfortunes?" Korolev's lunar craft carried scientific instruments which would determine whether the moon had a magnetic field as well as taking other measurements. The capsule held information of its country of origin – there was a heat-resistant banner spelling the letters 'USSR' and a small container designed to explode on impact and scatter seventy-two pieces of steel on the surface of the moon, bearing the launch date and the Soviet hammer and sickle. Korolev was sure he had the trajectory correctly calculated, but using radio to confirm the craft's course would be difficult. He put a brain trust together to work out a solution – someone even suggested using a small nuclear bomb and measuring the time from the appearance of its flash. This was dismissed, however, as the moon had no atmosphere, and hence no dust cloud would form. Eventually, Korolev settled for a device that would create a glowing yellow cloud which astronomers could use to track the craft. To do this, a pack was devised in which napalm would ignite 2 pounds of sodium.

On 2 January, the R-7 was launched successfully and Luna 1 was the first spacecraft to leave the gravitational pull of the earth entirely, marking this victory with a colourful flag of yellow vapour seven hundred miles high as it passed over the Indian Ocean. It had been meant to crash on the moon, leaving its Soviet signature. It came within 3700 miles of its target, and then, with a failure of the control system, it fell into orbit around the sun. Although the R-7 had failed to hit the moon and leave the symbols there which would have marked the phantom Soviet presence, for Korolev the little 'moon ship' was a wonder. He renamed his probe Mechta – the Dream.

The mission was hailed by Khrushchev as a great accomplishment. Managing to ignore the ultimate failure of the enterprise, he could not resist boasting to the world of his Soviet 'scientists, designers, engineers and workers who achieved a new exploit of world-wide importance, having successfully launched a multi-stage cosmic rocket in the direction of the moon … Even the enemies of Socialism have been forced … to admit that this is one of the greatest achievements of the cosmic era.' Korolev, meanwhile, listened to Khrushchev's self-glorying rant from his anonymous place in the shadows.

A Soviet 'moon' was now shining on America. A Soviet 'planet' had the presumption to circle the sun. Von Braun's frustration at the new Soviet additions to the planetary system was growing while his own hopes for an American achievement never lost their lustre. As for Korolev, he told Nina: 'The best minds of humankind dreamt about it, but we are the ones who have the wonderful opportunity to lead the way. It's a great joy and incredible luck to be pioneers on this path.'

CHAPTER THIRTEEN

'We really are in a great hurry'

On 9 April 1959, NASA held a press conference to introduce their final selection of Mercury astronauts. When they appeared on stage at Dolley Madison House, NASA's headquarters in Washington, the press were clearly moved at the sight of these unknown young men, and roared their approval as though the 'significant seven' had just saved the world; heroes already. This was an all-American story about small-town boys making good. Genuine admiration was displayed; applause continued until arms ached and the electricity engendered by so much emotion was in danger of lifting the roof. The seven were unaccustomed and unprepared for such a show of feeling. Unprepared, too, for the photographers, who were flashing a thousand brilliant lights as they fought to get the most telling, the most winning picture, no matter who was injured in the process.

Eventually the press calmed down a couple of decibels and demanded to know what had gone into the making of these seven champions, who were looking back at them with such an easy charm. Clearly, these would-be warriors were from an original mould, made when the rules were first written. Had they not just competed with hundreds of their fellow men for the honour and the dubious reward of sitting astride a throbbing mountain of metal with all the gaseous fury of the fires of hell under it as it hurtled towards the unknown, at a speed faster than sight can comprehend? And when they were thrown out of the earth's orbit into silence and emptiness, what strange or fearful odyssey might await them? Who would be there to rely on? What possible reassuring voice would come from that eternal blackness and say: 'It's OK, boys. I've been there, it's easy'? They were on their own.

The press wanted to know what had gone into the making of these lionhearts who were committed to risking so much for the glory of America. Finding out was not easy. What does your wife think about what you are doing? But tough heroes don't waste time on small talk. They answered simply, keeping the replies short, giving nothing away. Their wives were all with them, it seemed. Then Marine Lieutenant-Colonel John Glenn, the oldest at thirty-seven, summed it up: 'I don't think any of us could have gone with something like this if we didn't have pretty good backing at home.'

Somehow their answers did not satisfy the press, who wanted to know what made them tick. Why were they doing it? Were they religious? Standard answers came back, somewhat disappointing. Navy Lieutenant-Commander Walter Schirra admitted to being Episcopalian. The handsome Air Force Captain Gordon Cooper was a Methodist. Navy Lieutenant Scott Carpenter sometimes went to church. When they had finished, the sober-minded John Glenn took over. 'I am a Presbyterian,' he said, 'and take my religion very seriously ... I have taught at Sunday school ... I was on the Board of Trustees at the last duty station ... We are very active in Church work and all things connected with the Church.' The press were interested. The other six astronauts shifted uncomfortably at the image emerging from one of their number. 'I believe we are placed here with certain talents and capabilities,' John Glenn continued. 'It is up to each of us to use those ... as best we can ... I think there is a power greater than any of us that will place opportunities in our way if we use our talents properly.' This was rather more what the press had been expecting: an adventurous spirit which, guided by religious conviction, was also facing the possibility of making the ultimate sacrifice.

For Navy Lieutenant-Commander Alan Shepard the religious emphasis was sounding the wrong note. He had a practical intelligence that could go immediately to the heart of the problem. His love of racing fast cars until they became all but airborne and the tyres burned out, he regarded as nothing more than just taking fresh air. He decided to put the record straight and get the focus back on to what they were doing. 'I don't mean to slight the religious angle, but the Mercury project is merely one step in the evolution of space travel,' he said, his face expressionless. 'I would like to discount the fact that this project is extremely hazardous. It is not a technical race. It is a step in the evolution of space travel.' He wanted it understood that they were simply doing a job – a job where the prospect of a most fearful annihilation was par for the course.

The questions flew for what seemed an eternity and the reporters found themselves thoroughly charmed as the astronauts replied with the sentiments of small-town Middle America with its farmland and prairies. They were clearly the sons of the early settlers, born holding the badge of courage. This was confirmed when someone asked the big question: did they believe they would be coming back from outer space? John Glenn, the mystic, gave them the answer they wanted to hear. 'I got on this project because it probably would be the nearest to heaven I will ever get and I wanted to make the most of it.'

Intrigued about the women behind such men, the press descended on their wives and families. All blemishes airbrushed out, the wives appeared on the cover of *Life* magazine, examples of American womanhood at its best, perfect partners for heroes. The astronauts were fêted and admired, role models for American boyhood. 'From a nation of 175 million, they stepped forward last week: seven men cut of the same stone as Columbus, Magellan, Daniel Boone, Orville and Wilbur Wright …' declared *Time* magazine appreciatively on 20 April, marvelling at their bravery in electing to be shot beyond the atmosphere to a height of 125 miles at a speed of 18,000 mph. 'Rarely were history's explorers and discoverers so clearly marked out in advance as men of destiny.' They were called on to give talks and speeches, visited the White House, held up as examples of the kind of men the free world was blessed with. What was there to say against them? They were going to kill the Minotaur, bring back the head of Medusa, slay the dragon.

In May – the month that Korolev finally won approval from the Soviet government to start building his Vostok – the American astronauts visited McDonnell in St Louis to see their space capsule for the first time. This proved to be something of a surprise. It was dishearteningly small, a mere 6 feet across. It looked like the sort of precarious equipment that might be found at the end of a centrifugal arm in an amusement park, with unbelievably thin walls giving little confidence that it would cope with the unknown eccentricities of being flung into space and hurled through fire. The interior was a miracle of ingenuity with equipment and parachutes skilfully packed away. But the 'one size' astronaut's couch designed to accommodate all body types was inches from the heat shield and the instrument panel was just 2 feet from the astronaut's face. The overall impression was one of claustrophobia with a small entrance hatchway and no window; just a tiny porthole and a periscope. It seemed unfeasible to embark on such a hazardous journey with no means of direct observation. The astronauts made a fuss and demanded a window, even though this would add to the weight of the capsule. They also had concerns that they themselves had no means of controlling the craft. 'All we are is guinea pigs,' they complained endlessly. 'They insisted on getting as much control of the vehicle as possible,' recalled Max Faget. This was a great worry to the engineers, who felt the risk of human error was greater than the risk of failure of automatic controls.

On 18 May 1959, the seven astronauts were gathered at Cape Canaveral to see their first launch of the Atlas rocket that was to carry them into orbit. To place the 4000-pound Mercury capsule into earth's orbit, the rocket had to reach a speed of 18,000 mph. Despite its continual teething problems, hopes were pinned on the air force's Atlas, which had a thrust of 367,000 pounds and was America's first intercontinental ballistic missile. As usual, the adoring press was in attendance as well as a sprinkling of VIPs, congressmen and captains of industry. The seven astronauts had a part to play: they needed to look impressive, to attract funds, make small talk, all of which Glenn and Shepard did superbly. The atmosphere was relaxed and convivial. In the background, flood-lit against the night sky, the mighty Atlas rocket waited for its moment.

Liftoff was faultless; the exhibition of such power was thrilling. The crowd watched, utterly absorbed by the spectacle. Slowly the rocket lifted as though resisting the titanic pressure. It was up, almost away; then, with a drunken lurch, it became unsteady, toppling sideways, its thin skin wrinkling like burning paper. As the crowd looked up, horrified, an almighty explosion ripped the whole thing apart, shooting out fiery debris, sending everyone running for cover. From the safety of the bunker, Shepard turned to Glenn: 'Well, I'm glad they got that one out of the way … I sure hope they fix that.'

As a step towards the more successful takeoff desired by the astronauts, testing the complete vehicle that would launch an American astronaut into space began in June 1959. The first team of engineers from the Space Task Group began to arrive at the Cape. Scott Simpkinson was in overall charge of the party whose job it was to launch the hybrid union of an Atlas missile married to a Mercury space capsule – otherwise known as 'Big Joe'. The capsule itself was a 'mock-up' of the real thing assembled purely for the purpose of testing the heat shield. They wanted to know how it would survive under the extreme conditions of re-entry as temperatures soared to over 3000 degrees F. What effect would this have on the temperatures inside the capsule?

The engineers from the Space Task Group were guests of the air force who quickly decided they were of small importance in the scheme of things and roped off some space in an old hangar, which they were to share with naval research. They found themselves in what amounted to a dirty shed with no facilities – not what they had expected when signing up for state-of-the-art space research, imagining a clean and sterile laboratory in which to assemble the craft. The office staff were limited to a storage corridor, their desks filling the tight space. If someone at the far end wanted to leave the area, everyone had to stand up and move from the room to allow the departing person to squeeze out. The conditions were unpleasant, with long working hours. Even at night the heat was unbearable and the mosquitoes so ravenous that in desperation workers resorted to spraying themselves and the mosquitoes with ammonia from the capsule's cooling system. It was clear that they were not at the smart end of space engineering, but obviously considered something of an anomaly. They decided to go along with this and cultivated a sense of individuality: Scott Simpkinson, bypassing air force bureaucracy, arranged with his supervisor for $50,000 petty cash with which to buy the tools he needed from the local hardware store, rather than filling in forms and dockets for a hammer or some screws which would be delivered several months later.

When they were finally ready to fit the capsule to the Atlas out at the launch pad, they had no specialized vehicle to transport the precious object, and were forced to use their ingenuity. Lashing an old mattress to a pick-up truck, they placed the Mercury capsule on it and made their way to the Atlas, waiting in splendid isolation. But on top of the gantry, all was not well. The capsule did not fit. It was half an inch too big in diameter. And time was not on their side; the test launch was imminent. Back in the workshop with the mosquitoes, they decided that the heat shield was too big, not the capsule. Simpkinson, never at a loss, queued up at the local hardware store again for the tools he needed to get a precision-crafted job done, gouging off that extra half-inch from the heat shield.

After delays not just with the capsule, but also with the Atlas rocket, 'Big Joe' was sent heavenward on 9 September 1959, weeks later than planned. At first it looked successful as it blazed away into the night sky like a great torch, but the rocket travelled too high and separated late from the capsule. Unable to separate at the right stage, the capsule used all its fuel trying to get into the correct position for re-entry. Since it was upside down, it was expected to burn up on re-entry. To everyone's amazement it righted itself on descent and dropped into the Atlantic five hundred miles off course. Despite the fact that the capsule had survived, the press notched up yet another failure to America, mystified that the Soviets never had a setback, unaware that in Russia failures were simply kept secret.

The astronauts, having watched the disintegration of the Atlas rocket at Cape Canaveral, were more than pleased when, in due course, NASA announced that the Atlas, with its haphazard record of success, would no longer be used to send the very first American into space. Initially, von Braun's Redstone rocket would be the basis of space exploration. Until work was done on boosters, however, the Redstone, with its more modest 70,000 pounds of thrust, did not have enough power to launch a Mercury capsule *into orbit* around the earth. But it could be used for a twenty-minute ballistic mission, literally to punch the Mercury capsule into space, perhaps reaching a maximum altitude of 130 miles before returning for re-entry. The first flight the astronauts would take would therefore be *suborbital.* This would not match a Soviet orbital flight – but it would get an American into space, defined as sixty miles above the earth.

Much work had to be done to 'man-rate' the rocket to make it safe for an astronaut. Von Braun wanted to make it more reliable with improved guidance and control and to design an abort system that would enable a man to be thrown clear of danger at any time between launch and the upper atmosphere. He had to make the Redstone's engines more reliable to prevent overacceleration; high acceleration could cause the pilots to black out, to pass out completely. There was also the problem of reducing the rocket's vibration, which might harm the human body and reduce concentration and alertness. He wanted 100 per cent reliability – no astronaut was to die in one of his rockets.

On 21 October 1959, Eisenhower at last announced plans to transfer von Braun's Huntsville team to NASA. To von Braun's great delight, he himself would become director of an entire NASA centre, as Huntsville was to be grandly renamed the Marshall Space Flight Center. More than four thousand of his army staff were also to become part of NASA, including the few remaining loyal members of the original team from Peenemünde. At last, after fifteen years, von Braun and his men were no longer in the army but part of the glamorous space industry, identified now with astronauts, not sergeant majors. It augured well, and that same month official approval was given for the Saturn project, which aimed to develop a booster rocket with enough power to reach the moon. But the first step was to beat the Soviets with a man in space. Everything would now depend on an old army missile – the Redstone.

• • •

In Moscow, Korolev had seen the many detailed press reports on the US astronauts. With Nina's help, he learned the names of the clean-cut American heroes and marvelled both at the coverage they received and the adulation showered on them. Since NASA had been formed and the American astronauts had pirouetted on to the world stage, he had received increasing support for a Soviet manned programme. During the spring of 1959, a number of meetings had been held between the Academy of Sciences, OKB-1, the Institute of Aviation Medicine and military officials to discuss recruitment of the volunteers. Korolev was quite clear about what he wanted: 'I think the candidate's age should be about 30, height below 67 inches and weight less than 150 pounds,' he declared cheerfully. 'Above all, he should be a man with a smile. They must be brave!'

In contrast to the great fanfare in America, in the Soviet Union the hunt for a spaceman was undertaken with cloak and dagger secrecy. During the summer, 'recruiting' teams began to arrive at air force bases across the Soviet Union – although exactly what position they were recruiting for was not something they felt obliged to reveal. Interviews were carried out with teams of pilots but no mention was made of space travel or space ships; bureaucracy was careful to make the quest obscure. Any query was dismissed with the minimal explanation that they were being given the opportunity for 'special flights'. Details of three thousand jet fighter pilots were scrutinized for candidates – and when almost all of them were rejected, they had no understanding of why, or from what, they were being dismissed.

Two hundred candidates were eventually chosen to go to Moscow to take part in a more rigorous elimination process under a programme enigmatically labelled 'Theme No. 6'. At the Scientific Research Aviation Hospital they faced similarly anarchic medics apparently bent on destruction of the mind and body, just as the American astronauts had. Although the centre was under the control of the air force, the doctors held sway. Dressed in hospital regulation khaki pyjamas, candidates had to endure tests that were designed to push physical and mental health to the limits and assess strength of character. Many dropped out as they met their nemesis in the low-pressure barometric chamber or the inhuman centrifuge for manipulating gravity. Just twenty candidates made the grade, in time becoming known as the first Soviet 'Star Squad'.

While this recruitment was underway, that September Khrushchev visited Eisenhower, in the midst of a paper chase of comment from the world's press. Before he left, Krushchev made it clear to Korolev that it would be to his advantage if he could pull off one of his clever stunts, enabling Khrushchev to put Eisenhower on the back foot with delicious insouciance. Korolev was in fact preparing another Luna launch and timed it for Khrushchev's visit to the United States. He badly needed a success; another probe to the moon in June had failed. Launched on 12 September 1959, Luna 2 impacted as planned on the moon's surface in the Mare Imbrium – the Sea of Serenity – three days later. The Jodrell Bank Radio Telescope in England tracked the progress of Luna 2 and confirmed this Soviet triumph. Another breathtaking first was marked up for the Soviets who for the first time had planted a man-made object on the moon and made it theirs.

When the world woke up to the news, Khrushchev was in Hollywood, visiting the set of a musical, making his disapproval of the scantily clad dancers clear. He was insufferable, so bloated with success as he lectured the press on the virtues of the Soviet socialist system that Eisenhower found it easy to refuse him a visit to Disneyland, arguing that security would be impossible. Yet again, America could only look on in awe now that, for the first time in history, a craft had visited another celestial body. All American attempts to send a probe to the moon had failed. The closest they had come was 37,000 miles with Pioneer 4. While NASA politely congratulated the Soviet Union on 'a truly great engineering achievement', Khrushchev, thrilled at his new-found political cachet, proudly presented the American President with a copy of Luna 2's ensign. Von Braun was dismayed that Soviet propulsion power appeared to be far in advance of America's; they had evidently used a multistage rocket of enormous power. In fact, the first stage alone had a thrust of 500,000 pounds.

What America didn't know was that Korolev was poised to produce his showstopper. Barely three weeks after Luna 2, in time for the second anniversary of Sputnik 1, Luna 3 soared towards the moon on a quest to see its hidden face: the side of the moon no human being had ever seen. The earth and the moon are locked in synchronous rotation. With an exquisite precision, the two bodies, each with their unique rhythm – the moon turning slowly on its axis once every twenty-eight days, the earth once every twenty-four hours – perform what can best be described as a celestial pavane, which in its timeless routine allows the moon only to show one face; the other is always concealed from view.

Launched on 4 October, Luna 3 contained radio equipment, a television system with a film-processing unit, solar cells to provide power and a covered opening for cameras in the upper part of the craft. To prevent it falling into orbit around the sun, it had a slower speed than Luna 1 and took three days to reach the moon. It entered a trajectory that prevented the craft from being held by the moon's gravity and passed within 3800 miles of the moon's south pole as it made its way beyond to the far side. On 7 October, bathed in the light of the sun, the probe photographed 70 per cent of the moon's secret face. With the film carefully shielded from radiation to prevent fogging, the photographs were processed on the craft ready to be radioed to the control centre on the ground as it returned towards the earth.

Radio communications, however, were working erratically, potentially jeopardizing the chance to obtain the photographs. Although the radio station was situated miles away at Ai-Petri in the Crimea, Korolev preferred to deal with the problem himself. Taking Mstislav Keldysh, the distinguished mathematician and computer expert, and other senior executives with him, he knew that if they left immediately they could be at the radio station at 4 p.m. when Luna 3 should be within range. After a hair-raising ride to the airport and a flight to Simferopol near the Crimean Mountains, the helicopter taking them on to Ai-Petri ran into bad weather. Visibility was zero and a detour was made to Yalta where a car would take them to their destination. They found a driver who seemed remarkably relaxed about being asked to squeeze seven city types in dark coats and homburgs into his Zim and drive in blizzard conditions – fast – the most authoritative one with the intense black eyes saying, 'Well, dear boy, show us what you and the Zim can do … We really are in a great hurry.' Looking neither to left nor right, he drove like the wind through blinding sleet as though the devil were behind him, his seven passengers anticipating a head-on crash at every curve in the road.

On arrival, Korolev swept in with more energy than the blizzard itself, intent on discovering the fault in communication: was it bad design or careless workmanship? He set about a charm offensive; if that didn't work, the staff at the radio station, accustomed to the peace of a sleepy backwater, would feel the edge of his tongue. Korolev knew there was no such thing as an unsolvable problem. He listed the nuts and bolts of what could go wrong with his usual energy, assuming the radio engineers felt a corresponding urgency. Could it be the aerials perhaps, or their alignment? The workers quietly underlined the point that the design was Korolev's; they were merely operators. This was not perhaps the wisest of answers to give a man who felt such passion for his creation roaming the skies, its cries unanswered. The cosy little nest of operators was poked until the truth emerged: they had been passing work not confirmed as correct before undertaking the next procedure.

The next people to come in for the Korolev treatment were at the centre in Moscow responsible for maintaining calculations of the spacecraft's trajectory. After a night with no sleep, the staff had camped out in the building for a catnap. When Korolev telephoned to discuss the trajectory, the operator informed him that the staff were not available: they were all asleep. She was not prepared to take the matter further as she didn't know who Korolev was – he sounded like a nobody kicking up a fuss. The lacklustre level of technical backup was intensely frustrating for Korolev, and his anonymity made matters even more stressful. His fabulous craft, his 'dream', was flying high, waving the flag for the Soviet Union, leaving the Americans with their mouths open at the wonder of what he had achieved, and back home people were asleep on the job.

In the end, the mission was a spectacular success. By 18 October, seventeen images had been transmitted back to earth from Luna 3, indistinct patches of dark and white from the moon's surface – its secret face exposed to the mechanical click of the camera revealing craggy mountains and pockmarked sites of fatal meteor flights. Never before had a celestial body been photographed in such detail. A tentative map of the far side of the moon could be produced revealing different features, the most striking of which Korolev named after Tsiolkovsky. It was a large crater estimated at more than ninety miles across.

Later that year, the Soviet government rewarded Korolev and Nina with a new house in Podlipki. It was a revelation, a tangible token of appreciation; sufficiently grand to give a feeling of wonder, not so grand as to overawe. They moved in as New Year approached. 'It had six rooms, a patio and basement floor,' says his daughter, Natasha. Pink marble surrounded the fireplace in the sitting room. Bay windows to the floor offered views to the far distance. In Korolev's study was a very large German desk upon which stood a portrait of Nina; he filled the house with pictures and books. Sometimes Korolev would say he was going to rest in the forest and could be found half-dreaming in a deep armchair contemplating a picture called 'Forest Scene'. In the summer, he sat under the shade of an old oak tree viewing the leafy garden and the larch trees around the border. His daughter said 'he loved the house'.

But best of all for Korolev, Khrushchev did a complete volte-face on the subject of space travel. Waking up at last to the fact that the

Americans would soon 'outstrip us', he called a meeting in January 1960. 'Your affairs are not well,' he acknowledged to Korolev, Glushko and Keldysh. 'You should quickly aim for space.' Many of Korolev's original ideas that he had put forward with Tikhonravov as early as 1958 now seemed to be under more serious consideration. Khrushchev even wanted a probe to place the Soviet flag on Mars by October that year when he was due to visit the United Nations.

Events moved forward swiftly. That same month a Cosmonaut Training Centre (TsPK) was formed, temporarily based in a number of specialist institutes and facilities in Moscow while the search was on for a more suitable site. In charge of the training was General Nikolai Kamanin, whose life was a clear testimony to Spartan denial. Golovanov somewhat unsympathetically recalled a man of 'terrifying evil, a malevolent person, a complete Stalinist bastard'. Lean and hardened through active service, he expected the same devoted austerity to communism from everyone under his command.

Like their American counterparts, Soviet doctors were fascinated by the possibility that a man might lose his sense of reality should something go wrong in the capsule. If the cosmonaut lost that psychological, umbilical radio link with earth, would he go mad? To this end they emphasized tests that would stretch sensory deprivation and isolation to the limit. The dreaded windowless cell of silence became a subtle weapon of terror where the slowly passing days and hours could transform alert young men into subdued and passive zombies. Night and day did not exist in the chamber. No stimulus was allowed; books and music were forbidden. If an inmate drifted off to sleep, a light would shine in his eyes. He could talk to his keepers, but no reply was forthcoming as they were left to linger in the silence of their concrete coffin.

Oxygen starvation tests, beloved of the doctors, were also part of the programme. The would-be star man was locked into a cell while air was pumped out of it. The doctors observed, taking notes while their patient repeatedly wrote his name on a piece of paper. As the oxygen oozed away, the name writing became a scribble and the patient would lose consciousness. Why, asked Korolev, do we have this test? In a space capsule, he explained, an individual would only ever face a situation in which he either did or did not have oxygen; there would be no prolonged state in between. But the doctors won. Equally detested was the centrifuge capsule where the hopeful star traveller was spun until his eyes could no longer focus and the heightened forces of gravity transformed flesh into the weight of stone. One young trainee recalled: 'My eyes wouldn't shut, breathing was a great effort, my face muscles were twisted, my heart rate speeded up and the blood in my veins felt as heavy as mercury.' Yet he was longing to be selected for the 'special flights' and was delighted when he saw his name was on the short list: Yuri Alexeyevich Gagarin.

Twenty-six-year-old Yuri Gagarin was from the deeply agricultural region of Smolensk, a hundred miles from Moscow, and had worked hard to win his opportunities. He had grown up on a collective farm where his father had looked after the stock and his mother milked the cows. In 1942, when Yuri was eight, their quiet lives had been transformed as the Germans arrived, four columns of them marching through the village. Yuri had survived the brutal German occupation, and worked hard to escape his country roots and become a jet pilot.

Now he understood that his application had brought him to the threshold of something extraordinary. During the last days of the recruitment tests, he had been asked how he felt about flying something new. He had wondered what could be 'newer' than the most up-to-date jets, only to receive the jaw-dropping reply that being sent around the world to view it from 150 miles up in a rocket was newer. To be the first man to wave goodbye to the world, to look down on it and see the whole sphere as no man had ever done: that would indeed be new. Yet he felt his chances of being chosen first were limited. There were others whom he thought surely stood a better chance, such as the cosmonaut Gherman Titov. Titov had a sharper intelligence, was better educated and had a more sophisticated manner. He had read the classics and could quote them effectively and had acquired the confident assurance that comes from being born into the intelligentsia; it was bred into him – an innate understanding of his worth. Gagarin had no such inheritance with which to impress, only the centuries of dogged endurance bred into his forebears and the hard-won knowledge that survival really was the reward of the fittest.

The crucial next step was to meet the Chief Designer. They had all heard about this legendary figure, the man with no identity, with no

CHAPTER FOURTEEN

'Why aren't you dead?'

On 18 June 1960, the successful candidates were summoned to a room in OKB-1 to meet the mysterious maker of the rockets, a man sometimes known as the 'Iron King'. Although his name was never mentioned, which gave him an invisible, almost mystical status, the Chief Designer could never be mistaken for a man without identity. On close acquaintance, everything about him was indeed larger than life, as though the years of being incognito allowed him to express himself more fully and completely. He walked towards them and, as he did so, the more observant among them registered how his presence affected those around him. He was not tall, but seemed so. His impressive head was large, his features quickly changed from neutral, intelligent enquiry to a smile that embraced everyone. What Korolev saw impressed him: his 'little eagles'.

He spoke with warmth as he introduced himself and learned their names and piloting experience. While they listened attentively, he presented them with a vision of things to come; of journeying into the immensity of space in a machine that could travel many thousands of miles an hour, of arriving at a space station, living there or perhaps moving on to Mars: of the whole vista of a starry universe waiting to be discovered, just as the New World had been in the fifteenth century by those exploring galleons. Some of them would make that journey, one of them quite soon, he hoped. 'Patriotism, courage, modesty, iron will, knowledge and love of people,' he said: 'cosmonauts must have these qualities.' Unlike his usually 'controlled presentation', Korolev spoke with passion, gesticulating wildly to illustrate his ideas and was 'noticeably carried away'.

The company felt inspired and asked questions on space travel but Korolev wanted to know more about his 'little eagles' and how they had arrived at this point in their lives. He looked at Gagarin, the slight young man with the unmarked face of youth and a warm smile that seemed to illuminate a generous spirit. 'Tell us about yourself, comrade,' he said. Gagarin spoke for several minutes and, as he talked, it became apparent that the long and tortuous path from farm and village school to jet pilot had taken more perseverance and steely determination than was apparent in his looks.

As a child during the German occupation he had witnessed events that had brought him a certain seriousness that an eight-year-old child should not possess. There had been heavy fighting in the nearby woods where he used to play. When the fighting was over, he and his older brother, Valentin, had stolen among the trees to see what had happened. They saw a Soviet colonel lying where he had fallen two days earlier among the drifting leaves and snow. He was still alive – just. They watched secretly as German officers approached and began to question him. The soldier pretended blindness and asked them to come nearer. He could hear the Germans arguing, while the Soviet colonel remained very quiet. The Germans moved nearer and bent over him, the better to hear him. That was when the wounded Soviet released the pin in a grenade he was holding under his back. Gagarin could still remember the wild leap of flame that killed the group, the noise of the explosion and the calls of the frightened birds.

Yuri knew of neighbours who had been shot. People were rounded up and locked in sheds that were then set alight. When the Gagarin family had been turned out of their house, now occupied by Nazi soldiers, they had dug a hole in the ground and made themselves a hovel in which to live. To help the war effort, Yuri, with his younger brother, Boris, had littered the roads with broken glass and watched the tyres burst on the German lorries. They had been noticed by a great bear of a German soldier called Albert who lived in their house. He grabbed the six-year-old Boris and hung him from a nearby tree by his scarf. Gagarin's mother rushed to the scene and it looked as though there would be a double tragedy when Albert reached for his gun. Thankfully, he was called away by his superior officer and Gagarin quickly undid Albert's work and the terrified child recovered. Far from being cowed by these experiences, however, Yuri had found ways to make his own small contribution towards Soviet revenge. Albert's job was to top up the flat batteries of the German lorries and tanks; every night in the secrecy of darkness Yuri had found a way to render them useless. Although he had no knowledge of chemistry, he found he had a genius for rearranging the liquids in the various cells of the batteries and sometimes he would add a little dirt just for good measure.

After the war, the young Gagarin had been intent on starting to earn a living. When he explained that he had worked his way up from training as a smelter in a vocational school, Korolev immediately responded, pointing out that he too was a graduate of a vocational school. Gagarin had then won a place at the Pilots School at Orenburg on the Ural River, graduating in 1957 – shortly after the launch of Sputnik – only to find himself assigned a place at the Nikel Air Base near Murmansk in the Arctic Circle. It was here that he had first come across the mysterious recruiting teams.

Korolev was evidently impressed with Gagarin. He talked to each cosmonaut in turn and seemed pleased with the team. Once introductions were complete, he took them to see their spacecraft: Vostok, meaning 'east'. They were led through OKB-1 to a large room of hospital-like cleanliness, where white-coated technicians were busy working on the craft. He introduced the cosmonauts to Oleg Ivanovsky, the chief constructor, and Konstantin Feoktistov, the project leader. A fleet of Vostoks in various stages of construction were aligned on both sides of the room.

The one nearest to completion was a sphere of silver sitting astride a cone-shaped base. This was a surprise for the cosmonauts. Where were the wings, they wondered. They examined the various silver balls with circumspection. The work looked impressive but where were the controls for the pilot? Was it a practical joke, part of the bizarre training? No, the Chief Designer said, the Vostok would be guided. The silver foil covering the sphere would protect them from radiation. Inside the sphere was a reclining chair for the pilot who would have a perfect view of the ceiling inches from his head. Near him was a Vzor, an instrument used when orienting the craft for re-entry. It was made up of distorting mirrors and lenses which gave a view of the earth's horizon, enabling the correct position for re-entry to be ascertained. It was difficult to comprehend: no wings, no control of power. Was this precarious silver ball really meant to carry men racing through the heavens?

Gagarin was willing to believe so. When Korolev asked who would like to sit in the ship, Gagarin was the first to step forward: 'Allow me,' he said. Korolev noted his natural sense of respect as he removed his shoes before climbing into the pristine silver sphere. The little circular cell was thickly threaded with an electrical spaghetti of wires. These were the unseen nerves that would work the gyroscope and the instruments. There were some controls and switches and he guessed the reclining chair was also an ejection seat. To Gagarin it seemed a marvel of ingenuity.

Korolev was heartened by his meeting with the cosmonauts and followed their progress closely. Both Titov and Gagarin were doing well and receiving excellent reports from the new training centre being built about twenty miles from Moscow, not far from OKB-1 – which eventually became known as 'Star City'. In Vostok simulators, the cosmonauts took it in turns to familiarize themselves with the craft: the roar of the engines on launch that were fed through loudspeakers; the correct pre-flight positions of the instrumentation; the orientation of the craft in relation to the globe. The training was intensive because the Soviets still hoped to be ready for a manned space mission by the end of the year. It was evident from the Western press that the Americans could not be ready for manned Mercury flights before January 1960. Korolev was determined to beat them.

There were still countless technical difficulties to overcome, notably the design of the retro rockets, which were to decelerate the spacecraft below orbital velocity and guide it on a trajectory to fall back through the atmosphere to earth. The safe return of the cosmonaut would depend on the exact firing of these rockets; any miscalculation and he might be condemned to a metal coffin as he hurtled out into space, or a fiery grave as he re-entered the atmosphere too steeply and burned up. Alexei Isayev, who was assigned the task of creating the rockets, was worried. 'You and Korolev are twisting my arm,' he confided in Chertok, who had now been promoted to become one of several deputy chief designers. 'The schedule is too tight and you want to put one more noose around my neck ... What if someone does not come back to earth because of me? The only thing I could do is shoot myself!'

Chertok persuaded Isayev at least to discuss the matter with Korolev one more time. Isayev was determined to step down from the responsibility. 'You wait and see,' he said. 'It will take me two minutes to ditch that job.' When he emerged from his conversation with Korolev, somewhat shaken, almost an hour later, he was still in charge of the retro rockets. He lit a cigarette. 'At least I did a deal with Korolev that you [Chertok] will handle the electronics!'

The first unmanned Vostok had been launched in May 1960, under the careful supervision of Marshal Nedelin, Chairman of the State Commission. At first the mission was a great success. The Vostok successfully went into orbit, sending clear signals from space. While they were preparing a statement for the press, according to Georgi Grechko, Korolev announced a competition to decide exactly what to call it. Various names were called out: spacecraft, cosmocraft, rocketcraft ... Korolev did not like any of them. 'There are sea ships, and river ships and now there'll be *space* ships,' he declared. The excitement was palpable. 'Comrades, do you know what we have just written,' said Lev Grishin, Deputy Chairman of the State Committee of Defence Technology, when he received the message in Moscow. 'We have used the word *spaceship*. It is a revolution! The hair on the back of my neck is standing up.'

The excitement, however, did not last. After sixty-four orbits they were ready to test the re-entry system. The retrorocket fired successfully, but a fault in the attitude control system meant that, rather than entering the earth's atmosphere, the craft skimmed off the upper layers of the atmosphere and, by the force of the earth's gravitational pull, was flung into an even higher orbit. With no retrorocket fuel left, the craft was stranded. Everyone was aware that any cosmonaut would have died in a slow and very public way when the oxygen ran out.

Nonetheless, two months later they were ready to try their ship with the first live passengers: two dogs called Chaika and Lisichka – the latter meaning 'little fox'. Korolev loved the red-haired dog, Lisichka, and she clearly returned his affection. Boris Chertok remembers that Korolev visited the dog just before her flight and, brushing the white-coated scientists aside, lifted her up and stroked her fondly. 'I wish so much for you to come back,' he was heard to say, before he turned slowly and left without speaking to anyone. 'I had been working with Korolev a long time,' Chertok recalled. 'I even had contradictory feelings about Korolev, yet that hot day in July 1960, I felt a lump in my throat and a feeling of pity for the first time. Maybe it was some kind of presentiment.'

Almost immediately after the launch on 28 July 1960 one of the booster engines burst into flames and barely thirty seconds after takeoff the rocket exploded. Everyone raced for the shelter. There seemed to be as much regret for the wretched fate of the red-haired dog as there was for the demise of the costly rocket. Following this setback, Tikhonravov modified the design of the ship so that, during the first critical minute of the launch, the cosmonaut would be able to eject from the Vostok and parachute to safety if there was a failure. TASS never released any information about the launch. Despite this failure, Korolev was determined to press ahead with the next dog launch in less than a month.

On 19 August, Belka and Strelka were carried into orbit. At first, the camera trained on them showed them looking lifeless, almost dead, only the monitoring equipment revealing that they were alive. As the flight continued they appeared a little more animated. The crucial stage came after a day, as the team prepared for re-entry. This time, using back-up facilities, the orientation system worked, Isayev's retro rockets fired and they entered the atmosphere as planned. Indeed, the return was so successful that Belka and Strelka were parachuted to safety just over six miles from the designated landing position in Kazakhstan.

It was a great triumph: the first creatures to return alive from space. A celebration organized at Marshal Nedelin's house was hijacked by the need for an impromptu press conference even though it was late in the evening. In an attempt to let TASS get the story before the Western press, Belka and Strelka were paraded before an appreciative audience. It was yet another setback for America. John F. Kennedy, then an aspiring presidential candidate, chided that 'the first canine passengers in space who safely returned were called Strelka and Belka ... not Rover and Fido'.

Following this success, in September the Central Committee of the Communist Party approved Korolev's formal request for a manned flight, Keldysh, Nedelin, Glushko, Ustinov and other senior figures all put their signatures to the document. Korolev still aimed to launch a manned Vostok by December. In the meantime, Khrushchev hoped more exciting events were on the way as he attended a United Nations conference in New York in October 1960. As usual he was relying on a splendid performance from Korolev who was planning to send two probes to Mars. Inflated with Soviet conquests in space, he made extravagant claims for Soviet rocket science. 'We are turning out rockets like sausages,' Khrushchev said, 'and will soon have a man in space.' With irritating grandiloquence he promised a gift of one of Strelka's puppies to the White House. But on 10 October, Mars 1, which would probably have had Khrushchev doing a *pas de deux* for the UN representatives had it succeeded in its journey, failed even to leave the earth's gravitational pull before simply falling back to earth. 'Mars keeps its secrets,' Korolev wrote to Nina the next day, 'and our work – which is not always successful – helps it to remain hidden. It is such a great pity that the result of such titanic work ... is lying on the ground in a thousand pieces scattered somewhere in Siberia.' Three days later a second Mars probe also failed.

Displeased at being unable to flaunt Soviet joyrides to Mars, Khrushchev returned to Moscow and made it very clear to Marshal Mitrofan Nedelin, his chief of missile deployment, that some success was overdue and Marshal Nedelin had better provide it. What had happened to Comrade Yangel's R-16, the wonder rocket with the wonder fuel that would equip the army with storable rockets and that could be hidden in silos, safe from the prying eyes of US spy planes? He expected a successful test launch that autumn. Failure was not an option.

Khrushchev's hard words weighed down the epaulettes on Marshal Nedelin's shoulders, increasing his awareness of how easily they could be removed should he be seen to be less than effective. He set out for Baikonur to organize the maiden launch of the R-16, due to lift off on 23 October. Much was riding on the back of the new rocket. The army was impatient for its strategic advantages. At last it would be possible for a build-up of stored weapons ready to fire. It was essential that the new design was seen to do well. Irritatingly, on 23 October, far from the new propellants making fuelling easier, there were considerable difficulties. Eventually, a leak was discovered and the launch was postponed while repairs were carried out overnight.

The correct safety procedure was to drain the rocket of fuel while a proper investigation took place. Nedelin would not hear of this, however. He ordered several teams to tighten valves and patch up things in spite of the fact that the rocket was fully fuelled and highly dangerous. No one was allowed to rest. Next morning, he signed the documentation showing the rocket was fit to fly and a launch time was set while small, last-minute complications kept two hundred operatives busy on the site. With Khrushchev's words still stinging, Marshal Nedelin wanted results. He ordered that a stool be placed 60 feet from the rocket so that he could personally supervise and see that no slacking took place. If fuel was leaking, then valves must be tightened and screws given an extra turn. 'What is there to be afraid of?' it is claimed he said. 'Am I not an officer?' Safety was ignored as men worked among corrosive, dripping fuel and toxic fumes. Nedelin sat there, surrounded by dignitaries, his disquieting presence forbidding a challenge, the rows of glittering medals confirming his exalted position, his dark eyes missing nothing.

And there was nothing actually for the eye to see as, thirty minutes before liftoff, a faulty signal shaped the destiny of the R-16. In the complex sequence of procedures prior to launch, a wrong signal was sent to the upper stages of the rocket. Its engines fired, immediately setting alight the highly flammable liquid in the second stage below, which exploded with volcanic force. In a second, the rocket became an immense incandescent torch. The upper stage fell into the blaze, creating a whirlwind of apocalyptic proportions consuming everything in its path. Those trying to escape were suddenly embraced by far-reaching fingers of flames. The surrounding tarmac – freshly laid bitumen – melted and ignited, becoming a floor of fire which trapped those running from the horror. 'Many got stuck in the sticky burning mass and perished in the flames,' recorded the journalist Aleksander Bolotin. The deputy minister in charge of the Soviet defence industry, Lev Grishin, who only a few months earlier had been celebrating the idea of a 'spaceship', happened to be just a few steps from the rocket, talking to Marshal Nedelin. He managed to run all the way across this flaming tarmac and leap across high ramps and railings, breaking both his legs with the effort. He was taken to a military hospital where, four days later, he died of his burns in the arms of his good friend, Glushko. Most of the others were overcome almost immediately by the toxic fumes.

'The most horrific fate of all befell those on the upper levels of the service platforms,' continued Bolotin. Hanging from their safety harnesses in the gantry, the dutiful technicians 'simply burst into flames like candles'. As for Marshal Nedelin, who was standing positioned to watch the show, the roaring, swirling flames reaching temperatures of 3000 degrees simply melted him and his entourage away. The head of the emergency rescue team reported that victims could only be identified by their rings and house keys. All that was left of the vaporized Nedelin was his medals. He was identified by his gold star of the Hero of the Soviet Union.

When Yangel told Khrushchev of the horrific tragedy, the Soviet premier looked grave, his eyes focused on an inner reckoning.

'Why aren't you dead?' he asked Yangel.

'I was smoking a cigarette in the bunker,' Yangel replied. To calm his nerves, at the critical moment thirty minutes before the scheduled launch, he had gone for a break. He asked Khrushchev to punish him and afterwards he was seen crying at the launch site and at home. 'I could never get the sight out of my head,' he later admitted.

In true Soviet style, the disaster was hushed up. The country mourned Marshal Nedelin's death in a plane crash. The worst was kept from the cosmonauts in Star City. Official documents – withheld at the time – state that ninety-two people lost their lives. Eyewitness reports put the total death toll at nearer 150.

Korolev felt justified in damning those fuels so corrosive that they ate through the metal they were stored in. And he mourned the fact that crucial ground staff at the launch pad had died; he had so wanted to make up ground after the defeat of the Mars missions. In spite of being tired, shocked by Nedelin's death and defeated by the failure of the Mars probes, Korolev could not let go. He was impelled to work until his vision had taken on the substance of reality. He could no more change his compulsion than he could change the colour of his eyes. 'I wish I could be at home with you, or go out somewhere,' he wrote to the ever-patient Nina, 'but I am afraid that these are just dreams.'

• • •

In spring 1960, the deafening sound of Wernher von Braun's Saturn engines thundered their way into existence at the Huntsville test site. It sounded like the end of the world as windows trembled and the walls of houses shuddered; dogs slunk into dark corners and cats hid under beds. But the Huntsville citizens were tolerant; if the German engineers wanted to make a noise like the gates of hell opening, that was all right; it was probably just the sound of money anyway. Those living in the countryside were less charmed as their cows ran dry, their bullocks stampeded, heifers could not be got into calf, hens refused to lay and all manner of unwelcome phenomena haunted the farmyard. The sound could be heard a hundred miles away, the farmers claimed as they asked for compensation. But it was the sound of music to the ears of von Braun – at last his dream of space was waking to life.

In September, the President himself came down to Alabama to inaugurate the newly formed Marshall Space Flight Center and von Braun, as the new director of a NASA space centre, showed him around. Eisenhower was aware that America should embrace certain aspects of space technology more fully. Only the previous May, the Soviets had brought down an American U2 spy plane. Pictures of Gary Powers, the pilot, being paraded by the Soviets had been world news. It seemed as though Eisenhower had given Khrushchev a free opportunity to crow over the backward Americans. And worse, American intelligence feared a 'missile gap', with the Soviets suspected of having twice the number of ICBMs.

While the President was at the Marshall Space Flight Center, von Braun took the opportunity to be his persuasive best. Eisenhower recognized that space offered important strategic advantages, not least to spy on one's enemies; he had already initiated a secret spy satellite programme and his administration had created NASA. In spite of this, Eisenhower had doubts about expanding the space programme still further. A 'space race' with the Soviets had political dimensions that troubled him. He was concerned that this could stir up Cold War tensions and undermine his already fragile relationship with the Soviets. There was also the enormous potential financial cost to consider, with no guarantee of success. Indeed one of his advisers, George Kistiakowsky, had been scathing about the Mercury programme. 'It will be the most expensive funeral a man ever had,' he had claimed.

Von Braun proudly displayed the work in progress on the Mercury-Redstone project and plans for the development of the massive Saturn rocket, not to mention the research, still at an early stage, for a colossus among rockets, having 1,500,000 pounds of thrust. Preparations for the Mercury-Redstone were advancing well. They had managed to reduce vibration and noise levels in the Mercury spacecraft by the introduction of dampeners – vibration-insulating material – between the rocket and the craft. They had also introduced an 'abort-sensing system', an electrical system which monitored the performance of the engine, the rocket's trajectory and other parameters. Should there be any danger, the abort system would activate the escape tower rockets, pulling the astronauts to safety. Despite his reservations about a 'space race', the President was duly impressed.

All was not proceeding quite so smoothly with some of von Braun's NASA colleagues. Von Braun, anxious about safety, was keen to carry out integration tests on the rocket and the Mercury capsule at the Marshall Space Flight Center in order to check every system before it was taken to the Cape. With the sheer number of different components that had to be coordinated – the capsule made by McDonnell, the rocket by Boeing and countless other companies involved in the massive effort – would it all fit together and work as planned? Max Faget opposed von Braun, insisting that there would not be enough time for testing at Marshall.

Difficulties had also arisen with Chris Kraft, one of Gilruth's core team from the Space Task Group in Langley. He was assigned the task of working out a basic flight plan and had proposed the development of a 'mission control centre' which would gather all the data and direct the flights from the ground. Von Braun had vehemently opposed the idea, favouring full pilot control. They had clashed openly at a party; everyone fell silent, watching. It had been left to Maria von Braun to break up the argument and gently lead her husband away.

There may have been strong feelings towards the Germans within NASA. Gilruth and Kraft, while working for NACA (the National Advisory Committee for Aeronautics), had fought the Nazis during the war. 'We knew enough [about von Braun's ... rocket factory in central Germany...] for it to generate strong emotions, bordering on loathing,' Kraft wrote later. Others within NASA disliked von Braun's 'star' status.

Unlike Korolev in Russia, von Braun was now a small part of a rapidly growing space industry in America – yet he was a better-known public figure than many of the NASA team.

There was another rather personal setback for von Braun later that September with the launch of a film about his life, *I Aim at the Stars*. Far from being the flattering biopic that he had envisaged, he found it critical and disappointing. Worse still, there were pickets outside cinemas in some European towns, notably London. The film's title became transformed into a popular joke: 'I aim at the stars and sometimes I hit London.'

While there may have been mixed reactions to von Braun, the astronauts themselves were continually fêted by the public as living examples of the best of America. Every day they were in American living rooms, filling TV screens – their fearlessness seeming positive proof that America was going to beat the Soviets. They opened events, made speeches and impressed the industrial aristocracy. The smallest detail of their exemplary lives was public property. There were glimpses into their perfect marriages; their religious preferences, their favourite vegetable, their sock size, the colour of their eyes and their tastes in ties were known to all.

Out of the limelight, when they weren't required to represent America at some important function, the astronauts were training at Cape Canaveral. Often this involved long hours acquainting themselves with the space capsule, learning procedures. It could actually be tedious. There were two distinct landscapes to their lives: training at Cape Canaveral and weekends with their families. Sometimes weekends with the family didn't work out – fortunate, then, for Al Shepard, whose other love, apart from flying, was racing cars. He was able to get his speed up on the local dirt-hard beach that went on forever. His enthusiasm infected the others – apart from John Glenn – and they would often tear along Cocoa Beach until blue distance met the sky.

It was implicitly understood that wives did not appear at Cape Canaveral or Cocoa Beach. They would not like it. Cape Canaveral was not a civilized place. Until the government had started laying concrete and building bunkers and sheds, no one had wanted it. It was a mosquito-ridden, featureless stretch of aridity, devoid of all charm. And Cocoa Beach itself was hardly a resort, merely a conjunction of water and sand. No boutiques, bijou clubs or handsome hotels vied for attention: just a scattering of unprofitable, clapped-out motels for people unfortunate enough to be passing through. But to the astronauts, with its racing beach, anonymous motels and lack of reporters, it was the perfect place to relax – apart from John Glenn, that is, who preferred to unwind at home with his family.

Like the bush telegraph, word spread. The astronauts were there, racing on the beach, drinking in the bars. Fun of a special kind was bouncing around like hot, loud music. It was a place where anything could happen and probably would. The racing was serious. Al Shepard would eat up the miles in a Corvette. Scott Carpenter favoured a Shelby Cobra. Schirra preferred a Maserati. The 'personality' cars were customized and souped up. Brunettes and newly minted blondes appeared, the kind that lived in bars, wore high heels and not much else – and collected astronauts. The kind that were irresistible – except to John Glenn, of course.

It was a hard life being an astronaut, the main problem being that there might not be much of it. They watched rockets, which, with a persistent un-American lack of cooperation, exploded quite regularly. How was an astronaut at the peak of his health, taut in mind and body, to come to terms with that? Clearly Cocoa Beach was put there to solve this problem. At Cocoa Beach the astronauts could 'live', sampling every possible permutation of being pleasurably alive. Plenty of time left for sitting on top of an exploding rocket. Except, of course, for John Glenn.

John Glenn was different, like a knight of old; austere, abstemious, with a sense of purpose. As he ran on the beach clocking up the miles, keeping fit, he felt the motel morals were unhelpful to the astronauts' image. He had heard that several of the high-heeled, long-haired lovelies had been quoted as saying 'three down four to go'. As the eldest of the astronauts, he called a meeting to remind the other six that they represented America. They stood for wholesome manliness, honest labour, American motherhood and apple pie. This message was not always well received. Al Shepard asked on whose authority he was standing in judgement. Why exactly was he casting himself as leader of the group? The implication, of course, was that the leader, the most serious-minded among them, might be the first man in space.

And everyone in America knew that that would have to be soon, or the Soviet Union would prevail. The press fanned the flames with frequent jibes that the US could lose the race. Everyone on the Mercury project was working with Herculean dedication. As many as 13,000 people at McDonnell were labouring day and night. Holidays and weekends off were but a dim memory as von Braun's team focused on bringing abstract plans to reality, struggling to complete tests on the Redstone. Everyone connected with space flight had got the message: a superhuman effort was needed to beat the Soviets. But 1960 was passing them by. The yellow and red leaves of autumn had fallen, to turn brown and wither, blown by chillier winds. The first suborbital flight with an astronaut originally set for October had been cancelled and cancelled again, while Khrushchev continued to boast about an imminent Soviet manned flight. Gilruth still aimed to test two unmanned Mercury-Redstone flights and, assuming all went well, launch an astronaut early in 1961. But this plan too was looking unlikely – the first US flight was now postponed until the spring of 1961 and its success rate was put by the air force at 75 per cent. The latest fiasco was the failed flight of the Mercury-Redstone on 21 November. This was a key step, the first launch of a Redstone rocket with a Mercury capsule attached. It became a moment of unbelievable tragicomedy.

The whole firmament of the rocket industry stars turned out for the big event. For von Braun, who had not been able to test the capsule and the rocket together at Huntsville, it was a critical test. The Redstone was meant to lift the 1-ton Mercury capsule 130 miles up into space and return it safely back to earth – a crucial stage in the process before an astronaut could attempt the fifteen-minute space flight. Gene Kranz recalls his immediate sense of disappointment when he first saw the Mercury-Redstone. The rocket, 'far from seeming graceful in form, something you could love and rely on', struck him as 'stark, awkward, crude, a large black and white stove pipe atop a simple cradle'. The Mercury capsule was equally disappointing. 'It squatted atop the rocket, black in colour and seemingly constructed of corrugated sheet metal.' It was hard to imagine that this was 'a rocket ship from a science fiction novel', upon which the pride of America rested.

Everyone took their places for the launch; von Braun and the Huntsville team could see everything from the bunker. Bob Gilruth and Chris Kraft, the Flight Director, were in the Mercury control centre at the Cape, which could plot radar information about the position of the capsule during critical phases of launch, orbit and re-entry from tracking stations around the world. The tense countdown reached zero. The engine ignited with customary fury, blazing out fire and smoke beneath the rocket. Then, faster than the eye could trace, it appeared to take off, racing away confidently, too fast for the camera to catch.

Yet in Mercury control, Chris Kraft and his team could clearly see that the rocket was still sitting on the pad, having barely lifted 4 inches, the Mercury capsule surmounting it. So what was whizzing through the heavens? Whatever it was that had shot skyward so dramatically began to make the return journey, after reaching a height of several thousand feet. And it was evidently going to fall quite nearby. People began to run in disarray as the unknown object landed on the beach with an ominous thud – and still the drama wasn't over.

The assembled VIPs and glitterati eyed the reluctant rocket waiting for the great moment on the pad, smoke still whipping from under it. As they looked, a loud bang came from the top of the Mercury capsule and out popped a small parachute, opening as in a magician's trick. Gently, softly, it fell to earth, pulling the main capsule parachute with it, draping the coy and unwilling rocket in bridal white.

The assembled company were now facing an exceedingly dangerous situation. Thousands of gallons of volatile fluid wrapped in thin steel were centre stage. The worry was, what would the rocket do next? Would the parachute, dancing about with the wind in it, drag the great cylinder, gorged with fuel, from its precarious upright position into an explosive dance of death? At the height of the emergency, von Braun's team lapsed into German. This was the last straw for Chris Kraft, who went over and yanked the headset from one offending German engineer: 'Speak to *me*, dammit,' he yelled and later turned to a friend to vent his fury. 'Those damn Germans still have not learned who they work for!'

A sort of frozen anxiety took over. Someone with limited understanding of the situation advised hiring a crack shot to shoot holes in the rocket and drain its tanks. Suggestions came and went and eventually a decision was taken to do nothing. The heat of the day would warm up the fuel which would vent through the escape valves. The clamouring press soon notched up yet another American failure. 'What if there was an astronaut in the capsule,' they asked. 'Would you still be sitting here waiting for the fuel to evaporate?' 'The United States' hopes of rocketing a man into space early next year were dealt a crushing blow today by the third straight failure of an unmanned space capsule launching,' declared the *New York Times* the next day. 'The failure may have cost this nation our last chance to beat the Soviet Union in the race to send a man aloft.'

The cause of the failure of the Mercury-Redstone proved to be trivial. A connection had shorted and set off a confusing list of instructions that effectively triggered the abort sequence. Even before the rocket left the pad, a spurious electrical glitch sent a signal to the engines that they had come to the end of their prescribed burn time, and they shut down. This, in turn, triggered a confused response from the capsule's escape tower which was released and shot several thousand feet high before falling back to earth. Later it was found that the faulty connection was caused by an engineer who had filed down an electrical contact by a mere quarter of an inch so that a recalcitrant plug would fit in its socket.

Despite the endless setbacks, behind the scenes plans were being finalized for the first manned flight. The question was – who should go first? As the astronauts eyed each other up, it was difficult to assess who would be 'the one' who made history. The press and public had more or less decided that John Glenn should be first. He was the serious-minded, church-going hero who from the cradle had set his feet on this course. It was inevitable. He would be the one with the cool head and the calm ability to ride those eternal skies and smile his wise smile at the adulation when the job was done. Bob Gilruth, however, asked the astronauts themselves to choose. He asked them to cast a vote as to who should go first, if they couldn't go themselves. Nobody chose John Glenn. Perhaps the lectures at Cocoa Beach had played their part. Perhaps he wasn't one of the inner circle, his sense of responsibility disqualifying him from the forbidden fun.

On 19 January 1961, Bob Gilruth announced his choice. They were all good men but Alan Shepard would be 'the one' to go on the first trip. Gus Grissom would be the second and John Glenn would be backup to them both. Alan Shepard, the man's man, and also the ladies' man – the fastest driver, the hardest drinker and all-round favourite – was to fly the twentieth century's ultimate airship. Bob Gilruth explained that the decision would not be made known to the public. They would no doubt continue to bet on John Glenn and he would have to smile his way through the charade as though they were right.

The next crucial stage was to test the Mercury-Redstone with a chimpanzee as passenger at the end of January – and if this was a success, the way would be clear for a manned flight. Several chimps had been in training for some months at the Holloman Air Force Base at White Sands. Chimp 'No. 65' was considered ideal astronaut material – he was not yet permitted a name since this might give him too much personal profile. Brought over from Africa when he was quite young, No. 65 had soon found himself in a school where there was a distinct emphasis on punishment. The tutors were obsessed with it. In their immaculate white coats, sporting a stick or length of rubber hose, their demands were law. Failures to fulfil their requirements or the odd attempt at escape were met with memorable hostility. In spite of these hazards, No. 65 was particularly clever at making it clear that the weightlessness training, the centrifuge and the 'increased gravity' trick he was not doing from choice.

The main emphasis in the chimps' education was on 'operant conditioning'. This system rewarded correct behaviour and punished mistakes. No. 65's efforts to please his tutors' demands for him to push certain combinations of buttons and levers correctly were rewarded with banana pellets. If he lost concentration and made a mistake, the inscrutable tutor was quick to shoot electric bolts through his sensitive feet. The astronauts' doctors were assuming that if a chimp could cope with the rigours of space, concentrate on a correct sequence of button pushing and come back sane, then so could a human astronaut.

The chance to make history came to the reluctant chimp on 31 January 1961. He was wired up with every kind of sensor on and inside his body. The electric clamps were attached to his feet. He was strapped to his seat and transferred to the Mercury capsule. The countdown began. With an air of boredom, the amber eyes of No. 65 took in what looked like the familiar surroundings of his training capsule. He appeared relaxed as he sat on top of thousands of gallons of highly inflammable fuel 90 feet up in the air – evidently another session with the banana pellets. He had learned how to avoid the shocks.

Emmwood—London *Daily Mail*
"... THEN, AT 900,000 FEET, YOU'LL GET THE FEELING THAT YOU MUST HAVE A BANANA!"

Off went the rocket at an insane rate, going too fast, using up fuel ahead of schedule. The escape tower fired as if it were an abort, driving the Mercury capsule further than intended, as though the devil were chasing it. Unknown to the chimp, there was a fault in the electrical equipment. The punishment/reward system was faulty. For No. 65 the trip was a chastening experience. He could cope with the weightlessness and the very high 17g's of gravity which flattened him like a cartoon character meeting a steamroller. But no matter how brilliant he was at pushing his buttons and levers – he achieved Olympic standard in trying to avoid the horrible zizzing in his feet – to no avail. He received electric shocks all the way.

The capsule went higher and further than was intended that day – No. 65 banged frantically on his levers all the way – before it was walloped down into the ocean 130 miles off course. He sat in his chair topsy-turvy, watching the water rise in the capsule. The demanding buttons and levers were now quiet but the water continued to rise. One and a half hours later, the navy collected the capsule, saving him from a watery grave – just. Since he had survived, No. 65 was officially permitted to have a name: Ham. Later, at his debut with the press, Ham made his complaints known. The one apple he had been given was obviously a paltry recompense for proving that a chimp could do an astronaut's job, even with his feet on fire.

Analysis soon showed that several things had gone wrong on Ham's flight. The fuel flow rate to the combustion chamber was too fast. Consequently the rocket rose too quickly, using up too much fuel and flying too high, 157 instead of 115 miles. The angle had been steep, sending it off course. A faulty valve had released the pressurized oxygen in the capsule. There was a vibration problem. And Ham himself had experienced 17g's – close to g-loc – the point at which he would lose consciousness and eventually die. Nonetheless, Bob Gilruth, Max Faget and other key members of the Space Task Group were pleased that Ham had survived and determined that the next launch, planned for 24 March, could be Al Shepard's. Von Braun was not so sure. He summoned his core team and each member was asked whether he thought it was safe to go for a manned flight. The odds of a successful mission, he explained, were between 88 and 98 per cent. If they delayed, the Soviets could win. If they went ahead and suffered a fatality, the entire enterprise could be destroyed.

Von Braun wanted a unanimous decision. Only one person objected – Kurt Debus, who played a key role in launch proceedings at the Cape. He argued that the vehicle should be proved safe beyond doubt, convinced there should be a perfect Mercury-Redstone flight before the life of an astronaut could be put at risk. Where would the space programme be if there was a fatal accident? Von Braun respected his decision and informed NASA authorities they had to have one more flight. Al Shepard was furious when he heard the news and urged Chris Kraft to overrule von Braun. But he did not. 'When it comes to rockets, Wernher is king!' Kraft told Shepard.

Secretly Kraft was fuming. 'We had a timid German fouling our plans from inside!' he raged. The manned flight was now set for 25 April. They were all aware that the decision could cost America the chance to win the race.

• • •

By November, with repercussions from the Nedelin disaster and technical problems with the Vostok, it was apparent that Korolev's first piloted launch, provisionally pencilled in for December 1960, had to be postponed until at least February. Although Korolev's work on the R-7 and the Vostok were not directly affected by Nedelin's death, delays did ensue as many of the now depleted design bureaus, such as Glushko's, were also involved in the space effort. And the problems continued. On 1 December, two dogs, Pchelka and Mushka, along with scientific instruments for studying radiation, successfully followed the orbit intended for a manned flight. After seventeen orbits, however, the retro engines were fired but they fired for too short a time, which meant the Vostok would not land on Soviet soil. The Soviet need for obscuring everything under a thick shroud of secrecy led to a decision to explode the capsule on re-entry using a mechanism already installed. The dogs were sacrificed rather than allow the capsule to fall into foreign hands.

The faults were corrected and on 22 December two more dogs were sent into space. This time the upper-stage booster engine failed. The capsule separated from the rocket to land in the deep snow and bitter cold of remote Siberia, where some days later it was found, the dogs still alive. Built into the capsule was an explosive device designed to detonate after sixty hours. As more than sixty hours had passed, the rescue team approached with caution, wallowing in the fresh snowdrifts. They successfully defused it and found two very cold dogs. The capsule arrived back at Baikonur in January with a list of faults that further deferred manned flight. The ejection seat had jammed, remaining in the capsule and causing damage. The two pods in the Vostok designed to separate on re-entry did not separate until the heat of re-entry burned the connection. The self-destruct system had also failed.

There were now more failures to list than successes. Certain that the Americans were close to launching a man into a suborbital flight, Korolev could not rest from the pace he had set himself. With the imminent approach of the first Soviet launch, the twenty short-listed cosmonauts were asked the same question that Gilruth had asked the American astronauts: who should be the first one to fly assuming that you cannot? Seventeen of them named Gagarin, commenting positively on his character and ability. General Kamanin narrowed the group down to a short list of six, including Gherman Titov and Yuri Gagarin.

According to Korolev's biographer Aleksandr Romanov, the cosmonauts, aware of the endless difficulties, decided to go and speak to Korolev, who met them somewhat anxiously, waiting to see what they would say. But far from suggesting that it was too dangerous to fly,

Gagarin reasoned that had there been a man on board, all the launches would have ended in success. If the automatic system had failed, Gagarin explained, he would have switched to manual. Korolev was heartened. He thanked them for their belief in him and their dedication, but insisted he would not send them into space until he was 100 per cent sure of the outcome.

It was hoped that two more automated tests in March would right the problems. The first one on 9 March collected a motley crew into the Vostok craft. As well as Blackie, the dog, there were eighty mice, guinea pigs and various reptiles. A man-sized dummy accompanied them reclining languidly in the ejection seat. He was made to look as lifelike as possible, complete with mouth, eyes, eyebrows, 'even eyelashes', recalled Mark Gallai, an acclaimed test pilot who was advising on cosmonaut training. In their enthusiasm, rather than waste space the scientists stuffed the dummy's hollow body and limbs with yet more mice and guinea pigs. Then for decency's sake they covered him with a white smock and gave him a name: Ivan Ivanovich. 'There really was something deathly unpleasant in the mannequin sitting in front of us,' continues Gallai. 'Probably it is not good to make a non-human so much like a human being.'

One of the aims of the mission was to establish whether the communications systems worked in the Vostok, and whether a human voice could be heard. An automatic recording would be placed inside the dummy. The Soviet mania for keeping everything secret now asserted itself. Radio stations around the world would pick up what was being said; they could not therefore have anything that sounded like a cosmonaut. A recording of a Russian man singing was suggested. But then listening Americans would assume that the Soviet Union had sent a man up into space, who had promptly gone off his head as the 'Song of the Volga Boatman' resonated around the world. Finally, a recording of a choir was deemed suitable. No one would think a choir had been sent into space. On 9 March, a nervous-looking Ivan Ivanovich, fidgeting constantly as his wriggling animal insides settled down, took his first jaunt into space and was so entranced by his journey that every so often he sang like a heavenly choir.

Korolev was delighted. It was a perfect flight. The only unanticipated stumbling block, according to Mark Gallai, was that Blackie turned out to be a 'foam-eating dog' who insisted on consuming all the foam padding around her in her space cabin – causing concern that she would become ill. Blackie, however, suffered no ill effects from her curious diet. Another trip was scheduled for Ivan Ivanovich for the end of March. If that was a success, the next mission would be manned.

With the Americans apparently so close to claiming victory, Korolev could hardly bear to waste a day. He had had his fill of dogs, snakes, guinea pigs and singing dummies. He wanted a man in space; flesh and blood and intelligence that would see and feel the wonder of it. He would quite willingly have gone himself. He felt so close to his original vision, almost touching the infinite. 'We are getting ready and believe in our work,' he told Nina. 'It is important that all should believe that everything is going to be fine and I myself should believe in that also …'

CHAPTER FIFTEEN

'Which one should be sent to die?'

'It's hard to decide which one should be sent to die,' General Kamanin noted in his diary, unable to decide between Yuri Gagarin and Gherman Titov. Gagarin impressed his tutors with his serious attitude to work. He was committed; his actions were measured and careful; he didn't make rash decisions. And he was charming, with an innate sense of tact; he wore a permanent welcoming smile, as though he stood in perennial sunlight. Wrestling with the decision, Kamanin noted: 'I have kept an eye on Gagarin and he did well today. Calmness, self-confidence, and knowledgeability were his main characteristics. I've not noticed a single inappropriate detail in his behaviour.' Yet he was equally impressed with Titov, who had also come through the training programme brilliantly. 'He does his exercise and training more accurately and doesn't waste his time on idle chatter. Titov has a stronger character.' More complex in character and not so obviously popular, he was known to question authority and challenge what he called 'silly questions'.

Korolev had no such doubts. Since their first meeting he had known Gagarin was his man: the peasant boy from Smolensk, a product of the Soviet system, the tough graduate of a welding school who had had the sensibility to remove his shoes before entering the hallowed ground of Korolev's beautiful Vostok. When Korolev had asked Gagarin the question he had asked all the trainee cosmonauts after the session in the pressure chamber – what had he been thinking about? – the usual reply was, 'all my life passes in front of me'. Gagarin's reply was: 'What have I been thinking about? I have been thinking about the future, Comrade Chief.' Korolev liked that answer. 'Bloody hell! Comrade Gagarin,' he had replied. 'One could only envy your future!' He later confided in a colleague, 'I like this boy, he is so communicative and so gentle.' Korolev's mind was made up. Gagarin would be first. Titov would be the backup. This decision also found favour with Khrushchev who liked the idea of a man, like himself, from the rural heartland. Titov was certain to be an excellent choice, but Gagarin was a man of the people. It was a time to show the world what the communist system could produce.

In the tense atmosphere of the final few weeks, the Soviet testing programme lost its first cosmonaut. In Moscow, at the Institute of Aviation and Space Medicine, Valentin Bondarenko, was coming towards the end of a two-week spell in the dreaded isolation cell of silence. As he was waiting for the session to end, he stripped off the irksome sensory equipment worn throughout the test and soothed his reddened skin with cotton wool dipped in alcohol. In seconds, the room, freshly topped up with oxygen, was ablaze as one of his pads landed on the cooker. He was, of course, locked in and as the cell was soundproof no one could hear his cries. He tried to put out the fire but his clothing caught alight. Eventually, those on duty came to the rescue but it took several minutes to open the door. He was terribly burned and had lost every inch of his skin and all his hair; his eyes had melted completely. He was taken to hospital but death came within a few hours. News of the accident was suppressed for twenty-five years.

The very next day, Korolev heard that von Braun had launched a successful unmanned flight. On 24 March, the additional Mercury-Redstone launch he had requested had gone perfectly to plan. Alan Shepard followed its path; its journey burned into the landscape of his mind. He could have been on that flight. He could have made history. The sounds of success would have rung in his ears for a lifetime. 'We had 'em by the short hairs and we gave it away,' he declared, now convinced they would lose the race. Von Braun was prepared to agree that the next launch would be manned. Shepard's chance would come soon and he could yet be the first man in space. The date was confirmed for 25 April. Thousands of miles away, Korolev was left to wonder whether this was von Braun's defining 'test' flight? Was this the one that would signal an American flight in space?

Two days later, in freezing March weather, he nervously sent the R-7 and its Vostok into orbit. This would decide his own fate. If it was a success, a manned flight would be next. Korolev watched as the second Ivan Ivanovich and his dog went into orbit. With his menagerie of animals and the tape of the heavenly choir, now with the addition of recipes for cabbage soup, to confuse any listening Western ears – he made the perfect flight. Dressed in a cosmonaut's orange suit, his dumb presence was witness to the soaring flight. His sightless eyes took in a God's-eye view of the world. His unhearing ears heard the retro engines fire to perfection. His unfeeling limbs felt the rush as he landed on a quiet edge of the woods in falling snow near a remote village.

Baffled villagers watched a man with a parachute fall from the sky. When he came to earth he remained motionless. He must desperately need help. They looked at his deathly face, his lumpen body, his wide-open eyes, but before they could do anything to help, soldiers appeared and drove him away – with casual disregard for the fact that the flying man appeared to have died.

When the military brought Ivan Ivanovich back from the snowy woods, the success of the flight was confirmed. On 28 March, Korolev presented his case to the State Commission in Moscow and requested the final go-ahead for a manned flight. Originally he had hoped that the first manned flight could be timed to coincide with the great Soviet May Day holiday on 1 May, but Khrushchev, weighing the record of success and failure, had objected. If the smiling face of Gagarin, riding his fiery chariot, was blown into a million pieces somewhere in the great black void of space, it would be an unhappy image for May Day, and for all the May Days to come. The Russian soul would be full of grief. Korolev decided to bring the date forward. On 3 April, the Presidium of the Central Committee approved the launch of Vostok 3A with a cosmonaut aboard. Yuri Gagarin said goodbye to his wife and two children on 5 April – without telling them where he was going – and flew with the other cosmonauts to Baikonur.

Arrangements were made for TASS to announce the event the moment Vostok was in orbit. Three sealed envelopes were sent to TASS. The first provided information in the event that Gagarin reached orbit successfully. The second was a short statement prepared in the event that he should die. The third was to facilitate his safe return if errors occurred in the trajectory and he landed outside the Soviet Union. Should something go wrong on the flight and the cosmonaut be forced to eject, rescue would be speeded up if the world was on alert. And with TASS blazing it over the news wire, no one could accuse Gagarin of being a spy.

Flying between Moscow and Baikonur for last-minute preparations, Korolev hardly slept, fitting in a few hours here and there between essential work. When he woke up, he seldom felt refreshed. His fatigue was entrenched, the result of endlessly pushing himself beyond endurance. Korolev knew he was not well. Yet who could rest with the day that had seemed out of reach all his life finally in sight? He knew of von Braun's planned April launch. That clever, articulate German was near, so near, to claiming victory. From his perennial position in the shadows, within a week Korolev would begin an unstoppable process. He would give the final command that would focus all his years of accumulated knowledge into one daring adventure, and the world, never aware of who had pulled the strings, would be astonished.

On 9 April, a reception for the six cosmonauts was held in an improbable summerhouse at Baikonur on the banks of the River Syr Darya. It had been built by the now-departed Marshal Nedelin for use between icy winters and burning summers. With its archways and columns and blue-painted pillars, it suggested a bygone aristocratic taste. But no ghosts from Tsarist Russia would have dared to be present on such a day. This was a formal occasion for the Communist Party to celebrate achievement. At least seventy people attended; the elite of the Soviet defence and space programme were there to acknowledge their young hero, a man tried, proven and hand-picked from among the ordinary Soviet people.

Starched white cloths covered the tables. Wares rarely seen at Baikonur, many of them in crystal and silver, were displayed as though they belonged there; delicious dishes were presented among the abundant piles of oranges and other fruit; wine and vodka flowed. The final improbability was the heavy perfume wafting in on the breeze from the desert thick with its spring flowering.

The Commander of Strategic Missile Forces, Marshal Moskalenko, gave a speech congratulating Gagarin on the 'high and important task given to you by the Motherland'. Korolev also spoke, unable to resist setting out his hopes for the future when he believed a clutch of cosmonauts would man a space ship. The press was busy snapping away as Yuri Gagarin stepped forward to receive his commission. Surrounded by the might of officialdom in splendidly decorated uniforms, he looked far too young for the weight of the task. He smiled his confident smile, resolved and ready to venture into the vast, uncharted emptiness and begin the most extraordinary journey in the history of the world.

On 11 April, Korolev walked solemnly beside the great R-7 rocket, adapted with an upper stage that would take Gagarin into the future. The rocket was slowly moved the two and half miles to the launch pad. It was mounted on a rail car and held very carefully in tension. If any small thing went wrong, Korolev had to know. Only he could interpret how the unevenness of the road might unsettle its complicated temperament. With every step he took, his ears were listening for the infinitesimally different sound that might signal delay. The launch was set for the next day, 12 April 1961. Whatever the outcome, the date would be burned in his memory. Although it was only April, the weather was warm and Korolev had removed his hat. He moved slowly at the base of his R-7, a dark, determined figure guarding the great obelisk, pale and gleaming in the hazy sunshine.

In the afternoon, Korolev and Gagarin went to the Vostok alone. From their position high up at the top of the pad, Korolev explained the procedures for the launch the following day. There was so much still for Korolev to impart and Gagarin listened with complete concentration. But as he listened, he watched the older man grow pale and hesitant. Korolev was ill. His heart was troubling him. He had to be helped down to rest. A doctor was summoned: he patched him up and gave him pills and warned him he had to rest. Korolev felt too weak to disobey.

Gagarin and Titov spent the night before the launch in a cottage near the pad. Gagarin was trying not to show his elation at being chosen for the first space flight; Titov was trying not to show his desolation at having been rejected. There was, of course, the remote possibility that Titov would take over if Gagarin, for some reason, could not go. The quiet cottage was meant to provide a good night's rest, but they discovered that the doctors, anxious to monitor their reflexes to the last, had placed sensors on their mattresses to register movement. As a consequence they dared not move in bed and hardly slept at all. Korolev, recovered enough to ignore his doctor's advice, came to say goodbye and wish them well. 'In another five years we shall be able to fly into space as we now fly to holiday homes,' he joked.

At 2 a.m., Korolev had finally finished work. He walked to the small wooden house he stayed in at Baikonur. The housekeeper, Elena Mikhailovna, brought him the strong tea and rusks he requested. She drew his attention to his insubstantial diet, telling him that his wife had called reminding him to eat well. Korolev replied that there were more important things on his mind just now. It was impossible to rid himself of the worries concerning the launch now that he was alone. There would be no sleep. He was responsible for Gagarin. The young man trusted him completely, but nothing was guaranteed.

In a conversation with Khrushchev on sending a man into space, the Soviet leader had demanded: 'Is there 100 per cent guarantee?' 'Everything possible has been done,' Korolev had replied, adding, 'of course something unexpected is always possible in this business.' That was the unpleasant truth. The last two launches with Ivan the dummy had been successful, but nagging at Korolev's peace of mind was the knowledge that, of the last sixteen launches of the R-7 using the same engines that would put Gagarin in space, eight had failed, several with failure of the upper-stage engines. If they did not succeed this time, Gagarin would be flung into the stormy seas around Cape Horn and to certain death. Knowing that Khrushchev's desired 100 per cent guarantee was a dream, the possible failure of the upper-stage engines haunted him.

There were so many uncertainties. The launch was the most dangerous time, particularly the first forty seconds. Then there was weightlessness: the doctors still took it very seriously, unsure what effect prolonged exposure to it might have. And there was nothing guaranteed about Isayev's retro engines. If they did not perform correctly would Gagarin be burned alive or doomed to wander space forever? There was a backup plan if the retro rockets failed. He could return to earth in a decaying orbit as the rocket gradually slowed down over the next eleven to twelve days, and food was supplied. Although strictly forbidden,

Korolev had given Gagarin the three-number code that would unlock the system to give him some manual control in an emergency. If Gagarin completely lost his reason during weightlessness, as some experts suggested would be the case, would an unbalanced act borne from loss of sanity put the mission in danger? Sleep evaded him as the endless worries took their turn in the spotlight of his mind. At five in the morning his wife, Nina, rescued him. She telephoned to wish him well. He lied and said his health was good and, yes, she had woken him from his sleep.

Korolev joined Titov and Gagarin for a real cosmonaut's breakfast – concentrated food squeezed from tubes. Gagarin was shocked to see how exhausted Korolev looked. He wanted to hug him as a son would hug his father, he wrote later, and impart something of his own strength and health. Titov and Gagarin tried to reassure him. 'Everything will be all right. Everything will be normal,' they said. An old woman who had once lived in the cottage where the cosmonauts spent the night was determined to present Gagarin with some flowers. Her own son had died as a pilot in the war, she said.

While being kitted out in his space suit, Gagarin was taken aback when some of the technical staff asked for his autograph. 'It was the first time he seemed momentarily at a loss,' reported one witness. 'Is it really necessary?' he asked. He was taken by bus to the launch gantry. Korolev was waiting there with Kamanin and Moskalenko. They stood dwarfed at the base of the huge rocket. 'After leaving the bus, everyone got a bit emotional and started hugging and kissing,' recalled Kamanin. 'After wishing him a nice journey some were saying goodbyes and even crying. We practically had to tear him away from the crowd by force.' At the last minute, Mishin recalls, there was a sudden concern that, if Gagarin landed in foreign territory, he might be seen as an enemy rather than recognized as the world's first cosmonaut. Paint was hurriedly fetched and the letters 'USSR' were daubed in gold on his helmet.

Oleg Ivanovsky and the technicians escorted Gagarin on the two-minute journey in the lift to the top of the Vostok and helped him into the ejection chair made to his specific measurements. His straps were tightened and life-support systems plugged in. In the bunker, lights confirmed his pressure suit and oxygen supply were connected. At 7 a.m. the communications system was switched on. The ground call signal for Gagarin was 'Kedr' (cedar). The call signal for Korolev was 'Zarya-1' (dawn). Launch was set for 09.07 hours. The transcript from the Russian State Archive of Scientific and Technical Documentation shows Korolev trying to maintain a light-hearted tone with Gagarin while launch preparations and communications tests were underway:

Korolev: There are tubes of food waiting for you in storage – your lunch, dinner and breakfast.
Gagarin: That's clear ...
Korolev: Do you read me?
Gagarin: I read you ...
Korolev: There are sixty-three of them. You will get fat.
[Gagarin laughs.]
Korolev: You can eat everything when you arrive today ...
Gagarin: Nope – the main thing is that there is some sausage – to have with *samogon* [home-brewed vodka].
[Everyone laughs.]
A little later:
Korolev: Are you busy right now?
Gagarin: ... I am not very busy. What do you need?
Korolev: I found the sequel to the 'Lilies of the Valley'. Do you get it?
Gagarin [laughs]: I get it. I get it. 'In the Thicket of Bulrushes'.
Korolev: We'll sing it tonight.
There is a pause.
Gagarin: The buggers, they will kill me with their communication tests.

In the main room of the bunker, Korolev, although immersed in technicalities, was clearly under strain. In front of him was a red telephone which would only be used if the launch was a disaster. He would communicate the signal that would throw Gagarin clear from what might be a blazing inferno. Two other people in the bunker knew the password. The other essential thing he had insisted on was a telemetry system that would mark in patterns of 'fives' if the upper-stage engines worked as planned. If the engine failed, a pattern of 'twos' would be shown.

Korolev: This is it. Good luck.
Gagarin: Thank you.
Korolev: Good luck, dear.
Gagarin: See you …
Korolev: Good luck. See you.
Gagarin: Today in Kuybyshev.
Korolev: It's a deal.
[Gagarin laughs.]

On the gantry, it was time for Ivanovsky to say goodbye to Gagarin and close down the hatch of the Vostok. Yuri was still whistling quietly to himself – the tune of 'The Motherland Hears, the Motherland Knows' – and showing no sign of anxiety. Ivanovsky felt strongly that it was wrong to keep the three-number code for manual control from the cosmonaut. Heaven only knew what dangers Yuri was going to find. He needed all the help he could get.

'Yuri,' he whispered so that the intercom would not pick it up. 'The three numbers are …'

'Yes, I know. Korolev told me, and Kaminin,' Yuri laughed. 'Thank you anyway'.

It took some time to screw down the hatch and when it was completed, an angry Korolev on the gantry phone asked why the hatch was not sealed. Ivanovsky assured him that it was. Korolev assured him that it was not. There was no confirming light in the bunker. Ivanovsky 'turned cold'. He was convinced he had carried out the procedure correctly. Using colourful language, Korolev ordered him to repeat the process.

Korolev: Yuri Alexeyevich. It looks like after the hatch was closed one of the contact wires did not show. It got pinched and this is why we will probably take off the hatch and put it back on later. Do you read me?'
Gagarin: I read you. The hatch is open. I am checking the alarms.

The heavy hatch was removed and sealed once more using the thirty screw bolts.

If there were a fault this time, the launch would be delayed. Ivanovsky could see a little of Yuri's face from a mirror on his sleeve.

He was smiling, in spite of the strain of waiting, arguably even greater than it might have been since witnessing a terrible event a few days back when a rocket had exploded on takeoff. Yuri appeared calm; he would not allow negative thoughts to encroach. The explosion he had witnessed inadvertently with the other cosmonauts had been that of a combat missile. It had been most terrifying. He would rather not have seen so graphically what a launch-pad disaster looked like. But he was a cosmonaut. He carried on singing softly. Korolev now confirmed that the hatch was airtight.

In the bunker, the atmosphere was growing tense as the time drew near to liftoff. Korolev was thankful that he was alone with his assistants. The remainder of the commission, including Glushko, were in another room. He briefly spoke again to Gagarin, asking him how he was. Gagarin had requested music. There were only minutes left now before the final moment. The propulsion system and the life support had been checked. Every detail connected with the massive rocket had been checked and rechecked, the gyros, the fuel. Korolev grew more agitated, listening to the conversation of the launch technicians and frequently butting in. He took some tranquillizers, which made no visible difference.

Gagarin began whistling a Soviet song called 'Lilies of the Valley'. His pulse rate and blood pressure were normal. Mishin was concerned. Over the radio he could hear Gagarin had lapsed into singing a frivolous version of the song that he had taught him during training. Everyone could hear it.

> Today you bought me not a bouquet of red roses
> But a bottle of Stolichnaya vodka
> We'll hide in the bulrushes and
> We'll get drunk out of our skulls
> So why do we need these goddamn lilies of the valley?

With a few minutes to launch, Korolev was controlling his voice with effort, forcing it to sound normal. 'We could tell by the sound of his voice – heavy and broken – that the Chief Designer was more agitated than anyone else who was there,' observed Mark Gallai. 'He hid it well enough … but I was aware of his heavy breath and the beat of the blue vein in his neck.' Korolev looked ill. Soon would come the moment of no return. If the takeoff stalled, he was ready with the abort code. He gave the order for launch. The button was pressed. Ignition began, unleashing the sounds of a giant orchestra tuning up.

'Ignition is being given, Kedr. I am Zarya-1'

'I read you. Ignition is being given.'

'The preliminary phase.'

'I read you.'

'An interim phase.'

'Got it.'

'Complete takeoff.'

'*Poyekhali!*' Gagarin shouted. 'Let's go!'

Below the rocket was an inferno of white and orange flame sucking in and spitting out. The vibration was so great the bunker seemed to shake, almost a part of the launch. Lying in his citadel with walls of sheet metal, perched on top of the R-7 now glutted with fuel, Gagarin became aware of the subtle movement. The rocket shivered like an object in the wind as the gantry fell away. He had a sense of the struggle as the inert machine fought against the surge lifting it. He listened, trying to understand the sounds. Was the inferno under him going to work or would he suddenly find himself shot through the hatch in a bid to escape the roasting flames. The jangled noise of an orchestra out of tune grew and grew. The sounds were difficult to identify. Slowly power was lifting him. At 09.07 precisely, Lieutenant Yuri Gagarin began to make history.

The orchestral cacophony was turning into a roar. It felt as if the rocket was swaying. Gagarin's pulse rate was 157.

Twenty seconds into the flight:

'Zarya-1, I am Kedr. All is going well. There is a faint noise in the cabin. I am feeling well. I can feel the overload, the vibration. All is well.'

'I am Zarya-1. We are all wishing you a good flight. Is everything OK?'

'Thank you. Bye bye. See you soon, dear friends. Bye, see you soon.'

Two minutes into the flight, the four booster rockets fell away. Gagarin began to feel the g-forces pressing him down. Almost a minute passed. On time, the nose cone fell away. Suddenly, he could see the world below him spread out like a map.

'Kedr. I can see rivers and the folds of the terrain. I can tell them apart, well ... I can see the earth. Visibility is fine.'

'I am Zarya-1. How are you feeling? I am Zarya – over.'

'I am Kedr. I am feeling fine.'

The g-load was increasing, pushing Gagarin back into his seat, pressing hard on his facial muscles and making speech difficult.

In the bunker, Korolev was going through hell, his worst fears realized. The telemetry system had been delivering a series of 'fives' but now it changed, issuing a series of 'threes'. 'Twos' would mean disaster. The whole venture annihilated; Gagarin in unspeakable danger. Should he eject? In a moment so intense, seemingly outside measurable time, the many lights and dials demanding attention hardly existed as Korolev listened to the telemeter. Another 'three'. Another 'three', and then a 'five'. The 'fives' were back. Korolev felt sick with relief. He was shaking. He wanted to hear Gagarin.

'I am Zarya-1. All is well. Everything is working.'

'I hear you. I can feel it working. I am watching the earth.'

Gagarin lost contact with Zarya-1 and Tyura-Tam at around seven minutes into the flight and communications were transferred to Zarya-2 at Kolpashevo and at Zarya-3 in Yelizovo. 'At the moment of communication transfer, there were a few unpleasant seconds. The cosmonaut did not hear us and we did not hear him,' recalled one engineer. 'Korolev, who was standing next to me, was very worried. When he picked up the microphone, his hands were trembling. His voice changed, his face changed colour – changed so much we could not recognize him!'

About ten minutes into the flight, the upper-stage engines had done their job and were relinquished. Travelling at 28,000 mph, Vostok had left the earth's gravity and the protective halo of the atmosphere. Gagarin began to experience weightlessness, the familiar substance and weight of his body feeling like cotton wool. TASS was informed of the launch and directed to open the envelope that told of a successful manned space flight. The Vostok was turning slowly, flying at five miles per second. Gagarin had no feeling of speed. His view of the earth changed constantly as the capsule rotated. The hazy pictures of him coming across the television monitor showed him alert and well.

'I am Zarya. All is going well. Can you hear me? How are you feeling?'

'I am Kedr. I can hear you fine. I am feeling fine. The flight is going well. I am watching the earth … I am feeling well.'

As the Vostok was controlled from the bunker, there was little for Gagarin to do but observe the great display outside his window and monitor the equipment. His flight took him over Siberia and into the Arctic Circle. He found weightlessness far from being a worry, was actually enjoyable, but he was not ready for objects moving about and lost his pencil as it floated out of reach.

'The feeling of weightlessness is interesting. Everything is floating [joyfully]. Everything is floating! Beautiful. Interesting.'

He described the beauty of the earth, the radiant intense blue of the sky and the rainbow colours surrounding the earth. As he crossed America it was dark, but the sky was bright with a myriad brilliant stars. In the South Atlantic he noticed the sea was grey.

'I can see the earth's horizon. It is such a pretty halo … It is very beautiful … I can see the stars floating by through the Vzor. It is a very beautiful spectacle. The flight is continuing through the shadow of the earth. I am watching a little star in the illuminator. It is going from left to right. The star has disappeared. It is disappearing, disappearing … I am watching the earth, flying over the sea …'

Meanwhile at base there was growing concern because the expected announcement from TASS had not been made. In the telephone room, Gallai knew they were anxious to have the report before Gagarin attempted to land. Eventually, fifty minutes into the flight, 'when it seemed it was not possible to wait any longer', the music on the radio station was interrupted:

> Attention, Attention. All radio stations of the Soviet Union are making an announcement. It will be made in a few minutes …
>
> The world's first satellite ship Vostok with a human on board was launched into an orbit around the earth from the Soviet Union. The pilot cosmonaut of the space ship satellite Vostok is a citizen of the Soviet Union of Soviet Socialist Republics, Major of Aviation: Yuri Alexeyevich Gagarin.

After an hour, as the ship completed one orbit around the earth, it was starting to circle back towards the Soviet Union. The order came from base to align the Vostok to the correct attitude for re-entry into the earth's atmosphere. The retrorocket fired at 10.25. Gagarin noted and recorded this success. But the success was only partial. The rocket fired for forty seconds. Then came a sickening jolt and the Vostok was sent spinning. Gagarin noted: 'The rate of rotation was almost 30 degrees per second. I was like an entire *corps de ballet*: head, then feet, head then feet, rotating rapidly. Everything was spinning around. Now I see Africa, next the horizon, then the sky.'

When Isayev's retro rockets fired, Gagarin's descent capsule should have separated from the instrument module which was strapped to its back. This did not occur. They were still joined by cables, turning wildly over and over like a pair of cuff links at something like 30 degrees per second. Gagarin could feel the heat coating burning through and ominous cracking sounds. It was becoming very hot. After firing the retro rockets, he had no communication with bunker control. They did not know of his ordeal. He understood that his descent capsule had failed to separate and knew there was nothing ground control could do about it. He thought that if he could stay conscious, he had a reasonable chance of ejecting safely and landing on target. The major obstacle was getting back into earth's atmosphere under such impossible gyrations. Using code he transmitted the message that the retro rockets had fired successfully.

The g-load began increasing. About ten minutes of the tempestuous, tumbling, turning descent passed before the heat of re-entry burned through the cables, separating the two units. With loads as heavy as 8 g's, Gagarin was on the point of losing consciousness. His vision was blurred and grey. He made a huge effort to remain alert. There was a strange purple light in the cabin, which he thought must mean he was re-entering the earth's atmosphere. The capsule was still revolving, but less madly now.

Korolev was desperate for information. Although he had the report confirming the retro fire, experience told him that much could still go wrong. There was nothing he could do but wait. He wanted so much to see Gagarin standing in front of him unharmed, with that indelible trademark smile. While he waited for news, he telephoned Khrushchev, informing him of the historic flight. The telephone wires were hardly robust enough to transmit the Soviet leader's excitement. 'Is he alive, is he alive?' he wanted to know.

Yuri Gagarin became the first person in space on 12 April 1961.

LEFT: Crowds pack the streets of Moscow to celebrate Gagarin's triumph.

BELOW: Vostok 1 landing capsule. Gagarin ejected himself and parachuted to safety.

Yuri Gagarin with Chief Designer Sergei Korolev.

Wernher von Braun alongside his massive F-1 engines, which could consume 40,000 gallons of fuel a minute.

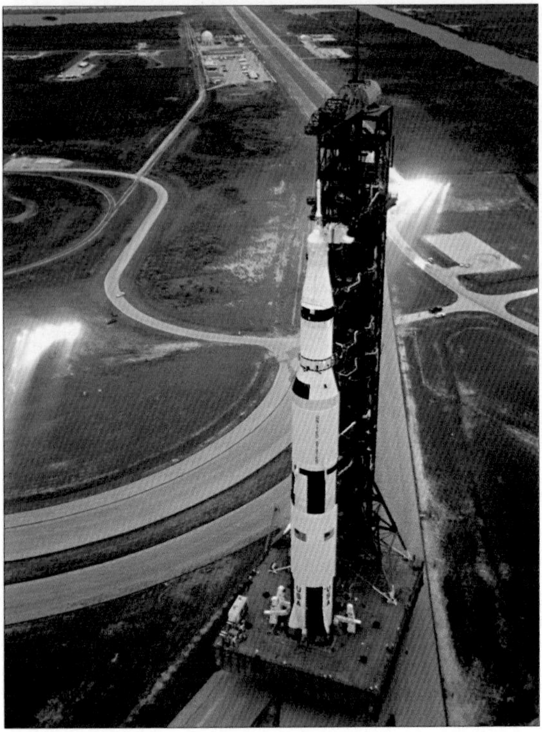

ABOVE: A section of the giant Saturn V rocket being transported to Cape Kennedy.

LEFT: The Saturn V – over 36 stories high – was moved on a giant crawler to the launch pad.

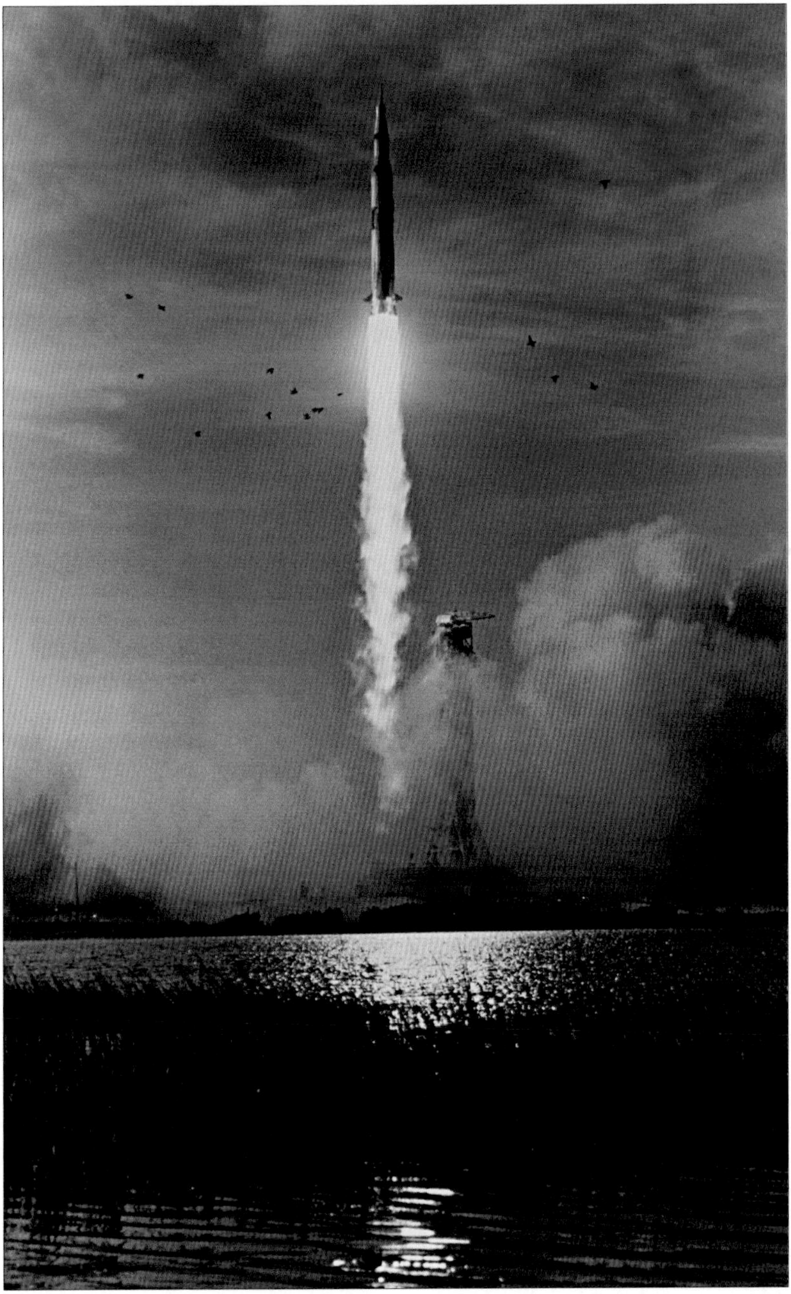

Apollo 8 sets out for the moon, 21 December 1968.

The Earth viewed from Apollo 8 orbiting the moon.

16 July 1969: Apollo 11 lift-off at the Cape.

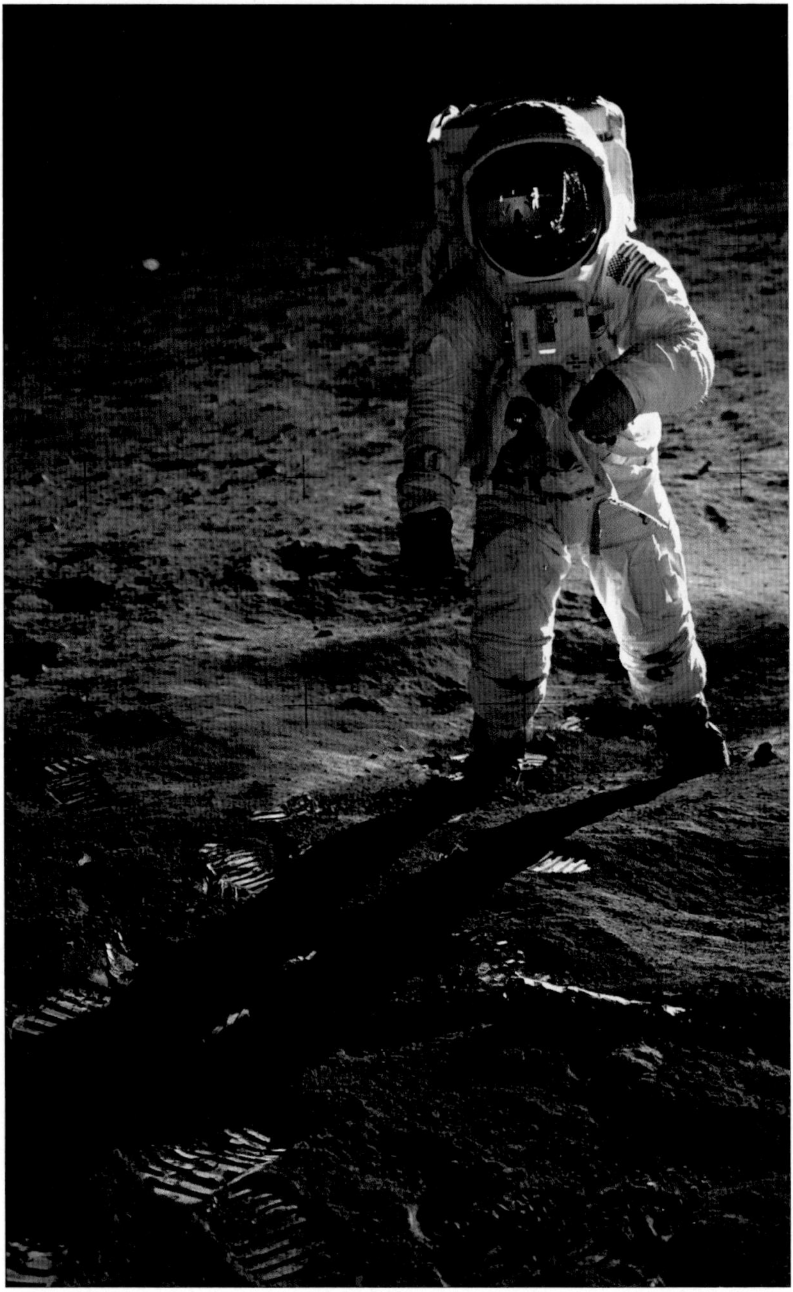

Buzz Aldrin walking on the moon.

In the turning capsule, Gagarin became aware of the rays of the sun, shooting through the porthole, nearly blinding him. He was in the earth's atmosphere. Soon he would be ejected. This could be a problem with the craft turning as fast as it was. Then a series of events suddenly unfolded. At about 20,000 feet, the main parachute opened, and the hatch above him instantly shot away. He felt vulnerable in the open cabin. The sound of it falling through the air was unnerving. Before he had time to ponder his position, he was shot out through the hatch in his ejection seat, thankfully avoiding the turning capsule and its parachute. When the rockets in the ejection seat had used up their fuel, he slowed then started to fall. From a great height he saw the land beneath him. Its features were growing larger very quickly. His parachute opened, then almost immediately the seat which had carried him through space fell away.

Now he hung in the sky making a slow descent. The Volga River and the features of his homeland were far below. He had parachuted when on training in this area. He knew it well. His journey in space had taken him nearly two hours. And in that time he had seen the world as no other human had ever seen it – the long, curving arc of the horizon, the beautiful transition between the turquoise atmosphere and the blackness of space. Now Korolev's incredible skill was casting him back down on familiar territory where his home and his wife were waiting, and where his footprints would mark a slower step.

He landed in soft ground near a village and struggled for some time with a valve before he could breathe fresh air. A woman and child were walking towards him but hesitated on seeing his strange shape. The child ran away. The woman was unsure. 'It's all right,' he yelled. 'I'm a Soviet, a friend.' Some workers arrived and a man on a tractor. They listened wide-eyed to his story of travelling in space. He told them his name. They were suddenly excited. 'You are on the radio,' they said. 'Your journey is being transmitted at this very moment.'

At Baikonur, before Korolev and the team could leave for the landing area, an impromptu celebration was hurriedly organized. The first toast was 'For Success'. Korolev downed the champagne and in time-honoured tradition threw the glass on the floor. Everyone was about to follow suit when one of the officials stepped forward and objected: 'The Chief Designer is allowed to do that, but we comrades can't. Who will answer to the glasses division?'

The champagne soon created a sense of gaiety. The deathly tension that had gripped the bunker for the whole of the flight was dissolved. Korolev was euphoric. He hardly needed champagne; he was intoxicated with the wonder of what had just happened. No one had ever seen him so totally unrestrained and free from anxiety. First they flew to see the capsule. Korolev examined it minutely, touching it, almost memorizing the marks of its journey as though it really was part of a dream that might fade. He simply could not be torn away.

Finally, they flew to be reunited with Gagarin. When Korolev saw him, he appeared like Lazarus returned from the dead. Suddenly shy, he did not know what to say. Emotionally, he was still adjusting to the enormity of the event. Gagarin came to the rescue.

'Lieutenant Gagarin reporting. All is well, Sergei Pavlovich, things are just fine.'

Korolev was moved, too moved to speak immediately, then slowly, emphasizing Gagarin's on-flight promotion, he said: 'Well done, Major Gagarin. You are a true folk hero.'

Moscow news was transmitting the story worldwide. The Soviets were celebrating having put one over on Kennedy and the Americans. There was a genuine interest and excitement over what seemed a miraculous endeavour. A son of the Soviet Union was a world-class hero.

At 5.30 a.m. in America, a news reporter decided the Soviet space story was hot news. He wondered what NASA's reaction would be. Ringing the press office in Florida, his call was answered by an overworked PR man trying to catch up on a few hours' sleep after dealing with publicity for Alan Shepard's flight in May. From his camping mattress on the floor, and with indignation in his voice, he shouted down the phone: 'Go away. We are all asleep.'

Next morning's headlines greeted America:

SOVIETS PUT MAN IN SPACE.
SPOKESMAN SAYS U.S. ASLEEP

The fourteenth of April was declared a national holiday. There would be celebrations in Moscow. The Russians had taken Gagarin to their hearts and were fiercely proud of his achievements, of his lonely flight around the world. In their hundreds of thousands, the people came to see their hero. There were crowds everywhere: at the airport, in the streets, lining the route to Moscow, jammed into Red Square. The media, busy unfolding Gagarin's story, had in the process turned the country boy into a glamorous and disarming figure, yet one who was of the people. The masses in Red Square were enraptured as Gagarin stood on the balcony with Khrushchev and Brezhnev on either side. When Khrushchev, grinning from ear to ear, bathing in the reflected glory, hugged the hero, the crowds roared until their throats were sore. Their applause was unceasing. This was a day for recognition and reward.

But not for Korolev. He was just someone in the crowd, undistinguished, unrecognized, unable to wear the medals of his awards, given no place at the high table and left to fend for himself when the fan belt broke on his car. He could not reach Red Square and eventually abandoned his efforts. He would have been so cheered to know that his daughter, Natasha, was part of the crowd that day. She was sworn to secrecy and unable to speak when her companions said they would love to know the identity of the Chief Designer. Nonetheless, she wrote later, her 'heart was full of pride'.

• • •

History was repeating itself. Once again America was up in arms. Once again, it had been shamefully beaten by its Cold War enemy, a country that many viewed as backward, barely able to feed itself, its industry obsolete. The Soviet success 'has cost the nation heavily in prestige' announced the *New York Times*. 'Only Presidential emphasis and direction will chart an American pathway to the stars.' John F. Kennedy, the new President, felt the defeat keenly. When he had entered the White House in January that year, he had such an irresistible air of success about him, and to the American people he seemed to embody the promise of a new golden age. The good-looking President and his glamorous entourage: what could they not do? 'How can we catch up?' was his continual question. 'There is nothing more important.'

The Shepard launch was scheduled for May but only three days after Gagarin's flight Kennedy became embroiled in an embarrassing fiasco with America's near-neighbour Cuba. The military took control of the Cape, potentially derailing NASA's schedule. The island of Cuba, a mere ninety miles off the coast of Florida, was under the communist rule of Fidel Castro, who received backing from Khrushchev. Communist Cuba was seen as a danger to America and Kennedy had inherited a plan from Eisenhower's administration to help Cuban exiles invade the island and overthrow Castro's administration. A brigade of more than 1400 fighters had been secretly trained and equipped by the CIA in Guatemala. Air cover was also promised by the CIA during the attack.

As the day approached, Kennedy had second thoughts, fearing that the American involvement against the Cuban government would be recognized. When the brigade landed at the Bay of Pigs on 17 April, the promised American air cover failed to materialize. Meanwhile, Castro quickly mobilized his own military, strafing the invading forces. Ships providing vital supplies were also destroyed. Within a matter of days, all the attackers were either dead or had been taken prisoner by Castro's men, leaving a messy American fiasco for Khrushchev's jibes. The Cuban exiles were betrayed and the Kennedy administration was embarrassed as the American role in the attack became all too evident. The stage was set for further confrontation between America and the Soviet Union. It was, declared one historian, 'the perfect failure'.

Kennedy needed a success; a plan to focus minds; something astonishing. Gagarin's flight around the world had been breathtaking as much for what it promised in the future as for the flight itself. Man was no longer earthbound; he could be his own legend. He could fly to the stars. Kennedy wanted to link his administration to the same dream. On 20 April, as a victorious Fidel Castro announced that he had destroyed 'in less than seventy-two hours the army the US imperialist government had organized for many months', Kennedy ordered his vice president, Lyndon Johnson, to find out what space mission the Americans stood the best chance of winning.

Meanwhile, interest countrywide was focused on Shepard's imminent launch into space. Its success was imperative. After several days of bad weather when low cloud hung over the Cape, the date was finally set for 5 May 1961. Tired of hearing about Soviet success, Shepard wanted to stop Khrushchev in mid-sentence from boasting about the space programme and remind him that this was a two-horse race. After an early breakfast, the doctors applied the usual monitoring sensors and by 4 a.m. he was in his space suit.

In the early morning chill under the garish lights, the great tower stood on the launch pad, mysteriously wreathed in mists of escaping vapour from its oxygen tank, almost seeming to breathe. Shepard was seated and strapped into the Mercury capsule. Goodbyes and good lucks over, he was all alone in what felt like a sarcophagus. His fate was in the hands of others. It was a time when fear could rise through a man, when that hollow feeling in the stomach could take over and turn him to jelly. But Alan Shepard didn't do fear. With considerable effort of will, he deliberately shut it out and concentrated on the controls.

Time passed – three hours in fact – and Alan Shepard, a man who could move mountains in his own mind, found he was in a situation his mind could not control. His bladder was full. It needed emptying, but there was nothing provided for this basic function. He explained his need to Mercury control and requested it be dealt with urgently. Von Braun was appalled. It couldn't be done. There wasn't the time, came back the order from the control room. He would just have to hang on in there. Shepard replied that he could be another couple of hours in the capsule. Hanging on was not possible. Von Braun says 'no' came the reply. Shepard, desperate, said he would do it in his suit. The medics were horrified at what that might do to their sensing equipment. It would short-circuit and Shepard would be fried. Unrepentant, Shepard said all they had to do was turn the power off. They did. Alan Shepard, sophisticate, King of Cocoa Beach, America's sweetheart, lying across his great tower of latent might, 'let go'.

Four hours after he had entered the capsule, countdown began. Six, five, four ... Shepard could hear Deke Slayton's voice at the control. That was an unexpected bonus: he would be in touch with his friend for the trip. Now he could feel the surge of power. Three, two ... Against the giant roar of the engines, no one heard him whisper: 'Deke and the man upstairs will watch over me. So don't screw up, Shepard. Don't screw up.'

One. Zero. Ignition. 'You're on your way, Jose,' Deke Slayton called out to Shepard. The huge thunder of 'liftoff' alerted everyone at the Cape to the successful launch. Three hundred reporters were on site, capturing the thrill, focusing on the power carrying America's hope. The world and his wife left what they were doing to look skywards and wonder, as the shining streak and tail of fire disappeared into the heavens. Everyone who could was watching on TV. Vicariously they also took the flight: they marvelled at the weightlessness, were horrified at the speed of 5000 mph, gasped at the revelation that Shepard could see the world from space. They listened with apprehension as the discarded Redstone rocket wallowed out of control, torn to shreds and spat out by the destructive maw of re-entry, and worried about Shepard's own re-entry through hostile red-hot air. And then he was down. Shepard was in the water. It was perfect; an American hero at last.

The country celebrated. Shepard was flown by helicopter to the White House where the golden Kennedys came forward, appropriately charming, sprinkling gold dust in their wake. With the Distinguished Service Medal glittering on his chest, Shepard was driven through the thick snow of a tickertape welcome in the capital. He was indeed the keeper of a great romantic dream. Although he had not done a full orbit, as had Gagarin – merely a fifteen-minute suborbital flight into space – the country was space mad. The idea of simply catching up with the Soviet Union was ignominious. No, America must set the standard and win.

Behind the scenes, Kennedy had wasted no time. Von Braun was one of many who had been consulted by Lyndon Johnson. 'We have an excellent chance of beating the Soviets to the first landing of a crew on the Moon,' he had advised unhesitatingly. On 25 May 1961, Kennedy's voice rang out confidently in Congress:

> I believe this nation should commit itself to achieving the goal, before the decade is out, of landing a man on the moon and returning him safely to Earth. No single space project in this period will be more impressive to mankind, or more important for the long range exploration of space and none will be so difficult or expensive to accomplish.

To von Braun, Kennedy's speech sounded too good to be true. Originally his dream of a flight to the moon had been a well-guarded secret, not discussed in Hitler's Germany. In 1941, he had confided to a friend, 'Oh yes, we shall get to the moon – but of course, I dare not tell Hitler yet.' Now it was out in the open, clearly delineated, with the whole country involved and excited. The endless dead years in White Sands where his motives had been questioned and thwarted were completely thrown off. All he could see ahead was a wonderful opportunity. It was *his* Redstone that had carried the first American into space. It would be *his* new Saturn rocket that would bridge the 238,000 miles to the moon if they were to have a chance of winning. And there was no time to lose. The Soviets were clearly intent on reaching the moon first. The sense of urgency was overwhelming. America must win. The President was with them. The vision that he had articulated as a prisoner in Germany in 1945 was now the vision of the American President. At last it seemed his time had come.

Although Kennedy's speech was not reported in the Soviet Union, Sergei Korolev knew of it through the Western press and in his mind took up the challenge. Funds for space projects were tight and in constant competition with funds for defence, but his determination could not be deflected. The new breed of rockets that the army required would be powerful enough to get to the moon. It was surely only a matter of time before some smiling Russian youth, like Gagarin, was planting the Soviet flag on the surface of the moon.

PRAVDA

31 December 1961

The Soviet land has become the gateway to the universe!

PROFESSOR K. SERGEEV

The march of time is swift. Only a year has passed, but how much has happened in the past year!

The bright dawn of Communism is lighting the way for the people of the Socialist world and all progressive peoples. The dream is becoming a reality, a realisable plan. Future generations of Soviet people will live under Communism!

Soviet science, technology and industry have achieved significant successes in the past year, culminating in the first space flight of Yuri Gagarin and Gherman Titov on 'Vostok 1' and 'Vostok 2'. The vast importance of these flights cannot be overemphasised, for they herald a new era of space flight for mankind.

The flight of Yuri Gagarin is unparalleled in history. The average speed during the flight was about 28,000km/hour, the altitude was 327km, the distance greater than 40,000km around the earth's globe, all in only 108 minutes!

Gherman Titov flew a distance almost equal to that of the earth to the moon and back in 25 hours and 18 minutes. This exceptional and lengthy flight was executed according to schedule. In conditions that he described as wholly comfortable, the cosmonaut completed a series of scientific experiments. He maintained a continual link with the USSR via radio, sent his greetings from space to the people of all continents, ate, slept and even did gymnastic exercises.

'Vostok 2' moved swiftly around its orbit, passing designated regions within a second's precision. Day became night every half hour. Meanwhile, on Soviet land, several dozen observation points kept a close record of the smallest occurrences taking place on board the 'Vostok 2'. While Gherman Titov slept, his heartbeat could be heard by everyone on our planet. He slept for 35 minutes longer than had been planned, but the medical specialists said: 'It's fine, he is sleeping well, let him sleep a little longer!'

The Soviet land and the whole world gave the hero cosmonauts a rapturous reception. They are the first, but as Khrushchev said in his speech in Red Square, addressing Gagarin and Titov: 'There is little doubt that your family of cosmonauts will grow and prosper.'

Soviet sputnik-ships will again set off to explore the distant regions of space from the shores of the universe, our Motherland. Each flight and return will be the cause of great celebration for the Soviet people and the whole of progressive mankind, a victory for reason and progress!

The most important result of the flights of the 'Vostok 1' and 'Vostok 2' is that they have fully confirmed the feasibility of sending man into space; proved the capability of our complicated rocket technology to send a sputnik-ship into orbit, and it has also solved the equally difficult question of the sputnik-ship's return to earth and landing. The question of sending man into space is, without doubt, one of the most complex, giving rise to a new set of scientific questions.

The attentiveness and daily concern of our Communist Party, the inspirational, selfless work of Soviet people, the cream of Soviet science – these are the great strengths which have made possible the unparalleled flight of our Soviet cosmonauts.

Previously uninvestigated regions of space will undoubtedly yield practical solutions to a whole range of problems in science and agriculture.

In the near future we can expect to see the invention of equipment for satellite stations that will transmit radio and television messages; provide vehicles and planes for navigation through space; monitor the weather and perhaps in the future exert an active influence on the weather.

The satellites and piloted orbital vehicles will carry out scientific studies of the earth. They will study those regions of space adjacent to the earth's atmosphere and phenomena related to the activity of the sun and the distant realms of the universe.

Finally, satellites and space rockets will provide a practical solution to problems of communication and passenger travel. Indeed, it was only about 10–15 years ago that flight on jet-planes was experimental and was accessible only to qualified pilots. Soviet jet-planes now provide the quickest, most convenient and reliable form of transport.

We can expect that passenger flight through space to any point on the earth's surface will take from one to two hours. This is still only a prediction. This technology will, of course, not appear immediately and it will take a while to become affordable, convenient and reliable. A great deal more work will have to be done before we can achieve this.

One of the most fascinating problems to have excited humanity for centuries is the question of flight to the other planets and the distant regions of the universe; at first to regions nearest to earth, such as the moon, the earth's eternal companion, which now bears the symbol of the USSR on its surface, and then to the planets of the solar system nearest to the earth – Mercury, the thickly cloud-enshrouded Venus, mysterious Mars, distant Jupiter and the four other planets.

These are the probable interplanetary routes for Soviet explorers. And after that: the massive suns and the worlds of the other galaxies.

1961 has come to an end. This year has seen great leaps forward for the Soviet people. It was the year of the 22nd Party Congress, which established the programme for building Communism; a year of triumphal achievements in Soviet science and outstanding displays of bravery by our pilots, who have paved the first road into space.

[Sergei Pavlovich Korolev's article in *Pravda* under pseudonym]

PART FIVE

The Race for the Moon

'Within 15 years we will be living and working on the moon … It will be somewhat like an Arctic camp. The first work will be the pouring of concrete around the first rocket to land humans on the Moon.'

Wernher von Braun interviewed by Victor Riesel
for the Hall Syndicate, 21 January 1959

'I just need another ten years …'

Sergei Pavlovich Korolev, 1966

Neptune's unearthly music. Of course he hid these ambitions behind a practical façade. His original 1958 report, 'On the Prospects of Mastering Outer Space', had set out a full schedule in a grand design that acknowledged no limits: flights to the moon, a solar-powered space station, manned exploration of Mars and Venus and the creation of a permanent space station on the moon. But it was always understood that military interests came first. To kindle official interest, he emphasized the part a space station could play in detecting movement of enemy missiles and nuclear activity. A crew of four or five would be in an unparalleled position to monitor enemy movement.

But Korolev's motivation could not be completely hidden from those who worked with him. His biographer Mikhail Rebrov reports that on a bleak day at Baikonur he suddenly turned on one beleaguered colleague, the Chief Designer Nikolai Pilyugin, and exclaimed: 'Forget the telemetry – do you want to fly to the moon?' Nikolai Pilyugin brought him swiftly back to earth. 'I want to fly to Moscow,' he retorted. When Korolev persisted, Pilyugin put him down: 'You are a dreamer, Sergei. When the time comes we aren't even going to be around.' Korolev would not hear of that. 'I don't know about you, but I want to live to see it.' Pilyugin just shrugged; it was 'on the edge of the impossible,' he said. When Korolev explained that he had been dreaming of this since his childhood, Pilyugin merely retorted that he had only 'dreamed about a big slice of rye bread'. Korolev finally conceded, 'so did I', and they both laughed. But it was nonetheless true that the grinding workload that Korolev carried for years, wantonly spending his energy at an evermore extravagant rate, was all for this ever-beckoning mirage.

Unlike von Braun, who now had the resources of the wealthiest nation on earth behind him, Korolev was still thwarted while military needs took precedence over space exploration. He had begun development work on different designs for a new and more powerful spaceship, the Soyuz – or Union – as early as 1957. One of the key aims of the Soyuz, Kamanin wrote, was to allow for docking, or 'union', in space – which Korolev saw as a crucial stage in space exploration. If docking could be achieved, individual spacecraft, launched separately, could be docked together to build interplanetary spacecraft and an orbital space station. This would serve as a staging post for crews and pave the way for long-distance flights. Soyuz was also designed to accommodate a three-man crew, which was seen as the ideal number for a lunar landing mission. To achieve all this, the Soyuz would have much more manoeuvrability than the Vostok; it could change orbit and adjust attitude with great precision. It comprised three main compartments: the service module with equipment at the rear, the main re-entry vehicle in the middle compartment and docking facilities at the front of the craft.

In addition, Korolev had also embarked on an ambitious design for a much larger rocket: the N-1. This was to be a giant; his initial designs allowed for a rocket that could be 295 feet high and carry a payload of 40 tons. Never losing sight of the fact that he was meant to be producing rockets for defence purposes, but, as always, with a secret plan for a space vehicle in his back pocket, he deliberately left a certain ambiguity about the final payload. It was presented to his military masters as a 'universal launch vehicle' and he took every opportunity for stressing the N-1's capability for use in creating a space station for military reconnaissance.

Korolev saw the N-1 as his masterpiece and the revolutionary design embodied unique features. He was preoccupied with obtaining more power yet determined to have a simpler design than the R-7 with its clusters of engines and strap-on boosters. The N-1 would have three stages, with a simpler engine layout and radical new spherical propellant tanks that would be considerably lighter than the conventional ones.

Yet although Korolev's name had been touched with magic since Gagarin's launch, and Khrushchev revelled in Soviet victories in space, he was not prepared to devote serious funds to a moon race. There was no firm commitment for an ambitious programme; for Khrushchev space was principally a vehicle for showmanship. To help foster the illusion that they had conjured up an entire space programme, Soviet intelligence services grandly renamed the bleak launch site out at Tyura Tam: it became the 'Baikonur Cosmodrome' – after the town of Baikonur, 220 miles to the north. The aim was to promote the impression that they had a vast cosmodrome at their disposal and this name was used in all public announcements. In reality, the army remained uninterested in a large space programme. The life of the N-1 was, for now, confined to the drawing board, and, with it, Korolev's hopes of space competition with America.

Korolev himself was planning another manned launch for August. Gherman Titov would be in orbit more than twenty-four hours, making seventeen trips around the world. This would put Shepard's suborbital flight of fifteen minutes into perspective. But when Khrushchev learned of the Titov flight, he had plans of his own. He let Korolev understand that Titov must be on his way *before* 13 August. Korolev had no idea why this date was so important but Titov was successfully launched a week earlier.

It wasn't a comfortable flight. Once in orbit, Titov felt miserably sick, as though he were flying upside down. He couldn't eat, drink or work his instrument panel. Doctors on the ground were concerned. They could see from his sensors that he was in severe discomfort, with vertigo and nausea. On the twelfth orbit he recovered and took a film of the earth. But his heating system broke down, he nearly froze and the re-entry rockets behaved as they had done for Gagarin. He found himself spinning uncontrollably until the heat of re-entry burned through the connection joining him to the instrument section. Once again, his sheer luck continued. At 150 feet from the ground, dangling from his parachute, he could see that he was about to smash into a passing train. He braced himself, fearing the worst, but a gust of wind providentially moved him backwards to the safety of a ploughed field.

The Americans were left wondering at the Soviet Union's brilliance. On 13 August, Khrushchev began building the Berlin Wall – and Korolev now understood the significance of the August date. Residents of Berlin woke up to find Soviet soldiers erecting a massive barbed-wire barricade across the city. Khrushchev had gambled on Korolev's success at a time when he wanted to emphasize Soviet superiority. West Berlin had been a gaping hole in the Iron Curtain that stretched across Europe. More than a thousand East Germans had fled into West Berlin the previous month alone; more than two and half million had left since 1949. Khrushchev had become obsessed with stemming the tide of refugees and blocking the link with the West. While the wall was regarded in the West as a powerful symbol of Soviet repression, Khrushchev was delighted. His rotund figure was seen around the world – the all-singing, all-dancing Soviet leader juggling space technology in one hand and Soviet policy in the other. He made sure *Pravda* reported accordingly.

And a few months later, in October 1961, under Khrushchev's leadership, the Soviet Union exploded the largest nuclear weapon the world had ever seen. Nicknamed the 'Tzar Bomba', or 'King of Bombs', it was estimated at in excess of 50 million tons of TNT – this one explosion, in fact, surpassed the total amount of explosives utilized in the six years of the Second World War. In a letter to Americans published in *Life* magazine, President Kennedy encouraged Americans to build fall-out shelters. In the Soviet Union, the urgency of military needs continued to dominate the missile programme. In spite of Korolev's repeated achievements, Khrushchev made it clear that he was not the only designer of large rockets. Mikhail Yangel and Vladimir Chelomei continued to carry considerable weight in the field of ICBM design.

Khrushchev still favoured Chelomei, the astute designer who had employed his son, Sergei. Korolev never bothered with social climbing; he felt his work should speak for itself. But Chelomei knew only too well that a certain amount of social clambering could be a beneficial exercise. He seized any opportunity to wine and dine Khrushchev and son in style – on the occasion of Sergei Khrushchev's graduation, for example, where everyone was well and truly drunk. Money and prestige came rolling in. The exquisitely attired Chelomei, looking a little like a high-class art dealer, sporting his avant-garde ties, would sit in his elegant office behind the desk it took him two months to design and intellectualize at length on matters scientific. A man of culture with an equally cultured circle of friends, he was a brilliant conversationalist. And Khrushchev, with his bucolic roots, sometimes looking as though he were pushing an invisible pig to market, was charmed. Chelomei's design studio with its emphasis on military hardware grew more and more successful with plans underway for new ICBMs, two space launch vehicles – the UR-200 and the UR-500 – not to mention a space plane, and even a craft for flying to other planets. All the while, he was attracting funds that should have gone to Korolev.

Korolev was incensed. With available funds split between different groups, work was held up on his N-1 rocket in spite of his willingness to experiment with variations in order to attract the military. He anguished over its halting progress like a mother over a sick child. To achieve the kind of thrust that was needed for the N-1, he was looking for an engine design that would give over 150 tons of thrust, with more than twenty engines clustered together in the first stage alone. But the relationship between Korolev and Glushko had 'gone sour', reported Chertok, and this was having a detrimental effect on the N-1's development. Korolev needed this pre-eminent engine designer on his team and, deciding to eat humble pie, went to visit Glushko.

The meeting started fairly well and both men made an effort, but the poisonous infection from past wounds soon festered into accusations and harsh words. Korolev had hoped to persuade Glushko to design new engines using the cryogenic fluids he favoured, but Glushko would have none of it. He accused Korolev of endangering progress with his inability to move forward, pointing out mistakes in the past which he claimed had held up progress. Korolev shouted him down with the Nedelin disaster and the part unreliable propellants had played in it. Both men were shaking. The shouting could be heard 200 yards away.

Hissing with indignation, the incensed Glushko officially demanded that Korolev redesign the N-1 for use with storable fuels. The matter now reached official ears. The Ministry would investigate and evaluate the two fuel systems. At a meeting in early February 1962 at the Kremlin, once again the irreconcilable faced the unmoveable as the two men shouted each other down. Korolev accused Glushko of turning rockets into 'powder kegs'. Glushko accused Korolev of being 'Mr Clean', the man who wanted to travel in space without getting his hands dirty, and suggested he design 'steam engines'. Both men were now on fire with anger. Glushko insinuated that Korolev was behind the times: the military were using storable fuels. Senior officials looked on in disbelief at the hostile display as the two men, oblivious of where they were, continued their verbal battle which ended with an outburst from Korolev: 'If you don't want to design the engines, then don't do it. We'll manage without you.'

His continuing estrangement with Soviet Union's top engine designer was not helping Korolev. It was therefore with considerable optimism that he set out for the holiday destination of Pitsunda on the Black Sea later in February 1962 where Khrushchev had called a meeting of the defence command in his dacha to discuss future plans. Yangel, Chelomei and Korolev would present their work; Glushko would also be there. Chelomei began the session explaining his latest

ICBM designs, one of which, the UR-500, with military and space capabilities, echoed Korolev's N-1. His performance was impressive, compelling attention, and Khrushchev beamed as Chelomei won approval to develop the UR-500.

In marked contrast to Chelomei's sophistication, Korolev's presentation was awkward and when he announced that Glushko would not be designing the engines but that the relatively untried Nikolai Dmitrievich Kuznetsov would have this responsibility, Khrushchev nearly fell out of his chair. Kuznetsov, who ran the design bureau OKB-276 at Kuybyshev, was acclaimed in the field of aircraft engine design but could not match Glushko's reputation as the foremost rocket engine designer of the Soviet Union. Khrushchev asked why Glushko was not involved. Korolev replied that Glushko had refused. Glushko immediately locked horns with Korolev. The eminent gathering looked on in disbelief as the two protagonists, angry and inflamed, shoulders hunched, heads down, once more yelled accusations at each other, now sharpened by a growing hatred.

All this was too much for Khrushchev. Korolev was calmly asking for endorsement of a massive new rocket powered by the unproven Kuznetsov. Khrushchev asked Ustinov to assess the situation and plans for the N-1 were put on hold while costs for the N-1 were re-evaluated with engines by Glushko using his preferred fuels rather than engines designed by Kuznetsov. Meanwhile, Chelomei's UR-500, which used storable fuels and Glushko's engines, combined with all his other projects, in effect gave Chelomei the lead role in space. Yangel, too, received approval for parallel space projects, while Korolev's N-1, with all the dreams it carried, was still little more than designs on a drawing board.

The rift between Glushko and Korolev now deepened into an unbridgeable chasm; they would never work together again.

• • •

While Korolev's time was divided between myriad different projects, at the Marshall Space Flight Center in America von Braun was fully focused on how to get to the moon. To launch a craft and travel the 240,000 miles would require carrying a significant weight of fuel. A rocket with over 10 million pounds of thrust would be needed. The Saturn I did not have the power for a lunar landing, even with its 1.3 million pounds of thrust. In January 1962, NASA gave approval for von Braun's team to develop an even larger rocket, which became known as the Saturn V, a giant, the first stage of which alone would generate 7.5 million pounds of thrust. Longer than a football pitch, weighing more than a light cruiser, with a diameter greater than the combined width of three large removal vans, and with its five clustered F-1 engines, it was claimed it could develop more power than a string of Volkswagens placed from New York to Seattle. And if that would not do the job, von Braun visualized a rocket even bigger than Saturn V, with eight clustered engines, which he called the Nova. That, he reasoned, would get Americans to the moon. The big uncertainty was, would it get them there first?

Everything would depend on the development of new engines. The first-stage engines, called the F-1, were to be twice the size of any previous rocket engine. To produce the required amount of thrust they had to be capable of burning 40,000 gallons of fuel a minute. The programme manager for the Saturn was von Braun's long-standing colleague, Arthur Rudolph. They had worked together since developing V-2s in Germany over thirty years before. In developing the Saturn engines, they now faced the ultimate challenge. The larger the engine the greater the chance of creating combustion instability that could destroy the rocket. The slightest disturbance, such as a small pocket of unmixed fuel, would create pressure waves that could lead to a damaging explosion and wreck everything. In seeking to scale up to create the largest rocket engine ever made, von Braun could not guarantee that it was even technically feasible.

Equally challenging was the work in progress to use liquid hydrogen as a fuel for the upper stages of the Saturn V. It had the advantage of being feather light, but it was mind-bogglingly cold and could shatter metal as it touched it because of the thermal shock. It could find the deepest, most hidden flaws in the welds of metal structures and snap them apart. North American Aviation, builders of the Saturn's crucial upper stage, had taken on the gigantic technical challenge of designing sophisticated tanks capable of using liquid hydrogen. But for von Braun the thought of the Soviets on the moon was a big incentive to make it work. The idea of passports surrendered to Soviet customs, of concrete spread over its pristine surface as ugly, utilitarian buildings sprang up, made him vow that that prize must never fall to the Russians.

Even assuming he could create a rocket with enough thrust, it was still uncertain *how* to approach a journey to the moon – the kind of craft that would be needed for each stage of the journey. Von Braun had always visualized creating a space station in earth's orbit, which would serve as a staging post to the moon. 'From this platform, a trip to the moon itself will be just a step,' he had enthused in *Collier's* as early as 1952. The space ship, with accommodation 'comparable to that on a modern submarine', could also serve as a launching vehicle for further exploration of the solar system. 'Space taxis' would carry sections of the lunar module up from earth by rocket for assembly into a lunar craft. And all this was just a first step in a grand vision of colonizing space.

Von Braun's fantastic science fiction vision had become modified and simplified over the years. He no longer imagined that an entire space station needed to be built in near-earth orbit – but the principle remained the same: components for the lunar vehicle would be flown into space separately and assembled in earth orbit. This approach became known as 'earth orbit rendezvous', or EOR. The main advantage was that components for a lunar craft were taken into space in sections and therefore a smaller launch vehicle was required. But there were also pitfalls. A rendezvous in space had quite simply never happened before. The Mercury capsule could do nothing so complex, so a new capsule with the power and the ability to manoeuvre would be necessary.

Von Braun and other supporters of EOR found themselves facing opposition from John Houbolt, a relatively unknown engineer from Langley Research Center. As early as 1960, he came up with a radical new idea: 'lunar orbit rendezvous', or LOR. Houbolt's bold concept envisaged that a booster rocket would launch two craft combined in one vehicle on a trajectory to lunar orbit. One would serve as the command module, orbiting the moon, while the second would fly directly to the moon, landing on the lunar surface. On return from the moon's surface, the second craft would dock with the command module in lunar orbit. This approach, Houbolt argued, would simplify development and greatly reduce the weight and complexity of the lunar craft.

When he had first proposed this at a NASA meeting, he was greeted with derision. His scheme was seen as unworkable, even dangerous. 'His figures lie,' announced Max Faget. 'He does not know what he is talking about.' Certainly Houbolt could not say how a rescue would be effected should there be an accident in lunar orbit, 240,000 miles from the earth. The plain truth was that it would not be possible and men would be condemned to certain death, whereas a space station in earth orbit would be within range of rescue. All senior NASA engineers, including von Braun, were convinced that Houbolt's far-fetched idea did not make sense. It called for a skill and manoeuvrability that were not yet in NASA's repertoire.

The other unknown factor causing delays and adding to design difficulties was the lack of information regarding the moon's surface. Was it, as some suggested, layers deep in dust or was it firm enough to take a landing craft? Would some million-dollar space capsule carrying astronauts land on the moon's angelic-looking surface only to sink fathoms deep in primeval dust undisturbed since time began? Caldwell Johnson, who was considering what kind of landing equipment might be needed, found no one could give a definitive answer about the moon's surface. 'How in hell are we gonna design landing gear if the moon's seas are nothing but pools of dust,' he despaired, 'and the mountains are nothing but blown-glass fairy castles?'

But one certainty was agreed upon: NASA had, at last, clocked up a victory. John Glenn had made his historic journey into space. He had not been the first in orbit, however. Another trip had been thought necessary before sending up a man to circle the earth on top of an Atlas rocket, and that honour had fallen to Enos the chimp who was as sharp at pressing the buttons and pulling the levers as Ham had been. He also had his own foolproof method of bringing a little joy into his life. Whenever his 'load' passed unreasonable levels, he had simply pulled down his clothing and begun masturbating. Much thought had been put into controlling Enos's little hobby, which he even preferred to banana pellets. A catheter was inserted into the offending member as a means of collecting urine and to help Enos concentrate on his piloting responsibilities.

Yet when the great day arrived, just as Ham had done, Enos experienced all manner of diabolical unfairness from the 'god' of the levers. On his flight through the eternal skies, utterly absorbed, concentrating on those complex levers, no matter how perfect his routine he was rewarded only with bolts of searing electricity shooting through his feet. Mean and moody when he met the press after his journey, he showed one or two of them what his teeth were for. And to everyone's embarrassment, in order to escape the endless boredom of the press conference, he turned to his favourable pastime, the one his name was famous for among his keepers: 'Enos the Penis'. As he pulled down his trousers, cameras clicked, flashing like diamonds, ensuring Enos's name would live in memory as much for his hobby as for his aeronautical achievements.

However, with John Glenn's flight in February 1962, NASA came of age. It was not without difficulties and it highlighted just what the input of the pilot's brain could do when automatic systems failed. Glenn was able to correct manually the craft's tendency to veer to the right, losing altitude, when the automatic system failed. But he really showed his mettle when ground control received warning lights telling them that the heat shield was loose. If this were the case, Glenn would be subjected to heat of around 4000 degrees F on re-entry to the earth's atmosphere. Ground control was sick with anguish. It was hard to see what could be done. Time was running out.

Max Faget advised mission control that they should leave the retro pack in place after the rockets had fired for re-entry. It might just hold the heat shield in place, unless, of course, one or more of the rockets had failed to fire, in which case NASA's first man in orbit would end up as a fiery sacrifice. America was watching. As the news spread, millions more tuned in across the globe.

It was time to fire the retro rockets. Three, two, one, zero … silence. Silence for five minutes. No signal could get through the ionized air surrounding the diving fireball capsule. Ground control continued calling …

Eventually they heard Glenn's voice reply: 'I hear you loud and clear.'

He was alive. By what miracle he stood in front of the cameras with that well-known smile the American public was not sure. But there was no doubt he was a hero, and he was going to get a hero's welcome, carried with countrywide euphoria, shoulder high, straight to the President to receive all his honours. He was the man who had kept his cool under the most terrifying threat of death, just as the American public knew he would. Four million lined the streets of New York on 23 February to catch a glimpse of him through a snow of ticker tape in the motorcade as he passed – to experience for themselves a moment when legend touches the everyday. Later, when the craft was examined, it was found the heat shield had not after all been loose. Faulty wiring had sent the wrong message to ground control. Glenn had spent what he thought must be his last moments while the fixed retro pack caught fire on re-entry, embracing the capsule in its fiery debris.

Although John Glenn's trip was a very public American triumph, it merely proved that America could put an astronaut into orbit. Behind the scenes, the stormy debate about how to put a man on the moon – LOR or EOR – was holding up progress. The argument had dominated NASA for more than a year. No craft could be built until there was agreement between all the different NASA centres about the basic principles. Max Faget was to design a powered lunar craft that could be manoeuvred by astronauts, but until the method of lunar approach was resolved he could not proceed. Precious time was being lost. 'If all the paperwork NASA generated were piled up, the stack would reach the moon long before a space craft ever did,' fumed one frustrated engineer.

Houbolt's seemingly far-fetched plan for a lunar orbit rendezvous was gaining supporters once the engineering was examined in detail. The lunar module would only be required to manoeuvre from a command module in lunar orbit to the moon's surface and provide some means of returning to the waiting command module. This had to be simpler than launching a large spacecraft from earth orbit to the moon and back again. As Faget scrutinized the proposal, he wondered whether the Saturn V would have the power to launch both the command and lunar modules to lunar orbit and began to consider what could be the minimum weight of both of these craft. He discussed his ideas with Gilruth and suddenly the scale of the engineering began to seem manageable. However, von Braun and his team at the Marshall Space Flight Center remained staunchly in favour of EOR.

The dispute continued one long day in April 1962 at a meeting in the Flight Space Center. Von Braun was there with his top assistants listening to yet another persuasive briefing on lunar orbit rendezvous from Gilruth's team at NASA's Manned Spacecraft Center in Houston. The staff at Marshall put forward the arguments for EOR. The day was spent in concentrated argument and nothing was settled. Finally, one Apollo programme contractor, who had originally been an advocate of EOR, rose to cut through the confusion: 'I've heard these good things about lunar orbit rendezvous and I'd like to hear what son-of-bitch thinks it isn't the right thing to do?' All eyes were on von Braun, who politely conceded that there were indeed advantages to LOR. A few weeks later, in the same place, the same arguments were aired once more. At the end of the day, to the astonishment of his team, von Braun rose to conclude the matter: 'It is the position of the Space Flight Center that we support the *lunar* orbit rendezvous plan.' He added that it 'offers the highest confidence factor of successful accomplishment within this decade'.

Von Braun had taken his time making up his mind. He had weighed all the evidence and now genuinely believed the lunar orbit was the right route to take. His mission was to get to the moon, not win arguments. But to his dismay, later that month von Braun found that his massive F-1 engines, each one the size of a room, were proving problematic. They were being tested at a specially built test stand at Edwards Air Force Base in California. In a matter of seconds, during a static test on 28 June, one engine was completely destroyed.

To get to the moon, everything was gambled on the success of the F-1 engines. Yet in scaling up a rocket engine to provide 1.5 million pounds of thrust, it was proving near impossible to prevent combustion instability as the propellants mixed. Kerosene had to be pumped to the combustion chamber at a rate of over 15,000 gallons a minute and the liquid oxygen at over 24,000 gallons a minute. Huge pumps operating at different temperatures pushed the fuel through 6300 holes in the injector plate. The two fuels met with exact precision at 3 tons per second in the combustion chamber where, in a few seconds, the temperature would rise to 5000 degrees F. To ensure a smooth and efficient burn, it was essential that the propellants were well mixed, otherwise pressure waves would be created within the combustion chamber and a destructive explosion would occur within a fraction of a second. Temperature, too, was important, to create a 'smooth flame front'. A little extra heat, a fluctuation of fuel flow, an imperfection in the injector plate: all these could cause an imbalance of pressure, sending a wall of flame zooming out of control in the combustion chamber.

Work on the problem of the F-1 engine's instability was a priority. Arthur Rudolph and von Braun discussed the crisis. Jerry Thomson, chief of liquid fuel engines at the Flight Space Center, was assigned to the problem and so began a long, slow process of trying to discover the cause of the instability. The flow rates were overhauled, the combustion chamber modified, the injector plate redesigned, to no avail. With more test failures during the autumn, 'it might just be too big to make it work,' concluded one cautious member of the President's Science Advisory Committee. There was increasing speculation as to whether an engine on this scale was even possible.

• • •

The Soviet space programme still had no coordinated plan to compete directly with the Americans in a moon landing. For military leaders, the number of rockets with nuclear warheads pointing at American cities was far more to the point. Glenn's trip, however, and its attendant ballyhoo, sent the message loud and clear to the Soviet Union: 'We are going to the moon; watch us.'

Khrushchev's immediate response was to push for Korolev to do something brilliant. Ustinov was on the telephone ordering results. Kamanin chose his most suitable cosmonauts. The time had come for Korolev to do another juggling act, which, if it worked, would leave NASA gaping, open-mouthed, and von Braun wondering just what he had seen. Korolev's plan was to launch a first manned Vostok on day one. Day two would see a second rocket launched at a calculated time and inclination that would ensure its entry into an orbit that would bring it close to the path of the first. At one stage, the craft would be so very near each other that uninformed onlookers might well assume that a deliberately manoeuvred link-up in space had been performed: something far beyond American expertise.

On the morning of the launch, Korolev did his best to hide his anxiety. Recent booster failures had unnerved the Chief Designer and had him ready to push the abort button at the slightest sign of a problem. But at 11.30 on 11 August 1962, Captain Andrian Nikolayev soared rapidly out of sight in a whirl of fire and fury and a perfect launch. As the day progressed, it was clear from the live pictures shown on Soviet TV that Nikolayev, now promoted to major, was enjoying weightlessness, eating sumptuously and waving to the world. He slept well and was ready to view the second Vostok carrying Major Pavel Popovich launch directly beneath him twenty-four hours into his flight at Baikonur. On the first orbit, the two Vostoks were a mere three miles apart. The Soviets announced nothing of how this had been achieved and the Western world assumed with Sir Bernard Lovell at the Jodrell Bank Observatory that the Soviet Union was massively ahead. 'I think that the Soviets are so far ahead in the technique of rocketry,' he declared admiringly, 'that the possibility of America catching up in the next decade is remote.'

Once again, Korolev had created a masterstroke. The jubilation in the Kremlin was matched by the celebration and partying in Moscow. More importantly for Korolev, that September he received the go-ahead to build his N-1 rocket. The design was revised to allow for a lifting capability of 75 tons. This would serve military purposes and fulfil Korolev's private ambitions for space. And Korolev had also won the fuel argument. The N-1 would be powered by the liquid oxygen and kerosene that he strongly believed were most suited to the task. Engine designer Nikolai Kuznetsov was developing the NK-15 engine with 153.4 tons of thrust for the first stage, and the NK-15V with 180 tons of thrust for the second stage. They could now proceed with planning the layout of engines across the different stages to create the most powerful Soviet rocket ever built.

With the backing at last of the Council of Ministers and the Central Committee of the Communist Party, a huge operation coordinated by Korolev and involving many bureaus was planned. Everyone of significance in the space and defence industry was involved, it seemed – except Glushko. With the first N-1 launch pencilled in for 1965, Korolev had just three years to accomplish the task. Although no specific space mission had been confirmed, Korolev could at last build the launch vehicle that would allow him to compete with von Braun.

CHAPTER SEVENTEEN

'Friends, before us is the moon'

In September 1962, President Kennedy went on a two-day tour to give a much-appreciated fillip to the space programme, dusting it with the usual Kennedy glamour. He marvelled at the giant complex taking shape at Cape Canaveral, was astonished by the size of the Saturn I rocket under development, proudly displayed by von Braun, grinned with delight inside the mock-up of the lunar module, and ended his trip with a rousing speech to an audience fifty thousand strong at Rice University, Houston. He re-affirmed America's intention to fight for the elusive grail of one day putting a man on the moon. 'We choose to go to the moon in this decade and to do the other things, not because they are easy, but because they are hard,' he said. 'Because that challenge is one that we are willing to accept, one we are unwilling to postpone, and one which we intend to *win*.'

Khrushchev might well have been eyeing the handsome Kennedy, standing in the sun, giving his rousing speech as cameras twinkled like stars. His cherubic smile of satisfaction might well have broadened to a grin as he thought of his nuclear missiles hiding quietly in Cuba. After the Bay of Pigs fiasco, Cuban leaders had turned to the Soviet Union for protection conveniently providing Khrushchev with a welcome opportunity. Khrushchev wanted to counter America's lead in developing and deploying missiles and aimed to strengthen his hand by creating a Soviet missile base in Cuba – just ninety miles off the coast of America.

In the summer of 1962, American spies on the island had become aware of unusual items being unloaded from Soviet ships. There were isolated reports of military equipment being transported on the island. The presence of Soviet engineers also did not escape notice. An air of threat steadily materialized. Finally, on 14 October, aerial reconnaissance from a U-2 spy plane confirmed the worst: intermediate-range Soviet missiles were in place on Cuba. American cities were potentially only three minutes away from Soviet attack. At a stroke the Soviet Union had dramatically increased its first-strike nuclear capability. The world held its breath. Khrushchev was taunting, daring. Kennedy was wondering how to avoid outright nuclear confrontation.

A fleet of Soviet cargo ships was moving towards Cuba. America threatened to invade the island. The hawks among Kennedy's military advisers urged bombing the missile sites. Kennedy played for time, worried that any bombing of Soviet engineers would bring the world tumbling towards World War Three. He imposed a naval blockade on the Atlantic sea lanes to prevent the arrival of more weapons and demanded the immediate removal of all missiles from Cuba. The Cold War had suddenly become fiercesomely hot. Khrushchev would not give the order to turn the cargo ships around. Kennedy prepared an invasion force. Khrushchev gave orders to his military to launch nuclear missiles if the Americans invaded. For thirteen days, the world stood poised on the brink of nuclear war.

Secret talks were held. There was a frantic exchange of letters. Finally, on 28 October, Khrushchev agreed to remove the missiles from Cuba in return for Kennedy secretly removing US missiles from near the Soviet Union's borders with Turkey. Both world leaders retreated to the safety of their positions as guardians of their own ideologies. The world breathed normally again.

Watching the escalation of hostilities throughout the autumn, von Braun was conscious of the need to overtake the Soviet Union in rocket design. As the crisis wore on, he and his team were battling with their own predicament: the continuing instability of the F-1 engines. The fifty engineers headed by Jerry Thomson had concentrated their efforts on the flow rate of the fuel, the injector plate and modifications to the combustion chamber, but they were no nearer to solving the problem. They could find no pattern to the instability and all they had for their pains were two more burned-out engines.

By January 1963, a radical plan was agreed: to accept the instability. It was argued that an engine of such size would always be subject to this problem. What was needed was the ability of the engine to right itself after instability and create a 'dynamic stability'. The aim of the team now was deliberately to create instability in the engine in order to study how to solve the problem. Ironically the method proposed for artificially creating instability was to place a bomb in the engine. That should certainly create instability.

• • •

Korolev and his team still devoured the news of NASA assiduously translated into Russian from the Western press. All too often there were interviews with von Braun, setting out plans for the lunar mission and elaborating on his total conviction that 'the Americans would soon have the lead role in space'. 'Unfortunately he is right,' Kamanin admitted with chagrin in his diary. 'We can do fascinating "tricks" but for the specialist it is now clear that we lag behind.' They knew that NASA leaders had decided on lunar orbit rendezvous using the massive Saturn V to launch the command and lunar modules to lunar orbit. A contract had been signed with Grumman Aircraft Engineering to build a lunar module that would descend to the moon's surface, while North American Aviation would build the Apollo command module that would wait in orbit around the moon. It seemed that Kennedy was causing money to flow like water towards the American space programme.

Korolev still had no comparable programme although his plans for the N-1 were now in development. Kuznetsov had formulated an ambitious design incorporating twenty-four NK-15 engines in the first stage of the N-1 alone, with eight NK-15V engines in the second stage and a further four NK-21 engines in the third stage. This would be a rocket that they hoped would eclipse even the Saturn V – but Korolev was held back by the lack of testing facilities for the engines, especially for firing all the engines simultaneously. He was suffering, too, from his estrangement from Glushko, the master engine designer. But money was the crucial factor. There was never enough of it. Korolev deplored the way he had to join the queue when money was handed out. Dispirited, he acknowledged that funds for his space programme were given in the spirit of the Soviet folk saying, 'Give earrings to all the sisters'. In other words, give some to Yangel, give some to Chelomei and give what was left to Korolev.

He was forced to make the fateful decision to waive the ground testing of the first-stage rocket engines firing together. Under the impossible circumstances of a crippling lack of funds and continued indecision on a direct space objective, he had no alternative but drastically to reduce the testing of each stage of the rocket. This meant the first flight would itself be the test, with no previous separate engine testing from which to learn from defects and mistakes. This would save time and fuel, too, which was in short supply. Only desperation could have brought Korolev – the man who liked to test every fault until it was ground to dust – to such a decision. Instead of testing each stage, he would now rely on a new control system, called the 'KORD system', which could detect a faulty engine and shut it down.

Khrushchev continued to be more preoccupied with rockets that would threaten the enemy rather than land unproductively on the moon. Soviet agriculture was failing to feed the country. How much grain could be bought for the cost of a rocket was the question on the premier's mind. There was one small positive step he could make, however: he could try and 'right' the schism between his top rocket men. In June 1963, Korolev and Glushko, with their wives, were invited to stay at Khrushchev's holiday dacha.

It sounded generous: a holiday with the great man himself in a June of cobalt skies, wine-dark sea and pale blue mountains in the distance. The wives were delighted. Nina hoped that Korolev would be able to rest; and how wonderful it would be to dispose of the rancour and see the two men friends again. But the two men in question were far beyond repairing the relationship. Forced to be polite, they hid their deep-felt antagonism, eyes never meeting.

Korolev did not waste the opportunity, however. Striking drawings of the N-1 rocket were exhibited, which illustrated the possibility of actually landing a Soviet cosmonaut on the moon. Korolev held Khrushchev's attention while he spun the possibility into reality outlining details of the spacecraft needed for a lunar landing – before the Americans. Khrushchev's son was there and he watched his father 'enthralled by Korolev's idea. But he could not forget Earthly concerns.

He enquired how much this project would cost ... [Korolev's] estimation was ten to twelve billion roubles. Father wavered.'

Within a few days, Korolev was back at Baikonur for his next amazing feat. This would be another 'first' ordered by Khrushchev himself: the first woman in space. Twenty-six-year-old Valentina Tereshkova was exactly the kind of woman Khrushchev approved of: an ordinary textile worker with no 'class' advantages whose father had worked on the land before he had died fighting the Germans. As before, the plan called for two cosmonauts to be in space simultaneously. Major Valentin Bykovsky circled the world for two days in his Vostok, then, on 15 June 1963, after a successful launch, Valentina Tereshkova, the first woman, was in orbit showing the world what equality meant in the Soviet Union. Soviet news pushed the story of the glorious Soviet Union, the land where socialism gave women the same opportunity as men. Decorated and medalled, Tereshkova's perfect Russian womanhood was paraded in a world tour. Her three-day mission was a significant propaganda coup and the world was suitably impressed.

In his office in the Marshall Space Flight Center in America, von Braun was unaware of the mysterious Chief Designer's financial worries. However, he was a little nearer to discovering the identity of his enigmatic rival. In the Soviet Union, on 3 November at a wedding of unprecedented fanfare hosted by the state in Hollywood style, the world's first woman cosmonaut, Valentina Tereshkova, married Cosmonaut Andrian Nikolayev. The Soviet publicity machine made sure the wedding bells were heard throughout the world. Everyone who was anyone in the space programme was there – for once even Glushko and Korolev were allowed to attend this public event. A few days later the *New York Times* had ascertained:

> Reports circulating in Moscow's western community last week have mentioned two rocket pioneers as likely key figures in the Soviet Space programme. Although the identities of the top scientists in these jobs remain an official secret, a number of unofficial reports have pointed to academicians Valentin Glushko, combustion engineer, and Sergei P. Korolev, a mechanical engineer.

A door had opened a little crack and a name had appeared, but that was it. There was more information on the bride's dress than on the mysterious figure who had appeared for an instant. Earlier Western intelligence had also identified the significance of the name 'Sergei Korolev'. An article in 1961 by Dr Gregory Tokaty-Tokaev, a senior member of the Soviet military who had defected to the West, had singled out Korolev as 'this leading rocket engineer of our times'. Tokaty-Tokaev saw Korolev as 'an excellent organizer ... and a highly imaginative and inventive engineer with tremendous concentration'. A few scant details of his background in GIRD in the 1930s and his role in developing the rockets for the Sputnik and Vostok programmes were mentioned. Glushko's role as 'one of the outstanding experts' in the development of rocket motors was also highlighted, but there was not much more information about the key players and how they had achieved their successes.

Von Braun and Arthur Rudolph were still preoccupied with the Saturn engines which were proving unworkable and beginning to hold up the Apollo programme. The team responsible for solving the instability problem hampering the Saturn engines was making no progress. Placing the highly insulated 3-inch bombs inside a roaring engine was not a job for the faint-hearted. Eventually they put the bomb in the engine before ignition and using heavily protected detonating wires could explode a bomb on demand and watch the chamber pressure move from 1150 to 4000 pounds per square inch. Now that they could create instability, all they had to do was rectify it. They were sure they could solve the problem. It would just take time.

Von Braun noted that everything connected with the Apollo programme was taking too long. Criticism was now beginning to grow in the press. The lavish amount of money being poured into NASA was being questioned. The last orbital flight had been in May 1963 and it looked very ordinary compared to the Soviet double act. Estimates for the first lunar landing were as late as 1971. A series of congressional hearings questioned the value of a lunar programme. 'Anybody spending $40 billion in a race to the moon for national prestige is nuts,' declared Eisenhower. The *New York Times* echoed popular sentiment with the headline:

LUNAR PROGRAMME IN CRISIS.

To give NASA a chance of staying on budget and beating the President's deadline, the new head of the Office of Manned Space Flight, George Mueller, came up with a radical solution: 'all-up' testing. This, when outlined to the usually cautious von Braun, was greeted with 'shock and incredulity'. Instead of taking a step-by-step approach, testing each stage before moving on, all features would be tested simultaneously. Rudolph explained very carefully just what this would mean; there were huge uncertainties and risks in 'all-up' testing for a rocket as complex as the Saturn V. Mueller was unmoved. In the very first Saturn V flight *every* stage would be live, he insisted, and the first manned flight would be brought forward from the seventh to the third launch. Von Braun, mindful that his caution had stopped the Americans winning an earlier stage of the race, reluctantly agreed to fly 'all-up' on the very first flight.

President Kennedy once again put in an appearance to help revive the flagging interest in space. At Cape Canaveral, astronauts Gus Grissom and Gordon Cooper showed him the plans for the growing moon port from the vantage point of a helicopter. The Saturn V rocket would be over 363 feet long and would require equally gigantic service structures. Too large to be constructed horizontally then raised to a vertical position like the old V-2, it would have to be assembled in an upright position. The vehicle assembly building would rise a sheer 525 feet, its huge doors 456 feet high, opening on to the flat Cape Canaveral landscape. It was to be so vast when completed that it was rumoured clouds would form inside at ceiling height. To secure this enormous hangar against the fury of autumn hurricanes, more than four thousand piles would pin it to the Florida bedrock.

President Kennedy smiled in amazement as he heard about plans underway for the construction's sites. The size of the vehicle assembly building would dwarf everything. Beneath them, workers were matchstick men, vehicles mere Dinky toys. It expressed so clearly American industry at its ebullient best. The date was 16 November 1963. Kennedy would never see the building finished.

He went on to Houston and then Dallas. To the wild, waving crowds he had the look of a young hero travelling in the open car. His wife, always a fashion plate, exquisitely dressed in a pink Givenchy suit and matching pillbox hat, sat beside him in the sun. The crowds were elated to see the Kennedys almost within touching distance. Then, in the time it takes to squeeze a trigger, the picture was changed forever. He was slumped, dying, against his wife. Her pink suit was red with his blood. John Fitzgerald Kennedy and his Camelot court were assigned to the dusty pages of history. In deep shock America mourned. The lament echoed around the world.

With the assassination of President Kennedy, the space programme had lost its most powerful ally. His vice president, Lyndon Johnson, now President, was increasingly preoccupied with the perceived threat of the spread of communism: the Berlin Wall, the Cuban crisis and now the growing fear of the 'red tide of Communism' flowing into American-supported South Vietnam. Kennedy had already sent some 16,000 'military advisers' to train the South Vietnamese to take on the Vietcong – the Vietnamese communists. The North Vietnamese were allied to China and the Soviet Union. With South Vietnam seen as 'the cornerstone of the free world in south-east Asia' it was a crisis that Johnson was determined to contain.

• • •

For Korolev underfunding was now becoming chronic and holding back his lunar programme ambitions. NASA was designing a new spacecraft, the Gemini, due to come on stream in 1964. Not only would this craft have the ability to manoeuvre, rendezvous and dock in earth's orbit but it could also change orbit and carry two people for two weeks, exactly the time it would take for a spacecraft to go to the moon, perform a landing and then return. Korolev's own spacecraft, the Soyuz, would not be available until 1965. Khrushchev realized that America, with its Gemini spacecraft, would soon put the Soviet Union into second place.

According to Korolev's deputy, Vasily Mishin, Khrushchev was on the telephone in early February 1964 ordering Korolev to upstage the two-man Gemini and 'launch three cosmonauts right away'. Kamanin confirmed that Korolev received orders to stop work on available Vostoks and use them to 'prepare and accomplish a flight of a three-person crew in 1964'. He noted in his diary that 'it was the first time I had seen Korolev in complete bewilderment. He was very distressed at the refusal to continue work on the Vostok and could not see a clear path on how to re-equip the ship for three in such a short time.'

Nevertheless, Korolev rose to the occasion. The Soyuz programme was put on hold while he was sidelined into this essentially unproductive mission to trick Americans into believing the Soviet space programme was more advanced than it was. Four Vostok flights planned for 1964 were also cancelled and the Vostoks duly cannibalized for the new Voskhod ('sunrise'). Feoktistov, one of the original designers of the Vostok, was horrified at having to sardine three men into the cramped space. He argued that it would make more sense to wait for the Soyuz to come on stream. As further encouragement to the design team, Korolev let it be understood that if they could come up with a satisfactory design, there would be a place for an engineer on the trip.

Feoktistov responded to the bait with the drastic proposals of eliminating the protective space suits with their air-conditioning systems and the ejection systems. The space travellers would wear normal clothing and pray that there would be no need for an ejection seat. Landing would be tricky, as the three men would have no alternative but to land in the capsule. This would be slowed by extra retro rockets and parachutes that would bring the speed down to 3 feet per second at landing. 'They would be cramped just sitting,' Vasily Mishin observed later, 'not to mention it was dangerous to fly.'

While work on the Voskhod was in progress, at his next meeting with Khrushchev, Korolev found that his plans for the N-1 and its possible career in space were listened to with a level of interest formerly missing. Khrushchev promised more money to the N-1 and its use in a future bid to land a Soviet cosmonaut on the moon. This, however, was not enough to ensure success. Army red tape was still causing delays. Korolev had a continual fight on his hands for funds. In fact that spring, the money ran out and work on the N-1 stopped entirely. The vast endeavour employing thousands ground to a halt as factories closed and offices emptied. Korolev bombarded senior military officials with letters spelling out the dire situation. On 25 May, he still had had no response, and drafted a strongly worded letter to the Secretary of the Central Committee for Defence Industries and Space, Leonid Brezhnev, pointing out that 'precious time' was 'unwisely and aimlessly being wasted':

> Two and half years have gone by since the end of the draft stage of the N-1 project, but at the moment work is totally unsatisfactory ... There are no firm deadlines, no essential organization, nor sufficient financial and material support. In brief, the initial sum of money set aside in 1964 for the building and launch of the N-1, which started as 11 million roubles, was unexpectedly reduced to 7 million roubles and then to 4 million roubles. Now finances are being refused altogether ... By May of this year we will have used up all our resources and the construction of the launch complex of the N-1 will have to stop altogether within a few days ... This completely unacceptable situation ... is mostly due to an underestimation of the significance of the N-1 not only for Soviet science and technology, but also for the preservation of our government's advantage in the realm of space, as the first socialist country in the world, the home of great revolutionary ideas, and the progressive nation which leads the world's socialist system.

Korolev was keenly aware how much the Soviet Union was lagging behind the Americans in the race to the moon and suggested that the Americans might even have the means to 'land people on its surface by 1967 – that is on the fiftieth anniversary of the first Soviet state on our planet!' He assured Brezhnev that, with 'due attention', Soviet explorers could beat them to it – even though there were just three years left in which to do so.

The endless campaigning for support on top of his work was taking its toll. Tikhonravov remembers that he and Korolev were at an area branch party meeting when he noticed Korolev was looking ill. During the interval, Korolev asked Tikhonravov's advice: 'What do you think, comrade?' he said. 'I'm not well. I just want to get home and into bed. Would I be missed?' Tikhonravov looked at him. His face was covered in sweat. It transpired that he was experiencing internal bleeding in the bowel as well as heart problems. Tikhonravov hugged him in concern and told him not to worry, that he would cover for him. When he recovered, a holiday was prescribed and he and Nina were flown to Czechoslovakia for three weeks in June.

On his return, Korolev was back in the thick of preparations for the Voskhod flight, which was set for August. The date passed, the month marked by a significant escalation of hostilities in Vietnam. President Johnson ordered American planes to bomb North Vietnamese torpedo-boat bases in retaliation for alleged North Vietnamese firing on US destroyers in the Gulf of Tonkin. Despite doubts about these reports, within days Congress had approved the Gulf of Tonkin Resolution that gave Johnson freedom to proceed directly with 'all necessary measures to repel armed attack'.

While the eyes of the world were focused on events in Vietnam, behind the scenes in Russia Khrushchev finally approved Korolev's lunar landing programme using the N-1 – a project which became known as the L-3. Flight testing was to start as early as 1966 to beat the Americans – but Khrushchev was still ambivalent about providing the necessary funds. Korolev continued to use all his 'battering abilities to try to raise money', observed Sergei Kryukov, who was in charge of overseeing the design of the N-1. 'It was constantly necessary to hurry, to persuade, to approve.' Yet although Korolev's team needed at least 45 million roubles for the project that year alone, they were only allocated half that sum. Kuznetsov fared even worse with his requests for funding for the engine development. He needed 50 million roubles, but was to receive less than 20 million. With funding for space still split between rival teams, Korolev continued to receive lacklustre support.

In September, Khrushchev was still reasonably confident that the space age was Russian. Surrounded by his political and military grandees, Brezhnev and Ustinov, he went to Baikonur to review the latest technology. Chelomei, Korolev and Yangel were demonstrating their current designs. Chelomei, with an air of the 'winner' about him, was exceedingly put out when the launch of his UR-200 failed and doubt was cast on his ability, especially by Ustinov. There existed a thinly veiled hostility between Chelomei and Ustinov, usually carefully hidden by both men from Khrushchev. The sheer disbelief that his work could attract criticism was assuaged somewhat when the assembled dignitaries examined the streamlined latent power of a full-size mock-up of the UR-500 and were suitably impressed. Then came the inspection of the enormous launch site. Chelomei continued to stun his audience into silence, producing plans for a colossus among rockets: the UR-700. He assured Khrushchev that this was the only rocket for a lunar launch, flagrantly criticizing Korolev and the N-1. Korolev, by contrast, ruefully noted how little time Khrushchev spent reviewing his work. 'I passed these days as though I was in some kind of toxic furnace,' he told Nina.

After Khrushchev's visit, Korolev welcomed the relief of long days working on his Voskhod project. September passed, while Korolev's engineers were battling against multitudinous defects. Bolts were out of alignment. The shape of the vehicle had altered imperceptibly during transport. One of the telemetry instruments was not working and would have to be replaced. The third-stage engine was showing minor problems. The date was reset for 11 October. When further problems beset the telemetry system, Korolev was beside himself with fury.

At last, on 12 October 1964, all systems were ready. The hopeful cosmonauts had had a mere four months' training. The crew commander, pilot Vladimir Komarov, Boris Yegorov, a medical doctor, and Konstantin Feoktistov, the designer, strolled casually to the craft informally dressed in woollen tracksuits and shoes. They inched their way into the oddly placed seats, where the feeling was always that somebody's elbow was in your face. Despite appearances, on the launch pad tempers were stretched to breaking point. If there should be a miscalculation of any kind, those on the ground would be condemned to watch while the three cosmonauts were consumed in the flames with no possibility of escape. Korolev's nerves were failing him. People could see him visibly shaking.

It was a perfect launch and the flight went to plan. During weightlessness, unpleasant symptoms were experienced by Feoktistov and Yegorov but they recovered well. Their orbit of the earth, reaching an altitude of over 250 miles, afforded the cosmonauts a most stupendous view and they wanted to prolong the flight beyond the schedule of sixteen orbits. Komarov pleaded with Korolev only to be reminded that this was not in the agreement. Komarov persisted that they had seen so many interesting things, they must continue. Korolev was firm: 'There are more things in heaven and earth, Horatio, than are dreamt of in your philosophy.' He insisted they began the descent.

There were still enough hazards to preoccupy Korolev. For this trip an extra retro rocket had been installed in case the main ones proved unreliable. He was also worried about the landing on hard ground, although several small rockets were used to slow the descent and minimize a bone-shattering impact. But to Korolev's utter relief, very soon the three men were sighted, waving at the rescue helicopter. Looking pale and drained by worry at the risks taken, Korolev could hardly believe that it was all over and that it had proved a success. 'I would never have thought it possible that the Voskhod could be made out of the Vostok and that three cosmonauts would fly into space,' Kamanin heard him say.

At Baikonur the party mood at the celebratory meal was exuberant. The toasts to the heroes were extravagant; the vodka flowed. Yet ironically, the rotund figure of Khrushchev was not there, nor would he ever be again. There had been a quick, cold revolution in the depths of the Kremlin. Toasts were now drunk to Alexei Kosygin and Leonid Brezhnev. Khrushchev himself was under house arrest.

Once again, America was put in its place by the Soviet Union, which seemed to be operating an advanced space programme. It was assumed that the Soviets had a multiseat space vehicle similar to Apollo. NASA's two-man Gemini programme had not yet been launched and Apollo was still at the design phase – while the Soviets were evidently roaring ahead. The headlines in America deferred to Soviet capability in space and some reports even warned that the first Soviet manned lunar mission could be as early as 1966.

Meanwhile, the great industrial beast that the space programme had become in America just kept going at snail's pace, testing engines, designing capsules, slowly working its way through the complexity of the quest. In public, NASA leaders always insisted they would beat Kennedy's deadline – with men on the moon before the end of the decade. Maybe they wouldn't be first, but they were confident they would get there.

Hundreds of different companies were now involved in the vast enterprise to produce the complete Apollo spacecraft. Sitting atop the three enormous booster sections of von Braun's Saturn V rocket would be three further sections: the Apollo command module, a cone-shaped spacecraft, about 13 x 11 feet, with the controls and room for the astronauts to relax. Behind this, the service module would supply power and oxygen. Then came the lunar module, the craft used to descend to the moon. While the lunar module was on its journey to the moon with two astronauts, the command and service modules would stay in lunar orbit with the third astronaut. On leaving the moon, a powered section of the lunar module would return the astronauts to the command module where they would dock. Under its own rocket power contained in the service module, only the command module would return to earth with its crew.

Engineers were working their way through completely new problems. How could they measure fuel in a tank in zero gravity? Would the heat shield be affected by the extreme cold of the lunar orbit? How to deal with computer overload? And the mighty Saturn engines had still to be tamed. It was clear the injector plate would have to be redesigned. In order to obtain the elusive stability, at least fifty modifications were tried before it was realized that designs were being repeated. The great American industrial space machine was lumbering confidently forward when suddenly the Soviets did it again. The latest Soviet star turn was to achieve the world's first space walk. Their success began to make it look as though the Americans might as well give up.

The first space walk, however, was not the result of advanced technology. It was another of Korolev's illusions, part of the Vostok package instigated by Khrushchev when he had been in power. The Voskhod 2 was another quick lash-up made from the remnants of the Vostoks modified to carry two cosmonauts, one of whom would leave the capsule and 'walk' in space. Leaving the capsule was achieved by adding a pressurized inflatable cylinder made of a rubber-based substance. Airlocks allowed the cosmonauts to leave and re-enter. A specially designed pressurized space suit would be worn for the event. Alexei Leonov, the cosmonaut selected to carry out the space walk, trained for every possible emergency. There was particular concern that he might lose consciousness in space – and rehearsals were even carried out for a dramatic space rescue by his fellow cosmonaut, Pavel Belyayev.

The flight was planned for March 1965. Before this happened, an unmanned flight was made to test the pressurization chamber and the space suit. The flight started well. The craft began its programme of automatically working airlocks, controlling pressure and depressurizing, but then it suddenly exploded. There had been a concern that the airlock would not function and now there was no way of retrieving the flight data to be sure. Korolev was worried. It was known that the Americans were finalizing plans for their first manned Gemini, to be followed soon afterwards by their first space walk, and the need to be ahead was paramount. Korolev went to see Alexei Leonov and Pavel Belyayev, and, according to Leonov, 'presented us with a stark choice'. There was no possibility of further testing, Korolev explained, because there were no more suitable vehicles apart from the one allocated to their mission. They could do further testing and delay the mission for a year – or take the risk. He stressed that there were grave risks, but that the decision was in their hands. Leonov and Belyayev did not hesitate. They assured Korolev they were 'ready to fly'.

Korolev, however, felt the strain of the decision. He confided in Nina:

> We are trying to accomplish all our work without hurry. Our chief motto is 'the safety of the crew comes first'. God grant us the strength and wisdom to always live up to this motto ... I personally always believe and hope for the best outcome even though all my efforts, my mind and my experiences are directed towards trying to foresee and outguess the worst that can happen – an ominous presence that stalks us every step into the unknown.

The pressure of the uncertainties of the mission, the need to beat the Americans and the sheer workload of it all conspired to bring Korolev down with pneumonia. The eighteenth of March, the day for the flight, arrived and no one could keep him away. Looking old and spent, he turned up in bitter, snowy weather still worried about the ring in the airlock and whether it would make the capsule spin. He regained something of his optimistic vitality on seeing the two cosmonauts, Alexei Leonov and Pavel Belyayev, and wished them a 'fair solar wind'.

Although Korolev suffered his usual apprehension at the launch, it went well and, once in orbit, Leonov began his preparations in the cramped capsule, strapping a support system on his back and pressurizing the small exit chamber. He crawled into the pressure chamber, attached himself to an 18-foot rope, depressurized and put his head out, the first man to view what exactly all that emptiness was made of.

As the hatch opened he caught his breath. Far, far below, the world turned slowly. Holding on to the spacecraft, in dazzling sunlight, he eased himself out completely. The shock of seeing the earth spread out below 'like a gigantic colourful map', was, he said, 'an extraordinary sensation. I had never felt anything quite like it before. I was free above the planet Earth and I saw it – saw it was rotating majestically below me. Suddenly in the silence, I heard the words: "Attention, attention! Man has entered open space."'

Slowly he removed his hands. He was on his own, held like a feather in still air, yet travelling at 17,400 mph. The Pacific Ocean glittered far beneath his feet. The vast panorama was shifting quickly as they moved. 'I'm feeling perfect,' he told base as live TV images of him were relayed back to an astonished world. 'Lenin once said the universe is endless in time and space,' he later recalled. 'It is the best description I had of those moments ... Nothing will ever compare to the exhilaration I felt.'

All too quickly his ten minutes were over and he became aware of Belyayev urging him to return. That was when a series of 'dire emergencies' began to unfold. When he reached the airlock, he couldn't find his feet or his hands. With a sense of shock he found he had no control. His space suit had unaccountably stretched. The fabric of his neatly covered hands and feet had ballooned. He was suspended in space, with his hands now useless and his feet no longer in the designed footwear. He was meant to ease himself back into the cramped airlock feet first so that he could close the hatch behind him. This was no longer possible but if he delayed any longer the flight would be in jeopardy and his life-support system would fail. There was no alternative. He would have to enter head first. He entered the shaft to find he could not move. His suit continued on its course of expansion.

Leonov was in serious trouble, stuck in the narrow shaft in his overblown, cumbersome suit. He spent twelve minutes struggling with it, trying to get the pressure lower by releasing a valve. He could feel his temperature rising as he struggled in the thin channel. By his feet, the hatch was still open. He would have to perform a somersault so his hands were near the hatch. In the tight, claustrophobic cylinder, surrounded by vast, black space, he slowly turned in the shaft with his unwelcome acres of space suit hindering him. He was exhausted when he finally closed the hatch and Belyayev could start to pressurize the cylinder. His pulse was racing. He could hear the blood thundering in his ears; but he was safe.

His safety, however, and that of Belyayev, was relative. The exit hatch for the space walk had not shut tight. This led to a drop in air pressure which the automatic system rectified by pumping oxygen until it was at levels of 45 per cent. In such an oxygen-rich environment, the tiniest spark could cause an inferno within seconds as it had done for the young cosmonaut Valentin Bondarenko. To add to the feeling that the flight was doomed, the automatic guidance system for re-entry had failed. They were in a flying chamber with faulty guidance and a door that did not shut in an atmosphere of almost pure oxygen. On the ground there was panic.

Korolev took charge. He still looked very ill but his mind was fixed on the problem. Belyayev would have to fire the retro rockets manually. In the control room, frantic calculations were made to work out the orientation for a successful re-entry. If the cosmonauts failed to get the orientation right, their fate was better not thought about.

The sums were transmitted. They had to orient the craft but it proved almost impossible. In order to get to the orientation porthole, Belyayev had to lie across both seats. Leonov got out of the way, under the seat, holding on to Belyayev to keep him steady. The task completed, they scrambled back to their seats before firing the engine. This took them forty-six seconds. They fired. With a terrible jolt the pace slowed and re-entry began. But as had happened so often before, the capsule did not separate from the instrument module. They were tied together like a pair of old boots floundering through space. The gravitational loads on the men reached 10 g's. The forty-six-second gap between fixing the orientation and firing the engines made a significant difference to their landing position. For four hours there was no communication. Korolev had no idea that they were, in fact, safe but very cold, having landed in an area of thick forest in deepest Siberia, where they spent two freezing nights, the craft suspended between two trees above great drifts of snow.

The celebrations were sweet after the struggle of recent weeks. The Americans had still not begun the Gemini programme and were increasingly embroiled in a messy conflict in Vietnam. Johnson had ordered Operation Rolling Thunder – bombing raids of North Vietnam. In March, the first American ground troops landed to protect the US air base at Da Nang. Criticism was rising of the US air force's use of Agent Orange, a herbicide containing dioxin that stripped vast tracts of dense vegetation bare to expose the Vietcong bases. As

American imperialism came under fire at home and abroad, Korolev's repeated iconic missions in space appeared to symbolize much that was good about the Soviet system. His triumphs could hardly be ignored by those at the top.

Korolev now began to benefit indirectly from Khrushchev's fall as the Brezhnev–Kosygin duo was determined to reverse Khrushchev's policies. Chelomei, as number one favourite, soon found the chill winds of reality blowing through his life. Everything in his growing empire was questioned, from his UR-500 to the size and thickness of the carpets in his house. His design bureau was diminished. His UR-200 was cancelled. The glitter and glory of his recent favouritism was swept out with the rubbish – almost. His powerful Proton rocket would still be used to launch a simplified Soyuz craft, known as the L-1, around the moon. The aim was to do this as early as 1967.

Meanwhile, Korolev, who had never lost sight of his original inspiration, sent endless begging letters to those at the top, and was at last rewarded. With Chelomei's loss of support, more funds would at last come Korolev's way for the N-1/L-3. The target was set for a manned landing as early as 1968. Korolev suddenly found himself in control of the first 500 million roubles, with the promise of more to come, and work could begin on the launch complex.

Like the Americans, Korolev favoured a lunar orbit rendezvous, but the N-1 would need gearing up to take a greater payload. Six more engines were to be added to the first-stage booster, giving thirty engines in total – twenty-four on the inner perimeter and six new engines centrally placed in a circle. The rocket would now have a lifting capability of over 95 tons, and a new guidance system would be housed in the third-stage equipment bay. Korolev also won approval for a complete schedule of work on the Soyuz from the Military-Industrial Commission. 'Don't give the moon to the Americans,' came the directive. 'We'll find all the resources we need.'

At long last Korolev was able to compete directly with the Americans to reach the moon. The infighting between different bureaus and government indecision had caused delays and cost the Soviets a high price in time and resources – which were still split between two parallel programmes. But Korolev was full of hope for the future. 'Friends,' he announced to his colleagues, 'before us is the moon. Let us all work together with the great goal of conquering the moon.' The year was 1965. He intended to be on the moon by 1968.

Von Braun, meanwhile, had become a target for popular criticism. At a time when there was continued opposition to the mounting cost of the Apollo programme, a popular satirical song from the period by Tom Lehrer summed up the ambivalence that was felt towards von Braun. Why should some 'Nazi Schmazi' spend 'twenty billion dollars of your money' to put 'some clown on the moon?'

Wernher von Braun

Gather round while I sing you of Wernher von Braun,
A man whose allegiance
Is ruled by expedience.
Call him a Nazi, he won't even frown.
'Ha, Nazi Schmazi,' says Wernher von Braun.

Don't say that he's hypocritical,
Say rather that he's apolitical.

'Once the rockets are up, who cares where they come down?
That's not my department,' says Wernher von Braun.

Some have harsh words for this man of renown,
But some think our attitude
Should be one of gratitude,
Like the widows and cripples in old London town
Who owe their large pensions to Wernher von Braun.

You too may be a big hero,
Once you've learned to count backwards to zero.
'In German oder English I know how to count down,
Und I'm learning Chinese,' says Wernher von Braun.

TOM LEHRER, satirical musician

CHAPTER EIGHTEEN

'I just need another ten years'

Less than a week after Leonov's remarkable feat, the Americans launched their first manned Gemini mission. Gus Grissom and John Young flew in Gemini 3, which was half as big again as the original Mercury capsule and much more advanced, with larger thruster rockets and enough fuel to test many more manoeuvres in space. This was swiftly followed on 3 June 1965 with Gemini 4, in which Edward White successfully undertook a twenty-minute space walk. Within two months, Gordon Cooper and Pete Conrad set a new endurance record with an eight-day mission in which they tested new fuel cells and radar equipment to guide rendezvous in space. They suffered no adverse health effects and successfully developed a routine of sleeping and exercising in space: an eight-day flight to the moon was beginning to look like a real possibility. The Americans announced plans later in the year for a two-week stay in space. They were catching up.

Despite his go-ahead to develop the N-1 for a lunar mission, Sergei Korolev had no immediate means of outshining the Americans until the Soyuz came on stream, and he felt the disappointment keenly. 'He spreads himself too thin and tries to keep everything under his control,' observed Nikolai Kamanin. For almost two years he had attempted to investigate the moon with unmanned probes. The Luna series of probes aimed to take measurements of the near-moon environment and ultimately to land on the moon's surface, record data and even bring back soil samples. But the first probes, designed to enter lunar orbit, were failures and this was fuel for attacks from Glushko who endlessly criticized him behind his back; still spitting venom, he was still advocating the complete redesign of the N-1 and lobbying for support among other chief designers whose betrayal Korolev felt acutely. Glushko teamed up with Chelomei and continued to promote his massive UR-700 as the lunar launch vehicle. Korolev was beside himself at the thought of funds being redirected yet again. He was already suffering enough production difficulties and delays. For Korolev, the constant pressure of work and his physical wellbeing seemed irrevocably entwined. The heavier his workload became, the more he buckled under it.

With low blood pressure and heart problems, he went to the clinic at Baikonur. The nurse thought he looked ill. 'You smell of Validol,' she said. He had been dosing himself with quite large doses because of chest pains. 'Oh, these country doctors,' he replied, asking her to keep quiet about it. From Baikonur he admitted to Nina that he was 'unusually, deeply tired', but tried to reassure her, 'the fatigue is in my nervous system'. He told her that he was trying to take rest. 'I am in a constant state of exhaustion and stress, but I can under no condition show that these things are getting to me. So I am trying to hold on with all the strength at my command ...'

His symptoms had already led to him having to spend a few days in hospital for tests in mid-1965. Nina Ivanovna had stayed with him. While he was there, like a breath of fresh air Gagarin and Nikolayev visited, coming from Star City with the latest news and gossip. Before leaving, in a typically generous gesture Gagarin gave Korolev his watch when Korolev complained that his had stopped. This touched Korolev. He hardly knew what to say and tried to give it back. Gagarin insisted. It was a gift. To recuperate, Korolev and Nina Ivanovna had a holiday in the Crimea. It had been full of memories of other summers as he had flown paragliders there as a young man.

Korolev's workload remained undiminished, all the more so because people came to him for every conceivable decision. Golovanov relates one apocryphal story: Korolev was trying to work out the nature of the moon's surface. He wanted to know if it was hard, or whether it was covered in a thick layer of dust, which could ruin any attempt at landing. After hours of debate it was clear the specialists could not resolve the matter. 'So, comrades, suppose we assume the Moon has a hard surface,' Korolev concluded, rising to leave. There was an immediate outcry. Who could possibly guarantee such a critical and unknowable decision? 'Who will take responsibility? Oh, this is what you are talking about.' Korolev grabbed some paper and a pen, his face screwed up in concentration. 'I will take it.' He wrote in large letters: 'The Moon is hard' and signed it 'S. Korolev'.

The American-planned two-week stay in orbit could not be allowed to pass unchallenged. The Soviet Union would make the Americans look amateur. If they were orbiting in space for two weeks, Korolev would send two men into space for three weeks using an elliptical orbit that would ensure they also collected an altitude record. The Voskhod had to be modified for the more complex trip but the life-support system particularly was causing delays. All this conflicted with what Korolev considered more important work on the Soyuz space capsule. Another disappointment was that the Voskhod programme for women was falling behind. No one could be found to design their space suits. As the year passed with the Americans notching up one success after another with Gemini missions almost every two months, the pressure on Korolev grew. He was asked to conjure up a landmark event for the following March to coincide with the next Communist Party Congress. A new wonder had to be produced, involving two Voskhods.

Weighed down by the continuing demands, Korolev continued to push himself. Kamanin was concerned at how old and tired he was looking. Nina quietly observed how tetchy he had become. He came home from work utterly depleted, exasperated by unimportant details, fussing if his slippers were in the wrong place or if she had inadvertently overstocked the fridge. His characteristic largesse of spirit was succumbing to anxiety. He hated it if Nina was not at home.

One Saturday she visited her sister. Korolev said he was tired and would relax. She must go, but 'put the phone near me,' he said. When she arrived, her sister passed on the message that Sergei had called. She rang him back.

'Why did you leave me all alone?' he asked.

'Do you want me to come back?' she replied.

'No, don't, I'm going to sleep.'

Twenty minutes later he called again. 'It's me. What are you doing there?' he asked. She returned home to find him asleep.

That autumn, Nina Ivanovna recalled that he would come home from work so drained that he would often say: 'I can't continue to work like this, you understand. I'm not going to continue working like this. I'm leaving!' Yet each day he pushed himself on, chasing that elusive goal when the Soviets could regain their lead. He confided in another colleague: 'I'll just reach sixty and that's all. I'll not stay a day longer. I'll go out and plant flowers.'

Work on the Soyuz expanded rapidly with more than three hundred different research organizations involved in its development. As soon as he had a full-scale model of the new ship, in the autumn of 1965, Korolev invited the leading cosmonauts and engineers to OKB-1 to see it. Korolev himself was late for the meeting, and when he arrived he looked harassed. Someone eagerly asked him to tell them all about it, but Gagarin was quick to silence him, worried at all the pressure on Korolev. 'We'll have time for that later,' he told his friend. But Korolev protested, 'I have twenty-five minutes.'

He proceeded to give a detailed explanation and soon warmed to his favourite theme. The Soyuz represented the culmination of a vision for which he had fought for years. It was a complex craft that would weigh 7 tons and comprised three modules. The capsule itself would hold crews of two or three and carried equipment to sustain flights of about two weeks. An important new feature was the solar panels that would give more power and therefore more time in space. The cosmonauts marvelled at the size of the craft. 'It must be larger than two Vostoks,' said Gagarin appreciatively. Komarov asked about the docking unit. Korolev explained how this would facilitate space exploration and conjured up a vision of a craft which would make flights from earth to space and back again to earth routine. The questions came in quick succession and Korolev, although clearly tired, responded to their enthusiasm. It was only when he had left the cosmonauts and returned alone to his office that he felt severe chest pains.

Ignoring his own wellbeing, in the coming weeks Korolev continued to supervise progress, sometimes travelling out to Star City to check up on the cosmonauts' work in the Soyuz simulator. According to Romanov, Korolev's biographer, once he arrived to find Vladimir Komarov in training. He took over from the instructor at the microphone.

'I am Dawn [Zarya]. Can you hear me?'

Komarov was surprised to hear Korolev's voice at the controls. There was a brief moment of silence before he replied with his usual professionalism: 'Good morning, Sergei Pavlovich. I feel great. Everything goes according to plan.'

The instructors were full of praise for Komarov. They told Korolev it was a pleasure to work with him: 'He was first and foremost an engineer.' Komarov was rapidly acquiring a reputation as one of their most talented cosmonauts.

In December the Americans claimed their two-week endurance record in Gemini 7, unchallenged by the Soviets. They also continued to perfect manoeuvres for orbital space docking as Gemini 6 met up with Gemini 7 – the craft just a few yards apart for several hours. Apart from facing up to the growing challenge from America, Glushko's criticism was also an endless source of irritation to Korolev. Making no allowances for his ailing colleague, Glushko seized every opportunity to put him down in public meetings. 'He thinks that he is the chief successor and descendant of Tsiolkovsky,' Korolev would say angrily, 'and that we are only making tin cans!'

Korolev started intestinal bleeding again and was forced to consult a doctor for reassurance. Yes, he had a problem but it was not serious. He had a polyp, a small growth in the rectum. He would need an operation. Korolev explained that he was too busy, half jokingly saying that he just needed another ten years, at least. The doctor persisted. It was not serious, but the problem must be tackled. After the operation he would be a new man. The minor operation was booked for January.

At last there was a reason to explain his tiredness, but it was not an excuse to slow down. Over New Year's Eve, Korolev could not see quite why, just because it was a holiday, he was obliged to stop work. Nina reminded him he had been invited to a party. Important officials such as Mstislav Keldysh would be there. Korolev confided to him, 'You know, I am going to hospital again and I have a bad feeling about it. I don't know whether I will leave it ever again.' Keldysh had drunk a little too much and afterwards he reflected with regret that he had not allayed Korolev's fears sufficiently.

In the New Year, Nina invited some of Korolev's colleagues in for a party. When the guests finally departed, Korolev urged Gagarin and

Leonov to stay. He felt the need to talk. The conversation continued for hours as they mulled over all aspects of their lives. It was four in the morning when Korolev finally began to unburden himself of the shock of his arrest in the 1930s, his mock trial and his experiences in Kolyma. He had not discussed this with the cosmonauts before. In fact he had never spoken of those years to anyone, but now he suddenly felt the need to re-examine what had happened.

Through the night over a bottle of cognac the two men listened as Korolev described what he had survived. He had been unconcerned with politics, happily wrapped up in his work and his young family when he had become drawn into the purges at the height of Stalin's terror. Once arrested, he had lived in fear of death, knowing full well that what was happening to him could easily be inflicted on his family. His torturers had a way of taking more than his life. He was beaten, yes, but after a brief respite, when broken bones were knitting and a sliver of hope appeared, they would begin again, taking him almost to the point of death. His mock trial was a travesty of justice. When Korolev had explained that he had committed no crime, the reply cut short any possible investigation: 'None of you swine have committed any crime! Ten years' hard labour. Go! Next!' Korolev was speaking of events that had happened twenty-five years earlier but his memory made it feel as if it were yesterday. His words made a deep impression on Gagarin and Leonov.

It was hard for the young cosmonauts to understand how Korolev had served the state so well when he had lived through such betrayal. No one was more truly Russian, more loyal to the state, more patriotic than Korolev. This was the man who had written to Nina Ivanovna from Kapustin Yar almost immediately after their marriage: 'Being apart is the battle order of our Motherland. We are going to go through it cheerfully and firmly. I miss you and I think about you.' All this when the state had taken years of his life in pointless imprisonment.

It was late on 4 January before Korolev was able to extricate himself from his office and even then he couldn't quite bring himself to leave. The door suddenly opened, recalled Chertok. 'He did not come into the room. He stopped at the door. He was wearing his overcoat and fur hat and looking at everyone with a sad, soft smile.' They began to wish him the very best of health. Chertok was struck with how strange it was not to see the 'powerful and authoritative army commander' but a 'tired, sad, dear SP'. He looked at everyone for a minute and then turned around and left, without closing the door.

The next morning he and Nina Ivanovna made their way to the Kremlin hospital where he was to undergo some tests. Almost absurdly for a distinguished man of science, he was fussing because he was unable to find his 'good luck' coins – two kopecks – by which he set store. Nina Ivanovna stayed all day, delaying the time when that air of authority prevalent in hospitals would reduce her husband to that anonymous creature 'the patient'.

While in hospital, he dealt constantly with the office. There was a crisis almost immediately. He had left his deputy, Vasily Mishin, in charge but Mishin was unused to dealing with all the different responsibilities that Korolev had juggled almost effortlessly. Quite apart from negotiating his way through the political bureaucracy and keeping abreast of the engineering challenges, the administrative task was formidable. The production of the N-1 alone involved more than five hundred different organizations, each committed to a different delivery schedule of key components. Within a matter of days, Mishin had had enough. He decided to resign. As soon as Korolev heard the news he called Mishin and urged him to stay. Korolev reminded him that ministers came and went with the seasons, but rocket scientists were there for the duration. Mishin tore up his resignation letter.

On 12 January it was Korolev's birthday but he expressly asked friends not to call. It would be too depressing. The previous year Nina had had to remind him it was his birthday and on the spur of the moment he had invited everyone he knew to celebrate it with him. All the guests had worn their 'Gold Stars'. Colours of red and gold, music and laughter, the warmth of his many friends: that memory could not be rekindled in the silent whiteness of a hospital room. Besides, he was bleeding quite badly. The surgeon who would perform the operation had removed a small sample of the polyp for tests and the bleeding would not stop.

The operation was set for 14 January. Korolev and Nina had said their goodbyes the evening before. They had agreed that she would not come to the hospital before the operation, which would be early in the morning. She would be there when he woke up.

In the morning, however, Korolev felt the need to hear Nina's voice. At 7.55 he rang her. The housekeeper answered the telephone. He wanted to speak to Nina. She wasn't in.

'Where is she?' he wanted to know.

'She left to go over to you,' the housekeeper explained.

He was reassured by this news, but there was a catch in his voice.

Nina Ivanovna had had a feeling that her husband needed her but she arrived too late. Korolev was being taken into the theatre on a stretcher. Something about the fact that he was taken in feet first struck her as wrong. A group of doctors followed. The doors were closed. The operation was due to last about three hours. She would wait.

Korolev's high status ensured an eminent surgeon would operate and it was the Minister of Health himself, Dr Boris Petrovsky, who was to remove the polyp endoscopically. When this was accomplished, Korolev haemorrhaged so severely that the flow could not be staunched. To allay the bleeding, Petrovsky cut the abdomen and discovered a tumour the size of a fist. In his memoirs he noted 'the presence of an immovable malignant tumour which had grown into the rectum and the pelvic wall. Using an electronic scalpel, we were able to extract this tumour with great difficulty and conducted a biopsy which confirmed the presence of a malignant tumour.'

All the while, Korolev was bleeding profusely. Alarmed, Petrovsky summoned a cancer specialist, Dr Aleksandr Vishnevskiy. There were now, said Golovanov, the two most famous surgeons in the country in the operating room. The size of the tumour was unexpected and tension was rising. They had to remove parts of his rectum in order to take out the entire pervasive tumour; still the bleeding would not stop.

Nina Ivanovna was desperate with worry. A three-hour operation had turned into eight hours. Someone was trying to calm her when the surgeons came towards her assuring her of success. Suddenly their names were called and they rushed back to the theatre.

For eight hours Korolev had been kept under general anaesthetic. For such a lengthy operation, he should have had a tube in his lungs, but he had omitted to tell his doctors that his jaw had been broken several times as a young man in Kolyma Gulag, and he could not open his mouth wide enough to allow a tube to be inserted. The doctors wondered about doing a tracheotomy. Instead it was decided simply to use a breathing mask. When this was removed his breathing was laboured. He was clearly in difficulty. Hurriedly, a tracheotomy was performed. His breathing returned to normal but his heart was weak, too weak – and it stopped beating half an hour after surgery. The doctors tried hard to revive him. They gave him adrenalin injections. They tried everything they knew. It was no use.

Nina Ivanovna was now in a state of shock, fearing the worst, not knowing what that could be. Petrovsky came out to speak to her: 'Be courageous,' he said. And then she knew. Her husband was dead. The news was difficult to understand. She heard the surgeon saying: 'I don't know how he lived or walked with that heart.'

The news struck like lightening. Korolev was dead. His friends and colleagues – Mishin, Chertok and so many who had worked with him for decades – heard the news and, like Nina Ivanovna, were dumbfounded. Suddenly a vacuum had been left by this man whose generous spirit had touched them all. Many could not adjust to the news. 'It has been three days since he died and I still don't want to believe that he is no longer amongst the living,' wrote Kamanin in his diary. 'He left us when his talent was in bloom.' Only Glushko received the news without charity. He was at a meeting when the Kremlin telephoned. 'Sergei Pavlovich is no longer with us,' he informed his colleagues. 'Now, where did we leave off ...?'

In the Kremlin, Brezhnev made a decision: to draw Korolev out of the shadows and bring him into the light. The Soviet Union had a hero. Mother Russia had produced a true son whose grandeur of vision must be known. It was time to reveal the Chief Designer's name. It would live forever in Russian history.

On 17 January, Russians learned the identity of Sergei Pavlovich Korolev and they took him to their hearts. They waited in the bitter cold to file past and steal a glimpse of this legend, lying so still, as though asleep among the massed flowers. All the chief designers and Korolev's deputies acted as guards of honour as he lay in state in the Hall of Columns in the House of Unions. So many people had volunteered for this honour that they could not all be accommodated. The whole of Russia, it seemed, had sent flowers: Korolev's factory workers, delegations from Kapustin Yar and Baikonur, Politburo members, ministers, generals, marshals and admirals. The scent hung in the air as hundreds of thousands paid their respects. Finally people 'had been shown a morsel of truth,' wrote Chertok, and learned 'to whom they had to pay respect for the greatest victory of mankind'. Everyone who filed passed his coffin felt they 'touched his historic achievement in some way'. At 8 p.m. the doors were due to close, but the queue of people had no end.

Everyone had seen Korolev's picture in *Pravda*, a strong, smiling, yet unfamiliar face, proudly wearing his medals. And they had read that 'he was warmly loved and respected by the many, many people who worked with him'; and that 'S. P. Korolev had lived giving all of his creative energy to the concern of the people and to progress'. And as they looked at the coffin draped in red, with wreaths from the highest in the land at his feet, the great white marble columns in the hall beribboned in black in the dull light of the shaded chandeliers, they understood that this must indeed have been a noble fellow Russian, whose passing should be marked and honoured.

The state funeral was on a massive scale: a huge drama with a cast of thousands laid out against the backdrop of the winter city under grey skies and falling snow. The serried ranks of sombre uniforms and crowds of people filled the scene. The urn with Korolev's ashes was carried by Gagarin, Keldysh and other leading officials to Red Square. Speeches for the twentieth century's practical visionary echoed across the square for the man who, since boyhood, had always gazed at the moon with wonder, dreaming of the stars.

There was one aspect to the ceremony that did not impress Kamanin. 'Korolev occupies a place in the Kremlin wall next to S. V. Kurashov [the Minister for Heath]. I was irritated by the fact that they were neighbours,' he wrote:

> It unnecessarily reminded me of the great guilt of our medicine in the premature death of Sergei Pavlovich. All of the orators at the funeral gathering thought that Korolev was a great scientist, but not the Chief Director of space studies. This is not true ... He was the Chief Designer of the spacecraft and not only in that post, but in essence as well. I will always place unlimited value on Korolev's talent.

Following her husband's death, Nina Ivanovna was swept along on the tide of interest and concern. She was the great man's wife and a certain dignity was expected. Private grieving would have to wait. Boris Chertok describes how she had to bear all the toasts and speeches teary eyed although she was 'in such a bad state … she could hardly make it'. Then suddenly all the noise and fuss was over. The silence of separation descended. There was no familiar voice with a touch of bemused petulance to telephone and say, 'why did you leave me all alone? When are you coming home?' She had no reason now to be at home.

His daughter, Natasha, was also trying to come to terms with her loss when Mishin came to see her with a small box which he had found in her father's desk. She opened it to find her school exercise books, photographs and letters from years before, treasured tokens from her missed childhood that he had kept with him at all times.

CHAPTER NINETEEN

'We're burning up!'

At the Marshall Space Flight Center, as he sat at his desk the strong features and intelligent eyes of Sergei Korolev looked von Braun squarely in the face from the pages of the American press. 'Mr Korolev was reputed to be the mysterious "Chief Designer" of the space programme,' stated the *New York Times*, revealing that 'during his lifetime, Soviet reference works described him as merely working on the "problems of mechanics"'. For the *Washington Post* he was the 'Soviet space scientist who designed almost everything Russia has put into orbit'. All the papers carried the same photograph – of a youthful Korolev in his early forties, his face relatively unlined, his eyes made more deep-set than they were in real life by the dark smudges of ink. For von Braun, the mystery figure behind the Soviet space programme was revealed too late. There would never be an exchange of views between them now.

The Kremlin's 'deep grief' at Korolev's passing and his loss to world science was reported at length; his death was evidently a 'heavy blow' to the Soviet space programme. He had seemed immensely capable, a man to rely on. Von Braun realized this was the man whose unseen presence had so often discomfited America with his astonishing firsts in space; the man behind the might of Soviet rocketry. If Sergei Korolev, from his place in the shadows, had not shared the same vision of space as von Braun, and pursued that vision with such energy, would America be fighting to get to the moon at all costs now? It was hard to know where his own ambitions would be without this Russian who had lived and died in secret. Was it because of Korolev that the American space programme was at last gathering strength?

Von Braun was especially pleased that the intensive testing of the Saturn engines had finally brought results. After many months, the day had arrived when the engine, even with a bomb placed in the combustion chamber, finally took the tumult in its stride and righted itself within a hundred milliseconds. They found that by changing the angle of the jets of fuel coming from the injector plate so that they met and united at a lower level in the combustion chamber, the bombs no longer caused a runaway instability, but one that was quickly rectified. They also implemented baffles – small, thin fence barriers – running from the ejector plate to the chamber, which ejected a smoother flow of fuel. No one was entirely certain why this produced results and there was no guarantee that it would always work. Consequently there was no actual day of celebration, just an increasing confidence that the engine was more reliable.

The Gemini missions continued to give valuable insight into the problems of manoeuvring, rendezvous and space walking. The Gemini VIII mission was to break new ground with the first complete test of docking in space – a crucial manoeuvre for any moon mission. The command pilot for the mission, Neil Armstrong, was known for his quick thinking and fearlessness in the face of danger. He had already proved himself to be a man of cool courage when he was twice almost shot down when flying over North Korea, but managed to nurse his craft back to base. His reputation was legendary: he and his plane, it seemed, were as one. He was to be accompanied by David Scott as pilot. Their task was to dock Gemini VIII in orbit with an Agena satellite and then David Scott would undertake a space walk. The Agena was launched first, on an Atlas rocket, at 10 a.m. on 16 March 1966. This was followed just over ninety minutes later by the astronauts in their Gemini capsule.

Six hours into the flight, the two men began to track the Agena ahead on the radar. Then they saw it – a shiny craft ahead of them, looking like a ghost ship lost in space, hardly appearing to move. They were about ninety miles apart with the Gemini capsule in an orbit seventeen miles below the Agena. Once the Agena was angled about 10 degrees above them, Armstrong pitched the Gemini's nose upwards and fired the aft thrusters, delicately manoeuvring the craft until they were at a distance of just over 100 feet. The Agena had a reputation for being unreliable. The one used on the last mission had exploded before it reached orbit. Armstrong took his time observing the target vehicle before moving in to a distance of about 3 feet.

'Go ahead and dock,' came the all-clear from mission control.

Very gently, at a speed of 3 inches per second, Armstrong eased the Gemini nose cone into the Agena docking collar.

'Flight, we are docked.' Armstrong spoke with confident assurance to ground control.

Everyone in mission control began to cheer, but their elation did not last long. They were about to face the worst American emergency in space yet.

The Gemini controls were turned off to save fuel and the joint craft was flown using the Agena controls. At first the mission seemed to be going well. Armstrong and Scott could see the shape of Africa below them, spread out like a map with its yellows, ochres and greens, reminding them that they were 185 miles above the earth. They had just passed over Madagascar when Scott became aware that the Gemini capsule had unaccountably *rolled* over 30 degrees. It looked as though the Agena was living up to its name. They switched it off hoping to regain control. They were back on the Gemini controls. All should now be well.

But it wasn't. The craft was rolling more than ever. Armstrong was concentrating on the controls in the hope of bringing stability to the madly tumbling craft. Whatever he tried, nothing made any difference and, worse, they were losing fuel. They decided they had no option but to disengage from the Agena, even though in the process it was possible that the two vehicles might collide with disastrous results – since there were about 4000 pounds of fuel on board. The stress on the two interlocked vehicles was extreme. If they didn't act quickly, there was a chance both craft would disintegrate.

To their amazement when they disengaged from the Agena they immediately fell into an even worse spin. This was fast approaching a full revolution every few seconds. Evidently the trouble was with the Gemini capsule, not the Agena. Over the western Pacific they just had time to relay a message to the tracking ship *Coastal Sentry Quebec*:

'We have a serious problem here ... tumbling end over end,' said Scott. 'Disengaged from the Agena.'

The message was relayed to mission control, but there was no time for a reply. They were on their own. The craft was now doing a complete spin every second. It was only a matter of time before they lost consciousness in the madly gyrating craft.

There was one slim chance left. They could try to regain control of the craft by switching off the manoeuvring thrusters and using the re-entry attitude control thrusters instead. The controls were somewhere above Armstrong's head on a panel of switches. Fighting the G forces, tunnel vision and the sickening strip of sunlight flashing every half-second through the windows with a burning brilliance, Armstrong concentrated on feeling for the right switch from memory. Agonizing seconds passed while he tried to make his fingers follow his brain's instructions, still trying to control the craft with the hand controller. He found the switch and pushed it down. The craft slowed. Within half a minute the gyrations ceased.

Once he had regained control, Armstrong coolly proceeded to test the thrusters in the orbital attitude and manoeuvre system to work out what was wrong. It soon became all too clear: thruster No. 8 had jammed in an on position, using up fuel and tumbling the craft. Although Armstrong and Scott were anxious to continue with their mission, they were ordered to return. With enormous relief, they heard all four retro rockets fire and miraculously, with no further adjustments, they splashed down in the Pacific Ocean five hundred miles east of Okinawa at 10.23 p.m.

Although Scott had not been able to carry out his space walk, the mission had proved beyond a doubt that space docking was possible – paving the way for docking in lunar orbit. The failure had nothing to do with the docking procedure itself. It was the Gemini craft that had let them down. Commander Neil Armstrong's clear thinking in the face of such a crisis was duly noted.

• • •

Korolev's death had left a huge void in the Soviet space programme. His second in command, Vasily Mishin, now found his name put forward by senior staff at OKB-1 as the new Chief Designer. Following in Korolev's footsteps was a daunting prospect and some felt that no one man could really replace him. Although Mishin was a brilliant engineer, it was said he did not have Korolev's overall grasp of the grand design or the political power to drive it through. And the sheer volume of work that Korolev had previously taken on was simply overwhelming.

The N-1 needed extensive testing but there were still not enough funds for building stands to ground test the first stage of thirty engines. To neglect this testing for such a huge and complex rocket was to assume that 'luck' was essentially a Russian commodity, under Mishin's personal control. But Mishin's luck had quite often been due to Korolev and Korolev was not there to guide him now. Mishin was finding it hard to achieve N-1's payload of 95 tons. Various techniques were used to lessen the weight of the rocket including reducing the initial earth orbit from 65 to 52 degrees, and cooling the propellants. Thrust was increased on all the engines by 2 per cent and a fourth stabilizer was added. The Soyuz capsule was still unfinished and the Soviets still had no experience at docking in space or transferring cosmonauts from one vehicle to another. Mishin also had to continue with the Voskhod programme. All eyes were on him as he took over Korolev's enormous workload and the impossible goal to beat America to the moon.

Yet the first foray into space, just over two weeks after Korolev's death, came not from Mishin, but another chief designer, Georgi Babakin, of the Lavochkin design bureau near Moscow, who had been put in charge of Korolev's robotic lunar programme. On 31 January he was ready to launch the lunar probe, Luna 9. When it was five thousand miles from the surface of the moon the attitude control thrusters aligned the vehicle and it began its descent. Sensors could detect when it was just 16 feet from the moon's surface and the engine was switched off. The craft made a perfect landing on the moon. This was another first for the Soviets. The moon was not made of improbable icing-sugar mountains of dust, but was indisputably firm, just as Korolev had said, and a Soviet craft was informing the world of this fact.

Mishin followed the lunar probe with a Voskhod flight, which took two dogs into orbit to carry out further tests on the life-support system over a prolonged period. They stayed in space a record twenty-two days and after the trip understandably lost some of their canine *joie de vivre*.

The physicians noted that the animals were weak and dehydrated; normal blood flow did not return until five days after the flight. This convinced Mishin that further experiments with the Voskhods were too risky, and a planned multicrewed space launch originally meant to coincide with the Communist Party Congress in late March was cancelled. There would be no more manned flights until the Soyuz was ready. This freed up staff to focus on completing work on the Soyuz.

On 10 July 1966, cosmonaut Vladimir Komarov announced optimistically to the press during a trip to Japan that the Soviets were on target to beat the Americans by one year. 'There is no need to hurry with a manned Moon trip; the important thing is to carry out everything safely,' he declared. 'But I can positively state that the Soviet Union will not be beaten by the United States in a race for a human being to go to the Moon.' A few days later, his comments were reported in the *New York Times.*

• • •

In August 1966, the first completed Apollo command module was delivered to NASA for testing prior to launch. Made of aluminium alloy to keep it light, the interior was spacious compared to the Mercury or Gemini capsules. Yet such was its complexity that during construction at least twenty thousand failures had slowed its progress. Gus Grissom was to be the captain for the first manned flight set for February 1967; Roger Chaffee and Ed White would accompany him. Grissom, however, was uneasy with the capsule design and initiated further changes. During the winter more than six hundred modifications were made, but that still didn't satisfy him. There were thousands of systems embedded in the craft and littering the floor of the capsule; it looked like electrical spaghetti. The patience of the astronauts was wearing wafer thin as endless countdowns prior to the real launch were interrupted by small failures putting the launch date back even further. So many changes were made on the hoof that the simulator model was always behind. Grissom hung a lemon on it.

The press had caught a whiff of trouble. What was wrong, they wanted to know. Grissom explained, 'I've got misgivings. We've had problems before, but these have been coming in bushelfuls. Frankly, I think this mission has a pretty damn slim chance of flying its full fourteen days.' Did this present a danger, the press wondered. Grissom replied: 'If we die, we want people to accept it. We're in a risky business, and we hope that if anything happens to us it will not delay the programme. The conquest of space is worth the risk of life.' He omitted to mention the pressure from the top for the team to 'get off their asses'. In an election year, President Johnson needed a significant step forward to impress the voters.

On 27 January 1967, a full ground test was scheduled to run through the countdown procedures. The spacecraft was in position on top of the Saturn rocket on Pad 34. The three astronauts were suited up inside. The double-hulled hatch was securely closed as if for space flight. The cabin was filled with oxygen.

They were over five hours into the countdown, having encountered the usual quota of annoying faults, when the communications system began to fail. Grissom couldn't hear flight control. How could he hear them from the moon, he demanded, if he couldn't hear them from a few miles away? Next, his microphone wasn't working properly. It was past six o'clock. Some of the engineers wanted to pack up for the night. They were overruled. The countdown had to be completed, all systems tested and the capsule passed as fit. Suddenly the decision of whether to continue or not was taken out of everyone's hands.

'Fire!' It was the unmistakable voice of Ed White.

A split second later, they heard Gus Grissom.

'We've got a fire in the cockpit.'

On the blockhouse monitors, Deke Slayton saw an intense brilliance from inside the craft. Frantic figures were dimly outlined against a whirling fury of flames. Grissom was trying to open the hatch. Chaffee, still strapped in his seat, refused to leave the communications.

'Get us out,' he cried. 'We've got a bad fire. We're burning up!' The flames were white. They filled the cabin.

The men working on the gantry could not at first comprehend what was happening. There was no fuel in the rocket, yet they could hear men screaming for help and the roar of flames. Someone pushed the emergency button. Then it became clear. The fire was *inside* the Apollo capsule.

Technicians ran forward to open the hatch but were thrown several feet as an explosion blew a hole in the side of the craft, shooting out flames. Flung back by the blast, they were forced to retreat as further blasts ripped from the capsule. More men appeared with fire extinguishers intent on killing the blaze. Rushing into the white-hot flame and thick black smoke, they attacked the fire, ignoring the choking fumes, determined to save the three men. But it took too long to put it out. The hatch was too hot to touch. Toxic flames disabled the rescuers.

Six minutes elapsed before they could open the hatch. While the capsule was still trembling with heat, the blackened interior dark with smoke and acrid fumes, they brought out the three bodies. As he rushed to the launch pad, Deke Stayton hung on to hope. An incomprehensible miracle was required but when he saw the three space suits melted and covered in fused electrical wires and the grim interior from which the men had struggled to escape, the haunting horror of it all was the only reality. The temperature inside the capsule had reached 2500 degrees F. Grissom, Chaffee and White had died of asphyxiation within a minute.

Over the following weeks, a team of nearly two thousand took the charred Apollo apart. Nothing was too small to warrant anything less than an exacting study. The many miles of wires weaving like intestines in the belly of the craft were especially suspect. The craft had been full of flammable substances, all steeped in oxygen. After the blaze, it was impossible to point the accusing finger with any real assurance at any one culpable system or piece of equipment.

The known facts were damning enough. The astronauts had been locked into the craft. Ninety seconds at least were required to unlock the hatch. For over five hours, pure oxygen had seeped into every absorbent surface. The atmosphere was pure oxygen under pressure. While the capsule was undergoing tests, bunches of electrical wires had been lying around, trodden on and tripped over. The wires under Grissom's seat were frayed. One small spark was all it took to transform the oxygen 'air' into an inferno. The three men had not the remotest chance of escaping. The explosion had ripped the capsule open seventeen seconds after the fire began, and the fearsome speed with which the flames took hold doomed anything they touched.

The loss was felt throughout the space industry. At a dinner in May to commemorate the sixth anniversary of Al Shepard's flight, the mood was sombre. Von Braun spoke movingly of Grissom, White and

Chaffee. Al Shepard spoke too. He was now reconciled to von Braun's policy of 'safe but sure'. Looking pointedly at von Braun, he declared his support for the decision to delay his flight into space. 'A different decision could have given us the first flight – but could so easily have ended in failure,' he conceded. 'Our total space effort today is second to none. I want every chance for this country to be first in everything it does. And yet, if we should lose the race to the moon, say for example by a month, we cannot be more than temporarily dismayed … We will be remembered in fact for how we did it, and not when we did it.'

The report on the Apollo tragedy comprised more than three thousand pages and was damning of NASA and North American Aviation. It was found that the next Apollo craft, still under construction, contained as many as 1400 faults. A duplicate of the Apollo craft was built, then deliberately fired in order to study it. Many recommendations were made and half a billion dollars spent on improving the design. The American space programme was temporarily halted. Months elapsed, which could only favour the Soviets.

• • •

In the Soviet Union, Chief Designer Vasily Mishin was aware that the Apollo 1 disaster left a most fortuitous gap in the American space programme. The Soviets had not produced a manned flight for two years but now had a chance to catch up. Mishin was hungry for further triumphs which he was sure the Soyuz capsule would provide. Rumours of the new mission spread, reaching the international press agency UPI, which promised 'the most spectacular space venture in history … involving in-flight transfer of crews between two ships'. To gain experience of orientating the craft and docking in space, the first manned test of the Soyuz was to be an elaborate double flight. On day one, a single cosmonaut would be launched in Soyuz 1. The next day, when the original craft flew over Baikonur, Soyuz 2, carrying three cosmonauts, would be launched. Once the ships had docked, two cosmonauts from the second ship would transfer to Soyuz 1 and return to earth.

Despite the optimism, the first Soyuz to be ground tested in May 1966 had produced more than two thousand defects – even the parachute system seemed to be less effective than the one used in Vostoks, failing to open with alarming regularity. There had been three unmanned tests of the Soyuz, all of which had given cause for concern. Launches had already been delayed several times but political pressure from above was increasingly insistent on a stunning space flight to coincide with the May Day celebrations. Mishin wanted a little more time for testing. The new Soviet leader, Leonid Brezhnev, conscious of the many successful Gemini missions, was impatient.

Cosmonauts Vladimir Komarov and Yuri Gagarin were both strong contenders to lead the mission. Komarov had earned his spurs on the three-man Voskhod mission, whereas, since his first flight in 1961, Gagarin had taken on more of a public relations role. For some time he had expressed his desire to lead another space flight but was constantly thwarted by Kamanin who decided he was too valuable to put at risk. He had, however, for the first Soyuz flight been appointed as Komarov's backup. The day for the launch was set for 23 April 1967 but, as the date approached, engineers continued to identify faults in the craft – more than two hundred. A report was duly drafted but the authorities would not listen and no one had the confidence to raise it directly with Brezhnev.

Kamanin made it clear to Vladimir Komarov that his safety was paramount and that the docking procedure would be delayed if problems arose. Such talk of possible 'problems' while in space confirmed a deep unease Komarov had about the flight and about Soyuz in particular. If he followed his sixth sense with its warning signals of danger and refused to go, he was worried they would send the backup pilot: 'That's Yura – and he'll die instead of me,' he confided to a friend. He felt he had to protect Gagarin. There was no alternative but to do his duty and take his allotted place in the lead Soyuz 1. He dedicated his flight to the fiftieth anniversary of the Bolshevik Revolution and ensured that he was the one strapped into the cosmonaut's couch when it took off at 3.35 a.m.

The launch went without a hitch but problems arose quickly once the craft was in orbit. 'I feel well,' Komarov reported. 'The parameters of the cabin are normal. The left solar battery has not opened ... Short wave communications are not working.'

The Soyuz drew much of its power from the sun. Once in orbit, two solar panels were supposed to extend on either side of the craft to gather enough power from the sun to run the ship. One had failed to open, causing an imbalance which meant that the remaining solar panel could not be orientated to take advantage of the sun's heat. With only battery power to keep the craft going, Komarov was not in immediate danger, but power was far below what was needed to run it for long-duration flight. With one solar panel out of action and the other incorrectly orientated, there was barely power for more than a twenty-four-hour flight.

At Baikonur, the senior officials from the state commission supervising the launch decided that Komarov might just manage to free the jammed panel if he changed the attitude of the Soyuz, enabling him to obtain full power from the sun. The operation did not work. Engineers now felt sure that Komarov's attempts at re-orientation to the sun would fail and that he would merely waste precious fuel in the attempt. This could put a successful retrofire for the return trip in doubt.

By his fifth orbit, the situation was looking grim. Whatever Komarov did, he could not orient the ship. He shut down the automatic control system and operated the ship manually. Control was sporadic. Soon he would be out of contact with the ground for a whole nine hours as he passed over America and the Atlantic Ocean. He was advised by ground control to get some sleep while they tried to resolve the situation. At the end of the thirteenth orbit, Komarov's voice was heard faintly at Baikonur. He was in serious trouble. He was advised to re-enter on the seventeenth orbit, using the automatic orientation system for stabilizing the Soyuz. But this system failed. At that point, ground control postponed re-entry to the eighteenth orbit. For this second attempt, he would have to attempt the impossible and re-orient the tumbling craft manually, controlling the ship in the darkness of the earth's shadow. The state commission recognized that Komarov's chances were limited. The most likely outcome would be a horrific burn-up as the craft experienced an incorrect re-entry or the more protracted disaster of being fired into a higher orbit from which there was no return. As he tumbled out of control there seemed little hope.

Rumour has long been rife that on the ground they decided the time had come for Komarov to have a last conversation with his wife. A large black car whisked Valentina Komarova to a radio point. She understood the situation: that they had little time together, that these could be their last words and that the tumult of emotions she was feeling must be suppressed. Twice he had been ill as his craft turned and tumbled. He forced his voice to sound calm and controlled. They said their most private goodbyes with the world listening as he tried to control his mad machine. And then she moved away. Alexei Kosygin, second in command to Brezhnev as Chairman of the USSR Council of Ministers, is alleged to have spoken to him next, informing him that he was a Soviet hero. Then suddenly it was time to act like one.

Re-entry was about to begin. Unbelievably, the retro rockets worked and slowed down the craft, but Komarov had no control over the machine. On the ground, radio outposts in Turkey could hear his cries of despair: 'This devil ship,' he shouted. 'Nothing I lay my hands on works properly.'

He struggled manually to make the craft obey him. With superb skill using the cabin gyroscopes, he succeeded in orienting the ship correctly on to the narrow re-entry channel. He was through. Gravity was winning the battle to pull the spacecraft back to earth. Soon the parachutes would open to slow him down and stop the dead weight of the craft falling like a meteor, spinning like a top. He heard the sound of the parachute mechanism – but he was still tearing through the sky. The parachute had not opened.

There was still a small hope. The reserve chute was yet to open. It failed, never filling with air, streaming behind, an orange banner mocking the enterprise as he made his headlong fall. Retro rockets should have fired before touchdown to cushion the landing. Ironically, they did work after the cosmonaut, aware now of his fate and with no reproach in his voice, slammed into the solid earth at 400 mph. That was when the retro rockets exploded, creating a funeral pyre in the deserted landscape.

Local farmers were soon on the scene. They covered the blazing remains of the capsule with soil in an attempt to put the fire out. By the time the search party arrived there was very little left but ashes. Later fragments from the exploded ship were buried with Komarov's hat in a mound at the crash site. A gun salute remembered the hero.

For Mishin and Gagarin, agonizing over the twenty-four-hour

horror flight with its grizzly conclusion, the only solution was to lose themselves in drink. But afterwards, according to Kamanin, Gagarin and other leading cosmonauts wanted to see Mishin himself named in the Soyuz crash report. They argued that Mishin was not conversant with the complexity of the Soyuz. Several of the cosmonauts felt that Mishin should have brought the craft down earlier, as soon as it was evident that Komarov was in trouble – rather than try to continue the mission – and felt that he had a 'weak knowledge' of the Soyuz. Kamanin was in no doubt. He thought Mishin sadly lacking in the leadership qualities that Korolev had had in such abundance. Where Korolev had inspired, Mishin usually just created friction. The unspoken understanding was that Korolev would not have let this happen.

'The road to the stars is steep and dangerous,' Gagarin later said to the press. 'But we are not afraid … Space flights cannot be stopped.' Neither America nor the Soviet Union had a spacecraft that worked. The tantalizing trip to the moon, so constant in its reappearance every month, was still the stuff of dreams.

CHAPTER TWENTY

'How can we get out of this mess?'

Nine months had passed since the Apollo 1 disaster, months in which the war in Vietnam had continued to escalate. Key strategic sites in North Vietnam had been obliterated in a series of air strikes yet this did not seem to diminish the Vietcong's power to retaliate. Almost five hundred thousand American troops were now in Vietnam. 'Search and destroy' missions were underway on the ground in a war which was both increasingly unpopular and unwinnable. Despite millions of tons of bombs being dropped on strategic targets, American military might had not been able to crush the spirit of the communist troops who continued to strike back.

If America was losing face in Vietnam with 'Uncle Sam's' vigour put in question, no one could doubt the superiority of the American rocket industry as the ultimate icon of power in the symbolic race for supremacy against the Soviets was wheeled out for all the world to see. On 26 August 1967, the giant doors of the vehicle assembly building at Cape Canaveral – now renamed Cape Kennedy – slid open to reveal the glistening white Saturn V. Reaching over thirty-six storeys high, with a height of 363 feet and engines capable of producing 7.5 million pounds of thrust, von Braun's Saturn V was an impressive cylinder of power as it was slowly moved, upright, on a gigantic crawler, to Pad 39A. It was so heavy some doubted it would ever get off the ground.

The unmanned launch was planned for early autumn, but myriad details connected with getting it fuelled and into position took longer than expected. The Director of Launch Operations, Rocco Petrone, came to understand that the Saturn V lived in its own time dimension. He was dealing with what he called the 'Saturn V minute', which he calculated was actually about five minutes. 'If I asked a guy how long something would take, he'd tell me 10 minutes and it would come up maybe in an hour,' he explained. 'Everything about the Saturn V was bigger. If you had to pick up a valve, you couldn't pick it up by hand. You had to get a fork-lift truck.'

In November, the rocket was ready for launch. It had been in position for some weeks, a new landmark on the flat landscape. By night, floodlights bathed it in brilliance. The wind was almost visible chasing leaves and debris around its summit. Launch was set for 7 a.m. on 8 November 1967 and by dawn the wind had eased. The morning was still. No one was allowed near the rocket. A safety zone of three and half miles was in operation. Four hundred and fifty technicians were absorbed in front of their screens at the control centre. The mission, Apollo 4, would not only test the Saturn V but also a redesigned Apollo command module – in which a raft of innovations had been introduced, including a new hatch. Queues of cars were lining the roads, full of people intent on seeing the launch and living a piece of history. All the leading figures from NASA were there to observe. En masse from Cocoa Beach the press were decamping in leisurely fashion to their observation posts, casually unconcerned about being late. Countdowns were never on time.

For von Braun and those close to him, the launch would be the culmination of years of work. As far back as the 1930s, on another continent, in another age, von Braun had dreamed of this legendary rocket with its amazing power. Now he sat with Arthur Rudolph, who had shared the long journey from Bleicherode, to watch the spectacle as they tried to lift a rocket with the weight of a Navy destroyer off the launch pad. Von Braun felt certain that Saturn V had the power to 'toss a jet aircraft into orbit – or even boost a Chevrolet clear out of the solar system'. Even so, there was no way of being completely certain of the F-1 engines, let alone all the new systems under 'all-up' testing on that day – fuels, pumps, electronic systems, control systems. They watched with concern as, with three minutes to go, computers took over.

Countdown was faultless. Fuelling was complete. Helium created pressure in the tanks that would bring the five powerful pumps into action. When the ignition flames were alight, 1 ton of fuel per second would arrive to be consumed by the huge engines, gently at first. Full power was delayed so that the rocket could adjust. Then the gargantuan appetite of the colossal engines converted power into a terrifying force, beyond human scale. And if one infinitesimal fault hidden among the many thousands of cooperating systems failed, the three-mile limit for safe viewing might not prove enough and the steel shutters on the control block designed to close in the event of an explosion might well prove inadequate. The four arms holding the rocket received the signal to withdraw. The giant, claw-like restraining bolts clasping the rocket at its base were released. On time and painfully slowly, the rocket began to rise. At first there was no sound, only tremors and reverberations that shook the roofs and windows. Then came the sound: deafening, like a physical force.

To the many anxious observers, the rocket was desperately slow to rise. It seemed to hover, indecisive, but then the white fury of flames under it, struggling to move the dead weight, became an inferno that nothing could resist. It was rising against its will but it was also beginning to tilt alarmingly. With horror the onlookers waited for the catastrophe. Now the rocket was gathering speed and it was obvious that it was meant to tilt. Soon it was against the blue sky, still trailing a plume of fire several hundred yards long. The volume of sound and sheer physical power was overwhelming. The rocket rent the atmosphere as though tearing a thunderous curtain. In the viewing rooms, watching the perfect performance, the crowds cheered madly. The CBS broadcaster Walter Cronkite found he was holding the glass window of his viewing room in place because it was rattling so violently. In the control centre, plaster fell from the ceiling.

Two and half minutes later, the first stage had done its job and was discarded. Now von Braun waited. The second stage, shorter by 57 feet with five equally massive engines giving one million pounds of thrust, should ignite and push the cone of metal into space. He felt more certain of the third stage. It was transporting a lookalike lunar module, with the weight and size of the real thing.

'Go, baby, go!' yelled von Braun. He watched anxiously as each stage of the booster did its job and then cut away until, eleven minutes after launch, the Apollo command and service module was in free orbit. What he had just witnessed was a man-made miracle: the synergy of thousands of systems into one powerful purpose. The column of fire could be seen more than 150 miles away. The elation he felt was unrepeatable.

• • •

In Russia, a sombre mood hung over the entire country. It was absorbed in mourning a hero. Yuri Gagarin was dead, killed in a plane crash. Kamanin had tried to protect him, Russia's bright star. After the Komarov disaster, Kamanin had informed Gagarin there would never be another space flight for him. He had bigger, brighter plans for Gagarin's career, although he was increasingly troubled by Gagarin's antics. 'There were many situations when Gagarin miraculously escaped big trouble,' he wrote. 'I was particularly concerned about his driving at high speeds ... The active lifestyle, endless meetings and drinking sessions were noticeably changing Yuri and slowly but steadily erasing his charming smile from his face.'

When news first broke of the crash on 27 March 1968, Gagarin's fate was uncertain. The plane was not found immediately. He had been on a training mission with a co-pilot; the weather had been snowy, visibility poor. Without warning, communications had ended abruptly half an hour into the flight. With the continued radio silence, helicopters were sent out to investigate. In the afternoon, a wrecked plane was found. Crisis meetings continued during the night. Those who knew Gagarin could not hide their emotions. 'Kamanin had his lips pressed tightly together,' recalled one cosmonaut. 'Kuznetsov was struggling to control his trembling chin. Leonov had his face to the wall.' In the morning, at the scene of the accident, Kamanin identified a piece of Gagarin's jacket. The two pilots had hit the ground at a speed of more 430 mph. The plane was scattered over a wide area, and it became painfully evident that their bodies were as fragmented as the aircraft.

Gagarin's death was felt as a personal loss to every Russian. As far as the eye could see, silent crowds waited patiently in the snow to file passed his coffin lying in state at the House of the Soviet Army.

Gagarin's mother, a simple countrywoman, overwhelmed by the event, waited to see her son in his coffin one last time before his cremation. She wanted to mourn in a way she understood. Officials refused her request. But she fussed and insisted and the officials gave way. The red velvet cloth was removed. She opened the coffin to find a plastic bag containing human fragments – all that could be found of her son.

Gagarin was irreplaceable. 'Gagarin's death will be my biggest loss for the rest of my life,' Kamanin confided in his diary. 'I know years will pass and new outstanding space explorers will come along but not a single one of them would be able to go as high as Yuri Gagarin's grand deed.' A bright star had burned out, leaving the sky dark. For the Soviets he had been a hopeful symbol of what was good in the all-too-dreary communist system of mediocrity and shortage. Russian grief could not be assuaged. Gagarin's face, recognized throughout the world, was so much a part of Soviet triumphs in space. Now he was dead and the Soviet Union was slowly losing its premier position to America.

For two years now there had been little but failed flights and disasters. At Baikonur, Vasily Mishin recognized it was crucial to restore the balance but he himself was struggling to make headway, all too often seeking the help of alcohol for this task. According to Kamanin, Mishin had been sent into rehabilitation for a drink problem at least once. In his diary he was unsparing in his criticism, describing Mishin as 'rude and always up for a fight' when under pressure, as well as 'hostile to any advice'. At times it seemed that Mishin was making 'so many mistakes' it was hard not to 'feel sorry for him'.

Earlier in the year, according to Kamanin, he, along with leading cosmonauts, including Gagarin and Leonov, had taken their worries to the First Deputy of Defence, Marshal Yakubovsky. The marshal had given them a 'very warm welcome', Kamanin recalled, and listened carefully to their concerns about 'the USSR falling behind America'. Many of their criticisms centred on Mishin, who was finding it difficult to walk in Korolev's shoes. Yakubovsky conceded that Mishin, although a brilliant engineer, was possibly a little lacking in charm; those who possessed an abundance of creative energy, he noticed, often had the drawback of a short fuse. He was inclined to the opinion that chief designers as a breed were not the easiest to deal with and promised to help as much as possible. But nothing happened and no action was taken against the struggling Mishin. Kamanin made a forecast: if Mishin were to stay as Chief Designer, 'things could only get worse'.

• • •

America now had the rocket to fly to the moon. And the Apollo spacecraft was phoenix-like, rising from the ashes, born anew. It performed flawlessly on the Apollo 4 mission, even on re-entry where temperatures reached almost 10,000 degrees F. The next mission, Apollo 5, would test the lunar module. By any standards, this was an extraordinary vehicle. It did not need to be an aerodynamic shape as it would be flying in a vacuum in space or in the moon's gravity – which is only one-sixth of that of the earth. Custom-built to land on an unknown surface with its big padded feet on long, extending legs, it could never be called streamlined. The overwhelming prerequisite was that it had to be light. Its metal skin was so thin; it was barely thicker than a couple of sheets of paper. It stood 23 feet high and came in two sections. The upper part carried ascent engines, a fuel tank and cockpit and would boost the two-man landing crew back to the command module orbiting the moon. The lower half, which would remain behind on takeoff, carried the descent engine, fuel and necessary equipment.

There was no detailed first-hand knowledge of the lunar surface, yet skilful handling and landing of the lunar craft were vital for success. Astronauts practised this in a full-scale mock-up of the cabin layout, with a computer camera-simulated view of a moon surface model seen through the window. In addition, staff at Langley Research Center devised an unlikely flight vehicle designed to give the commander and pilot real flight experience.

A strange craft with more than a hint of Heath Robinson about it was to be seen in the skies above Ellington Air Force Base, seventeen miles outside Houston, Texas. Called the 'Flying Bedstead' – and looking rather like one – its function was to give astronauts some experience of a lunar landing. It seemed to be only framework, with no decent covering hiding its complex interior. Reassuringly, four legs jutted out from its four corners. Less reassuringly, the pilot's control seat jutted outwards, surrounded by nothing more than fresh air. Its official name was the 'lunar landing training vehicle' and it was especially designed to remove the feeling of gravity, using a downward-thrusting jet engine and rocket to lift off. Separate manoeuvring thrusters, similar to those used on the lunar module, were also fitted to provide the hapless pilot with enough control in pitch, yaw and roll, to land successfully back on the ground, or, if he failed, to send it spinning off to destruction. Just in case, NASA had fitted an ejection seat.

Neil Armstrong was something of a virtuoso on the machine but one day in May 1968 he almost played his swansong. The 'Bedstead' decided to give a bucking bronco performance with an almost vertical 800 feet rise and an equally stomach-churning vertical drop. From his controls in the corner of the 'Bedstead', Armstrong held the machine steady, hovering near ground level. Then, without warning, up it went again 200 feet, bucking and veering to the left, rolling and turning over the runway, determined to throw off its passenger. The bed was wild. Ground control shouted at Armstrong to eject, but something in Armstrong's character responded to bucking bedsteads in a positive way. He was the boss; he would control the thing. Not until he was past knowing for sure that the bed had won would he eject. Seconds later it crashed on the runway. As for Armstrong, his parachute opened only just before his boots met the ground.

The inquest on the untimely demise of the 'Flying Bedstead' was a sobering event. On that windy day it had guzzled extra fuel and the inert helium supply used to push propellant through the fuel tanks was squandered, making the engine stutter repeatedly. But more sobering still was Armstrong's timing. Had he delayed his departure by two-fifths of a second, there would have been no time for his parachute to open. NASA officials hoped that if and when a lunar landing was made, Armstrong's timing would be less interesting.

In 1968, George Low, Program Chief, Manned Space Flight, and other senior NASA officials were becoming increasingly worried about delays in the manufacture and delivery of lunar modules – it was unlikely an operational craft would be ready for testing before the end of the year. The next mission was intended to test both the lunar and command modules in earth orbit. But it was becoming clear that if they waited for delivery of the lunar vehicle, they might not get their man on the moon by the end of the decade. It was looking more and more as though that prize would fall to the Soviet Union.

The pressure increased as the CIA made NASA aware of a Soviet programme to have a cosmonaut in orbit around the moon by Christmas 1968. It would be hard to claim that they had reached the moon first if the Soviets accomplished such a feat. In 1968, Mishin launched a series of modified unmanned Soyuz capsules called Zonds. In March, Zond 4 had returned from deep space to earth orbit, having tested re-entry from a lunar orbit, and in September a cargo of tortoises, flies and assorted plant life was also taken for a journey around the moon in Zond 5. In October there was a successful rendezvous in space with a manned Soyuz craft. Next up was the N-1 launch itself. It was wildly rumoured that the Soviets were on the brink of going one step further in December with a manned trip around the moon. A date was even specified in the Western press – 8 December. Although the Russians had had their share of failures, their ambitions to reach the moon were undiminished. They too were desperate to see their flag planted on its surface.

George Low was increasingly concerned that the Soviets might beat them to it; it was time to take some short cuts. He went to see Chris Kraft with a risky proposal. Rather than waiting for the lunar module and testing its flight worthiness in earth orbit, he suggested that the next flight should be a huge gamble. They needed to gain experience of navigation and communication on a manned flight around the moon. Why not send the first manned flight of the Saturn V straight into lunar orbit and around the back of the moon? The astronauts would gain the valuable experience of flying in lunar orbit using the Apollo command module. Kraft liked the idea, although it meant speeding up the development of the lunar navigation software.

Unfortunately they had come to this decision at a bad time. During the Apollo 6 mission on 4 April 1968, the second unmanned flight of the Saturn V had run into serious difficulties and revived all the old anxieties. During the initial stages of F-1 testing, vibrations had been so severe that the mechanical stress was approaching 10 g's. This 'pogo' effect in the first stage lasted a full ten seconds and made the command module above judder so much that any astronauts on board might have been in danger. During the second stage there was more erratic data and two engines failed. This made it impossible to put the command module into the right lunar orbit. At the time these failings scarcely made front-page news, eclipsed on the day by the assassination of civil rights campaigner Dr Martin Luther King in Memphis, Tennessee, which prompted a wave of civil unrest across America. Von Braun himself was worried that yet more engine faults were being found at this time. He had believed the problems of the F-1 engines were behind them.

With some foreboding, Kraft counselled the cautious von Braun. He had expected protests from him. It had been his decision to have 'one more launch' before Alan Shepard's flight that had lost America the title of 'first man in space'. It required more than a leap of faith to send astronauts on the very first manned Saturn trip straight into an untried and complex mission. Von Braun's team at Marshall had analysed the data from Apollo 6 and introduced further shock absorbers and vibration-absorbing dampeners to reduce the effect – but there was no guarantee of safety.

'Wernher, we need you to commit to your next Saturn V flight. It has to have men on it and it's going to the moon.' Kraft's appeal did not fall on deaf ears. In spite of recent problems, von Braun had confidence in his rocket. He didn't insist on yet another test flight. If it could achieve earth's orbit, then it ought to be able to go further to a lunar orbit. He did not hesitate. 'It's a great idea,' he said.

• • •

Had the Americans known the difficulties facing the Soviets they might not have taken such a risk. The original plans for the N-1 and its attendant services had been magnificent, as though conceived in the eye of Colossus. The assembly building, 150 feet high and 800 feet wide, was immense. The two launch pads, 1500 feet apart, would have service towers 500 feet in height. But by the time Mishin was ready to wheel out the first N-1 booster in May 1968, something in the Soviet system had conspired to undermine efficiency. The vast metal meteor was still not ready, with questions about its engines. Almost farcically, cracks were also found in the outer skin of the first stage and it had to be taken back to the assembly building for repair. Ground testing continued to run into problems and in September a bulldozer accidentally severed the main electric cable to the N-1 launch complex causing a two-month delay in testing. Repairs to the booster took far longer than expected. Delivery dates for equipment were invariably missed. A perverse unreliability was becoming the norm, presided over by Mishin who was still struggling with his own little problem with the bottle.

Any plans to modify the Zond missions for manned flight had to be revised in November, when the unmanned probe, Zond 6, raced back from a successful circumlunar mission only to develop a pressure failure, which killed the biological specimens on board. The fall in pressure jeopardized re-entry and also killed the dreams of a manned mission around the moon. Cosmonauts would not have survived. Mishin felt the strain acutely. 'Mishin did not look good this morning,' observed Kamanin on 15 November. 'His eyes were red, his hands were trembling, his face puffed up, he keeps drinking.' The following night, as the pressure plummeted still further in the landing apparatus, 'he could no longer stand the strain'. Kamanin was summoned in the small hours to find that Mishin had drunk so much and slept so little he was 'incapacitated'. He was firmly escorted straight from the control room to hospital. Zond 6 itself crashed into the ground as the parachute failed to deploy correctly.

• • •

On 21 December 1968, the mighty Saturn carrying Apollo 8 rose majestically into orbit. The three astronauts on board, Frank Borman, Jim Lovell and Bill Anders, were embarking on arguably the most dangerous manned mission yet undertaken: navigation into the moon's orbit. In 1966, Gemini 11 had powered 850 miles into space. No one had gone beyond that achievement. Now Apollo 8 had to travel a vast 234,000 miles into the unknown to navigate successfully into the grip of the moon's gravitational field. Precision and timing were essential. They must enter the moon's gravitational field while in control of the craft, using engine power to slow down to a speed of 3700 mph. This would take them into lunar orbit.

Mission control wanted to be sure about the engine – known as the service propulsion system – that would power Apollo 8 into lunar orbit before they made a decision. If all was well, it was feasible to go for 'trans lunar injection', or TLI, to power the craft on to a trajectory to the moon. Ground control liked what they saw on the screen monitors. The craft made a perfect orbit around the earth.

'Apollo 8, you are go for TLI,' said Mike Collins from mission control. They had the all-clear to power away from the earth, increasing their speed to 33,500 feet per second, beyond escape velocity – the speed required ultimately to break out of earth orbit.

On the third day, they had reached more than 200,000 miles out into space. At 38,000 miles from the lunar surface they began to fall towards the moon, pulled in by its gravitational field. Soon, the astronauts would be behind the moon and out of radio contact for more than twenty minutes. They would be on their own. Entry into lunar orbit required the Apollo engines to fire for exactly 247 seconds. This would take the craft to the lowest point of its orbit, seventy miles from the moon. If this failed, there would not be a second chance. If the engines did not fire for the exact required time, Apollo 8 would be unable to enter lunar orbit, but would sail on forever, lost in space. 'All the mathematicians in the world had looked at the calculations,' said one engineer, but still there was no way of knowing for sure that it would work.

In a few minutes the astronauts would know if this was a new beginning or a fearful end to their lives. The engines fired for the required 247 seconds at full blast. They had done it. They could now use the attitude control thrusters to align the craft. For the first time in the twin histories of earth and moon, a tiny speck of humanity was sailing confidently in the vastness of space behind the moon travelling at 5000 mph, viewing what looked like the world of the dead, a blanched landscape of peaks and troughs devoid of life. They were in lunar orbit, and could hardly believe it.

Ground control, meanwhile, did not yet know the outcome. Adding to the nail-biting suspense, they could not stop calling the ship although they knew they could not be heard:

'Apollo 8, come in. Apollo 8 ... Apollo 8 ...'

Agonizing minutes and seconds were to pass before they heard:

'Go ahead, Houston.'

Jim Lovell's voice sounded calm and clear as though it were coming from just round the corner. In mission control, grown men, sober men, whose days had been filled with equations and figures and bone-dry facts displayed on monitor screens, exploded into noisy delight. The relief was enormous. The thrill unbelievable. The huge complexity of their exquisite calculations had created a thing of wonder.

Kamanin recalls that on 24 December as he and some of the cosmonauts were travelling in a bus at the launch site, they suddenly rounded a corner and saw the bright crescent of the moon. 'Everyone grew silent for a minute.' They knew full well that Apollo 8 was at that minute circling the moon, and were 'filled with contradictory feelings'. It was impossible not to admire the courage of the American astronauts, yet also they 'felt hurt that it was not our men orbiting the moon'.

It was Christmas, a time when humanity traditionally paused to consider its collective myths and miracles: and now they had a real one. Some of their number were within a short distance of the unreachable moon, almost among the stars. The crew of Apollo 8 felt a sense of wonder. Looking out of their window, they felt themselves surrounded by the numinous. The view of the moon's surface beneath them was spectacular, millions of years of history embedded in its heavily cratered surface. They wanted to investigate the Sea of Tranquillity, a possible landing site for the following year. Jim Lovell, who was assigned to the task, reported that he could see no reason why it would not make an excellent landing site.

On Christmas Eve, they broadcast to half a billion people. 'This is Apollo 8 coming to you live from the moon,' began Borman. Each described the scene. 'The vast loneliness is awe-inspiring,' Lovell said. 'It makes you realize just what you have back there on Earth.' The moon 'looks a vast, lonely, forbidding place, an expanse of nothing … clouds of pumice stone'. And, in contrast to the dead and ghostly moon, Lovell saw the earth 'as a grand oasis in the big vastness of space'. Bill Anders sent earth a Christian message reading from Genesis: 'In the beginning God created the heaven and earth …' And Frank Borman was moved by the sight of the earth rising beyond the moon's bleak surface, a beautiful globe of swirling white and blue surrounded by the vast blackness of space: 'This is the most beautiful, heart-catching sight of my life.'

The euphoria continued, then on Christmas Day Apollo 8 completed its tenth and final orbit around the moon. It was time for the journey back into earth's orbit, and it was this journey, the 'trans earth injection', that unnerved NASA. A small miscalculation, a wrongly placed decimal point, an infinitesimal cough in the engine, would translate into disaster from which there was little hope of rescue. Everything now depended on the service propulsion system engine. Failure meant the Apollo 8 would not come back. When the oxygen ran out in about seven days' time, the men would die and the space craft with its phantom crew, alternately bathed in silver light and blackest night, would circle the moon forever.

Once again, the Apollo 8 astronauts were behind the moon and out of radio contact. Once again, NASA was sitting on the edge of its seat. The time came to operate the engine, which fired for the exact 304 seconds required, taking them out of lunar orbit and on their two-and-a-half-day trek through the darkness of space towards the brilliant, luminous earth. The computer was in charge. They felt the jolt and heard the roar of flame as they sped to the precise point that would allow re-entry. At 25,000 mph, with a flaming tail 120 miles long, they entered the narrow orb that would take them back to their lives on earth. The intense burn-up of re-entry inevitably slowed the craft. Then, still high above the Pacific, the ship's three parachutes opened like exotic flowers. Their triumphant return was a heady affair even by American standards: these were men who had touched the infinite and their epic journey was a harbinger of things to come.

• • •

The Apollo 8 circumlunar mission sent a shock wave throughout the Soviet space industry. America was leading the way, showered in glory, while the Soviet Union, with faltering steps, looked like the country cousin. As 1969 began, a meeting chaired by the Minister of General Machine Building, Sergei Afanasyev, was held at Baikonur with the aim of rescuing the situation. 'How can we get out of this mess?' he appealed to the chief designers in the room. The discussion centred on how to beat the Americans to the moon. It was acknowledged that it was a good thing that Mishin would soon be testing both the manned Soyuz space docking and the N-1.

Once again, Mishin confounded his critics. On 14 January 1969, Soyuz 4, with one man on board, was launched from a Baikonur covered in snow. The next day, Soyuz 5 followed, with three further crew members. After successfully docking, two of the cosmonauts transferred to Soyuz 4 and returned. Re-entry was somewhat more hair-raising for the remaining cosmonaut, Boris Volynov, left in Soyuz 5. The service module failed to separate from the main descent capsule and the craft turned over and over, out of control, with temperatures rising fast and smoke wafting inside. Volynov had the impression that the craft was disintegrating around him. He was convinced he was going to die. By a miracle he survived, only to find that the parachute was deployed too early and the cords became caught up, preventing it from opening. He hurtled through space, waiting for the sickening impact that would mark his final exit from the world. But incredibly the disastrous situation righted itself. The parachute straps took their correct positions and he was able to land safely. In spite of the hazards, the trip was a success, and the Soviet press were quick to claim that this represented the world's first orbital space station.

The N-1 launch was planned for February. By January, work was well underway to prepare the N-1 for its coming test. Waiting to be called into life, it lay in the assembly building attended by more than two thousand workers who had been drafted in from the services to wait on the behemoth and deliver it to the site on a transporter. On its slow journey to the pad, there was a ghostly, unoccupied presence just ahead of the rocket, where Korolev had once walked. Slowly it rose up vertically, a colossal tower held fast by forty-eight bolts at its base, pale against the leaden skies. As it stood alone, its size brooked no comparisons. This was Korolev's beloved N-1, 344 feet tall with a thrust of 4500 tons. It was his dream ticket to the moon and he had bet he would get there before the Americans.

The launch date set for 20 February 1969 was missed due to icy winds, blowing snow. The next day was bitterly cold, but bright. The occasion was toasted with a big bottle of champagne by Mishin, and high-powered visitors awaited that singular and exciting moment of takeoff. For Kuznetsov it was to be the first time that the layout of the engines would be fully tested firing together. 'Even if you had attended our Soyuz launches dozens of times, you couldn't help being excited,' Boris Chertok observed. 'The image of an N-1 launch is quite incomparable. All the surrounding area shakes, there is a storm of fire and you would have to be insensitive to be able to remain calm at such moments. You really want to help the rocket: "Go on! Go up! Take off!"'

The rocket blazed off with a fury of fire and thunder as all thirty engines sprang to life. The ground submitted to the terrible upheaval. The buildings absorbed the sickening shock waves. The windows trembled. The power of the engines and the white fire were awesome as the majestic column slowly moved upwards through the mass of billowing clouds at its base. The audience was enthralled.

Suddenly it became clear that something was wrong. Two engines shut down. The rocket kept going at reduced power. Then all the first-stage engines cut out. Still rising, with a tremendous surge of power it shuddered with enough violence to break fuel lines, bathing the interior in cascades of flame. Still rising, all the engines were now ablaze. The great star of fire travelled seventeen miles before exploding in unseen glory over the barren desert.

Solemn faces emerged from the bunker. 'I'm so sad I could cry,' Kamanin admitted in his diary. 'But it could have been worse. The rocket did take off and the launch pad is not damaged.' Their analysis showed that the KORD control system may have caused the initial shutdown of the two engines. Quite what triggered the huge fireball that engulfed the first stage and prompted all the engines to shut down was unclear but Mishin was reassuring. It was early days. More testing was needed; this was entirely normal. They had been through it all with the R-7. America had not won yet.

CHAPTER TWENTY-ONE

'One small step'

Deke Slayton was in charge of assessing the American astronauts and choosing the crews for various different missions, with the assistance of Alan Shepard, who had been grounded for medical reasons and was now chief of the Astronaut Office. They had made their decisions for the next Apollo trips, 9 and 10, which would test further docking in space and the flight of the untried lunar module. These astronauts were all currently in training for their missions but now they needed to confirm the crew for the 'bull's-eye' trip – the one that would land on the moon. They had established a rotation system for the crews: the backup crew of one mission would fly the third flight after that. On this basis, the backup team for Apollo 8 would fly Apollo 11. Neil Armstrong would be commander. His prowess and unique skills with the 'Flying Bedstead' would be essential when it came to making the critical lunar landing. Buzz Aldrin had proved himself brilliant at space walking – as had Mike Collins. They seemed the ideal team.

Deke Slayton summoned them to a meeting and got straight to the point. The three men practically looped the loop when they heard the news, even though the 'bull's-eye' date was as yet unknown. It might never happen, of course; in the space programme, so much could go wrong. Overnight, disasters had a way of changing schedules. But if and when it happened, the Apollo command and service module in lunar orbit would be manned by Mike Collins, who would stay in that orbit, while Neil Armstrong and Buzz Aldrin would attempt to land on the moon using the lunar module. Meanwhile, the emphasis would be on training, especially in the lunar landing simulator.

Buzz Aldrin was alight with excitement. The captain always stayed with his ship. That surely meant Aldrin would be the first man to set foot on the moon. Not necessarily so, replied Armstrong, who also had plans to make the first historic step. Both men in their mind's eye had the same vision of standing alone in the white moon world. Their name would be a living legend till the end of time.

The lunar simulator became a battleground. In too many sessions, Armstrong crashed. Aldrin challenged what he saw as irresponsible behaviour. Armstrong replied that he was quite deliberately pushing the machine in order to find its limitations. Aldrin continued to accuse; Armstrong continued in his own unique style, skimming disaster. The two men were caught up in a squabble that could threaten their lives, to the growing alarm of Gene Kranz who was to be Flight Director for their mission. The astronauts reined in their feelings but simmered like dormant volcanoes. Behind the scenes at NASA, a decision was made that Armstrong would be the first to step on the moon. Buzz Aldrin was bitterly disappointed. As difficulties persisted in the simulations, Kranz began to wonder if they would have enough time to rehearse the lunar landing.

The deadline was approaching all too fast. In March, Apollo 9 tested docking the command and lunar module in earth orbit and the lunar module was given a test run of more than one hundred miles. In May, Apollo 10 took the lunar module down to within nine miles of the moon's surface, and in a dry run for the real flight tested the ascent stage of the lunar module as they flew back to dock with the command module. Next up was Apollo 11 – the moon mission. This would carry the red, white and blue American flag that would be planted firmly on the moon's pale surface. The date was set for 16 July.

Surreptitiously, people began bringing the astronauts souvenirs to take to the moon. A flag was chosen, one with no manufacturer's label, to thwart advertising. And interested noises in the project were growing from a whisper to a steady clamour. Who were these men, so ready to gamble their lives, willing to risk the hellfire of rockets or the prospect of being lost in space for a day's glory? The world's press was avid for details. *Life* magazine arranged a family get-together for the three astronauts. From the comfort of their armchairs, people the world over could see them at last, nice, normal family men, with pretty wives and beautiful children, just doing a job like everyone else.

Just when the Americans seemed on the verge of winning, the CIA produced yet more worrying information for NASA. They had news that the Soviets were ready to launch a massive new rocket. All year rumours had been growing that the Soviet Union had ambitious plans for a sensational space venture that would dwarf the Apollo missions. Now American spy satellites had photographs showing a rocket bigger and more powerful that Saturn V on the launch pad at Baikonur. Von Braun believed that the Soviets could reach the moon first using its huge booster, if for any reason Apollo 11 missed its July deadline.

• • •

The Soviet Union was abuzz with talk of an imminent, astonishing space event. With conflicting reports it was difficult to distinguish rumour from fact. The cosmonauts themselves were in full training for a lunar mission. They were only too aware of America's imminent target to reach the moon with a launch date of 16 July for Apollo 11. 'The Soviet Union is also making preparation for a manned flight to the moon, just like the Apollo programme of the US,' cosmonaut Alexei Leonov declared. 'The Soviet Union will be able to send men to the moon this year … We are confident that pieces of rock picked from the surface of the moon by Soviet cosmonauts will be … on display.'

In April and May, Vasily Mishin's team was making frantic adjustments and improvements to the N-1 rocket. In spite of frenzied activity to meet the mid-June launch deadline, it was missed. The new launch day would be 3 July, which would still beat the Apollo launch. Baikonur buzzed with activity as staff laboured to bring the launch to fruition. Soldiers and technicians struggled in the July heat. Slowly order was created from chaos. The rocket stood on the pad, unbelievably tall and slender against the blue sky, waiting to orbit the moon. On 3 July, preparations lasted all day. All the prominent people in the space and military industry made their way to Baikonur, and every road leading to the site was congested with vehicles. Chief designers, manufacturers, cosmonauts and ministers mingled with the less elevated, all eager to witness the great event.

Most spectators were well away from the site. The cosmonauts had found an observation point, over four miles distant. Those in the bunkhouse were waiting in expectation. The countdown went ahead without problems and liftoff began at 11.18 p.m. The enormous firestorm of white flames under the rocket forced it upwards, bathing the night sky in shadowless brilliance, almost too bright to watch. In the glare, the rocket, so slender against the night sky, appeared to hesitate. Thick clouds of smoke billowed out. As the thirty engines arranged in two concentric circles reached full power, the ground shuddered and heaved. The rocket trembled. The noise was overwhelming.

At 600 feet, it was clear something was wrong. The upward flight could not be sustained. The rocket doubled over and fell back towards the pad where it exploded with almost the force of a nuclear bomb. Hot blast waves shot out in increasing circles and a mushroom cloud in lurid purples rose over the steppe as hot metal rained on the ruined launch pad. Those viewing the launch at some distance in the open felt the shock waves breathing hot air over them as the rocket, splintering into a thousand brilliant fireballs, fell like hailstones.

The light of the morning revealed carnage as thousands of dead birds and other wildlife littered the blackened launch site at Baikonur. The damage was extensive. It would not be possible to recover quickly from such a blow. Mishin was in despair. For months there had been no break from his unremitting workload. Three days after the disaster, he collapsed with chest pains.

'I was convinced that the rocket would not fly, but somewhere in the depths of my soul there glimmered some hope for success,' Kamanin admitted. 'We are desperate for success, especially now, when the Americans intend in a few days to land people on the moon.' Initial analysis showed that, once again, the KORD control system had turned off most of the engines a few seconds into the launch. The reasons for this were unclear although the evidence pointed towards the possibility that some loose component, perhaps one single shard of metal, had entered the oxygen pump of one of the engines, triggering the catastrophe. Other evidence pointed to an overall design flaw: the thirty engines of the N-1 were aligned in two circles, with low-pressure regions between them. If some of the engines failed, the low-pressure regions would not balance out, the forces in the rocket would no longer be symmetrical and it would become unstable. Whatever the reason, months of detailed investigation lay ahead.

Although it looked as though the Soviet Union was out of the race, an indomitable refusal to accept this remained. So many years of endeavour had gone into the space effort. So much money, so many hopes and dreams: there was still time to send a rocket to the moon. Within days of his heart problems, Vasily Mishin was back at work monitoring Georgi Babakin's next lunar attempt: the unmanned, robotic Luna 15. Once it reached lunar orbit, it was to descend to the moon's surface and a drill would bore into the ground collecting soil samples.

Three days before Apollo 11 was due to launch, a less powerful Proton rocket left the pad at Baikonur with its payload. There was still time to steal America's thunder. The Soviet robotic mission would bring back the fabled pale lunar soil before Apollo 11 returned – *if* it returned. There was no guarantee that the American undertaking would be successful. There was plenty that could go wrong. The Soviet team might still carry a bouquet of stars back to earth. At six o'clock on the blue and gold morning of 13 July, the rocket soared on an exact trajectory to the moon, staunching the flow of endless failures and taking with it so many wistful Russian hopes for success.

• • •

At Cape Kennedy, tension was mounting as last-minute preparations got underway. An argument erupted over the suitability of filming the moon landing. Chris Kraft and Max Faget argued vehemently in favour of filming and shouted all opposition down. The next concern at NASA came from the medical team who wanted to keep the astronauts protected from any chance of infection. When the astronauts met the press prior to the launch, elaborate safeguards ensured that a glass screen came between them and the offending, germ-laden journalists.

Finally, on Wednesday 16 July at 9.30 a.m., the moment arrived. The three astronauts said their most important goodbyes and their minds were now wholly given over to the purpose of the mission. As von

Braun watched the monitor, he could see the Saturn V lift off from Pad 39A. The ascent was a perfect, clean, upward sweep into the seamless blue; the long tail of white fire faded as the sound diminished. This was the fulfilment of his life's ambition. The world was watching. The Soviets were watching. At the BBC in London, commentators speculated on where the Americans would land and whether the Russian robotic mission would be in the way.

Four days out in space, with no worrying incidents so far, Neil Armstrong and Buzz Aldrin transferred from the command module to the lunar module, Eagle, which lay embryo-like above the Apollo third stage. This strange spidery craft with its fragile skin defied the normal rules of engineering. Now the time had come to prove it worked. The astronauts were due to try for their lunar descent on 20 July.

At Baikonur, the Soviets were worried. Problems had delayed the landing of their robotic mission. The mountainous terrain of the lunar landscape had made it impossible. A new orbit was calculated but they could not be completely certain exactly where the craft would land. At NASA, the concern in everyone's mind focused on whether the Soviet Luna 15 would cause problems for the American Apollo 11. The astronauts were informed of the situation. The world continued to watch with baited breath.

In mission control on 20 July, Flight Director Gene Kranz and his team were preparing themselves for the lunar landing. This was the one day when nothing must go wrong. To help everything along, Kranz had turned himself into his own mascot, wearing a handsome silver and white waistcoat in the moon's colours his wife had made for the occasion. On arrival, he had given his team a pep talk in language they understood:

> Hey gang, we're really gonna go and land on the moon today. This is no bullshit; we're going to go land on the moon. We're about to do something that no one has ever done. Be aware that there's a lot of stuff that we don't know about the environment … I trust you implicitly. But I'm also aware we're all human … We're working in an area of the unknown that has high risk. But we don't even think of tying this game, we think only to win … We're going to win … so let's go have at it, gang …

Kranz returned to his seat. He appeared calm, almost casual, like the rest of mission control, but the reality was they were all sitting on pins. When the lunar module came from behind the moon, the descent would begin. The lunar landing would present difficulties, maybe unsolvable problems. It would be up to Armstrong, the skilful champion of the 'Flying Bedstead', to find a safe landing spot. And if he couldn't, then it would be his decision to abort the mission, fire the ascent process and, hopefully, return to Mike Collins, who was orbiting the moon in the Apollo craft.

The Eagle was on course for the moon, descending backwards. The descent was timed to take advantage of sunlight. At 40,000 feet the landing radar began giving information to the computer on the altitude and speed of the Eagle. Suddenly, Aldrin heard the alarm buzzer. The code '1202' showed on his display panel. He reported this to mission control. Was there a problem? Mission control was uncertain.

The light was flashing. '1202. 1202,' Aldrin repeated. Information kept cutting out on the panel in front of him. It was looking serious, but only Houston would have the answer. Kranz didn't know if '1202' meant abort. He referred to Steve Bales in the Flight Dynamics Division. All eyes in the room were on the computer whiz kids. The computer systems and electrical circuits in the Apollo mission were multitudinous. Sometimes an overload of information led to an emergency warning light. But suppose it was a real emergency? Steve Bales had seconds to weigh up the difference.

'We're ... we're go on that, Flight,' twenty-six-year-old Bales stammered. He was sure it was computer overload.

'We're go on that alarm?' Kranz queried.

'If ... if it does not recur, we'll be go.'

Aldrin had his answer. Ignore the alarm.

At 7500 feet, the Eagle was descending at 50 mph. But a few tense moments later they heard Aldrin's voice questioning:

'1202 again?'

The alarm was flicking on and off. It seemed an eternity before Kranz heard Bales reply, 'Ignore,' as he did several more times.

At 3000 feet, Kranz informed Aldrin that it was 'go' for landing.

'Understand, go for landing at 3000 feet,' Aldrin replied. '1201 alarm.'

'1201 alarm?' Kranz repeated.

Bales had little time to work out whether this new alarm was a life-threatening situation and they should abort the mission. Kranz wanted an answer.

'Ignore,' said Bales. 'It's the same type. We're go, Flight.'

The nearer the astronauts got to the lunar surface, the more worrying it looked. The designated landing area was pitted and strewn with boulders. At 1000 feet, Armstrong had no alternative but to override the computer and take over. He needed to find an area of level ground out of the zone of rocks they were in. Accelerating, he moved the craft forward, searching for an ideal spot. Mission control did not know when Armstrong had taken over, but were aware that his fuel was limited.

At 300 feet, Armstrong thought he had a landing place but at 200 feet he changed his mind. Aldrin remembered Armstrong's flights on the 'Flying Bedstead' and how often he had taken it to the wire. Mission control was nervous.

'There ain't no gas stations on the moon,' a voice reminded them.

Now Armstrong was over a crater. A warning light came on. In Houston, the technician responsible for the radar sensors that detected the moon's surface fainted. No one noticed. There was only sixty seconds of fuel left. Twenty seconds of fuel was needed for takeoff if they couldn't land and had to abort. The control room listened in horrified silence. For these last seconds, as time seemed to stretch, the mission was in Armstrong's hands, reliant on his peculiar mastery of the gangling lunar craft.

'Thirty seconds of fuel left,' Houston warned.

Armstrong detected a flat area ahead. This was essential as the awkward craft had to have its four spindly legs all planted equally firmly on level land or eventual takeoff would be in jeopardy. Moon dust was thrown up as he lowered the lunar module with caution. There were still far too many boulders. The attention of both men was focused only on the rocks and fissures in the crumbling greyish plaster of Paris soil beneath them. At last Armstrong found what he needed: that miracle of a small flat area. Moon dust made a snowstorm as they descended.

'Contact light,' Aldrin said with relief.

They were down. Armstrong had handled the machine so skilfully there was no feeling of impact. They had twenty seconds of fuel left. It was 3.17 in the afternoon in Houston when they heard Armstrong's voice:

'Tranquillity base here. The Eagle has landed.'

'Roger, Tranquillity. You got a bunch of guys about to turn blue. We're breathing again. Thanks a lot.'

The men in their strange lunar vehicle looked in wonder at the crystalline world outside the window stretching as far as the horizon six miles away, where a sky the colour of night revealed a gently curving horizon. Outside was a dead world. For time immeasurable no sound had rung out across the white waste to echo among the boulders. No living creature had left its mark. Already, the disturbed dust of their landing had settled to hide evidence of their arrival. Only the fall of meteorites had broken the silence down the centuries. But the two men were enraptured with what they saw. This historical moment could never be duplicated and the whole world was waiting to hear what was really out there beyond the small periphery of the safety of the craft.

As Armstrong backed out of the hatch, automatic TV cameras on the lunar module switched on and six hundred million people watched the hazy figure on television. He stood in his strange white suit, his golden helmet meant to ward off the sun's rays, its reflective visor obscuring his features, turning him into 'Everyman'. His feet touched the moon dust and the world looked on while he tried a step in the new element. In the Soviet Union, the cosmonauts, crammed into a military reception centre with senior officials in Moscow, watched as he lived the moments they thought would be theirs. Then his voice narrowed the huge distance:

'That's one small step for man; one giant leap for mankind.'

The Soviet viewing room erupted with applause as they stared at the grainy images of Armstrong, looking like a moon creature in his pale, puffed-up suit in the pale new world so blanched of colour. 'Everyone forgot that we were all citizens of different countries on Earth,' Alexei Leonov wrote. 'That moment really united the human race.' A few hours later, the Soviet probe, Luna 15, crashed at almost 300 mph into the suitably named 'Sea of Crises' on the moon's surface.

America was the winner. The red, white and blue Stars and Stripes on the flag planted nearby proclaimed that they had legitimately won the glory. The years of dedicated work, never deviating from the original dream, had led to this brilliant conclusion. But woven into the American endeavour were also the dreams of men like Korolev, whose spirit no doubt, given the ghost of a chance, would be standing unseen by Armstrong's side. But the camera only registered one small anonymous figure standing on that bleak waste of cherished soil, waving at the blue earth, halfway to the stars.

It was a moment to savour, a moment for which Korolev had lived and died, a moment that for von Braun concentrated the long journey from the dark days of Hitler's Nazi Germany and the V-2 to this momentous culmination of a youthful vision. As he looked at Armstrong's figure on the TV monitor standing where he had dreamed he would stand, it was a timeless moment, rich with victory second to none in man's short history, and it left him without words.

EPILOGUE

With President Kennedy's mission accomplished and nine further Saturn Vs in production, it seemed there were now no limits for Wernher von Braun. An American flag marked the new frontier and the next step was to colonize it, just as he had always planned. A lunar base, a space shuttle, space station and even manned flights to Mars by 1983 were all under discussion. But for the American public, space had already begun to lose its lustre. 'We have run out of Moons,' admitted von Braun. He was transferred to Washington to crusade for further space missions, but the world had moved on. NASA's budget was slashed further and von Braun became deeply depressed. He left NASA in 1972 and was soon diagnosed with cancer. He died five years later.

Two years after this, in 1979, a former inmate of Camp Dora, Jean Michel, published *Dora*, his account of the camp, in America. 'I have been silent for a long time,' Michel wrote, but now, as an old man, he strongly felt the need to 'make the past speak to the future'. His book was a warning. People had to understand that there had been men 'who dared, in the middle of the twentieth century, to turn men into slaves and re-establish hell on earth', and that Dora, providing labour for V-2 production, was 'the Hell of all concentration camps'. In his powerful account of his experiences there, Michel now specifically linked men 'intimately involved' with the creation of Dora with the US space programme. Men today 'venerated and admired' for the conquest of space, he claimed, had 'drawn a veil' over 'an unspeakable sum of suffering, misery and death'.

Von Braun and his colleagues from Peenemünde had always given the impression that the SS had ordered them to use concentration camp labour and that they themselves had taken no part in organizing it. Spurred on in part by *Dora*, von Braun's charmed reputation would gradually change after his death, as a different view was to emerge.

Under mounting public pressure, that same year the US government set up the Office of Special Investigations to pursue Nazi war criminals in America. Their research turned up evidence against Arthur Rudolph, veteran of the Saturn V programme, which highlighted his involvement in the V-2 plant at Mittelwerk. Witness reports were found buried in the National Archives, which dated back from the 1947 Dora–Nordhausen War Crimes trial. These alleged that Rudolph had been responsible for liaising with the SS to supply slave labour and that he had also had some control over the prisoners' conditions including the pitiful food supply and levels of punishment. Seventy-five-year-old Rudolph was in retirement in California when investigators caught up with him in 1982. He found himself facing the prospect of a trial in the US. He denied the claims, renounced his US citizenship and returned to Germany.

In 1984, von Braun's own records, which had been classified during Project Paperclip after the war, were declassified by the American authorities. These also revealed telling new details. Quite apart from Wernher von Braun's Nazi and SS membership, the records showed that von Braun was not only aware of this slave labour but also, on occasion, had apparently taken part in arranging for it. For example, the minutes of a meeting held at Mittelwerk on 6 May 1944 showed that a number of engineers, including Rudolph, Dornberger and von Braun himself, had been present with the SS to sort out a delay in the manufacture of certain components. The recommendation of the meeting was that 1800 skilled French workers were to be taken to Dora as prisoners and put to work producing the key machinery.

Journalists and historians began to uncover some indisputable facts. Use of forced labour in the V-2 programme began as early as 1940, when around one thousand Polish prisoners of war had been sent to Peenemünde. Their numbers had grown steadily, with Russian, Polish and French prisoners systematically exploited, working eleven-hour days and six-day weeks. By the summer of 1943, 2500 concentration camp prisoners were allocated to the development of the V-2 at Peenemünde. While documents show that Dornberger and Rudolph were well aware of this and even had a hand in organizing it, eyewitness accounts also link von Braun. There is no evidence, either at Peenemünde or Mittelwerk, that von Braun and his colleagues opposed the use of slave labour or took steps to alleviate conditions for the prisoners.

Michael Neufeld, a curator at the National Air and Space Museum, Washington, DC, came across further damning evidence. He found a letter dated 15 August 1944 from von Braun to Albin Sawatzki, the director of production at Mittelwerk. In this, von Braun specifically admitted he went 'to Buchenwald ... to seek out more qualified detainees. I have arranged their transfer to the Mittelwerk ... as per your proposal'. Von Braun was, in effect, personally seeking out technically qualified slave labourers to join the concentration camp labour building his V-2. This, argues science historian Dennis Piszkiewicz, 'according to the principles established at Nuremberg, was a war crime'.

It seems beyond doubt that von Braun knew of the terrible conditions in which slave labour worked on his rocket programmes during the war. His quest to conquer space perhaps made him blind to the moral consequences of his work. Yet how could a man tolerate a situation in which his ambitions were being realized by the brutal mistreatment of thousands of men and women imprisoned in camps in which many died of hunger and disease? Was he acting under orders or serving his own interests and satisfying his own ambitions? To defy the Nazis would have been a brave and dangerous thing to do – but some in Germany did just that. Von Braun was not one of them. As he grew older the legacy of his past haunted him: 'Did we really do the right thing?,' von Braun is alleged to have asked one of his friends as he lay dying. 'Was it right what we were doing all those years?'

As for von Braun's American employers – they too were quite prepared to turn a blind eye to his past while he was useful to them, only to expose and betray him when he was not. Von Braun's energy, brilliance and ambition were cynically exploited first by the Germans and then by the Americans to serve their overriding political and military needs. Ultimately, perhaps, von Braun can be seen as a man with a naive enthusiasm for space who became a pawn in a wider political conflict beyond his control.

Meanwhile in the Soviet Union, Chief Designer Vasily Mishin struggled to cope with the responsibilities left to him after Korolev's death. The N-1 was launched for a third time in June 1971 – for just fifty-one seconds. In November 1972, the fourth launch lasted 107 seconds before six engines failed and the rocket, once again, erupted in a fireball. The manned Soviet moon programme was in disarray and Mishin was blamed. Everything he attempted seemed to be in the shadow of the US Apollo programme as the Americans astronauts returned several times to claim the moon. Though there were successful Soyuz missions, these were eclipsed by the tragedy of Soyuz 11 in which three cosmonauts died, prompting a day of national grieving and finally drawing to a close any hope of restoring national pride. Mishin was fired in 1974 – and even then his troubles were not quite over. By chance in 1985 he was interviewed by a journalist, M. A. Suslov, who was later arrested, allegedly for passing on secrets to the West. Mishin too fell under suspicion and was even threatened with a full trial for acting as the journalist's accomplice.

Valentin Glushko felt the humiliation of the Soviet lunar programme deeply. According to one witness, he kept a moon map on his wall. Every time a Soviet or American spacecraft hit it, he stuck a flag on his map. After Neil Armstrong had landed, he said bitterly, 'They should have listened to me.' In 1974, he had his chance. He replaced Mishin and took over as head of Korolev's former design bureau and immediately suspended Korolev's cherished N-1 project. Under his management, the Soviet space programme went on to achieve notable successes with the Salyut space stations and a joint venture with the Americans in July 1975 when Apollo docked with a Soyuz craft. The world watched as American astronauts and Soviet cosmonauts shook hands and smiled. Glushko lived long enough to see the development and launch of Soviet's own space shuttle, Buran, but he was plagued with ill health in his final years. His last request? That his ashes be preserved so that one day they could be carried to Venus.

Sergei Korolev never lived to see the moon landing for which he had devoted so much of his life – and a Soviet-manned ship never reached the moon. But Korolev's legacy is at last visible. While ironically von Braun's glamorous image from the fifties and sixties began to tarnish, for his Soviet rival the reverse occurred as Korolev grew in stature after his death, attaining near-legendary status, his life in the shadows revealed, his name a source of pride. In 1975, Korolev's home was turned into a museum, preserved exactly as he had always had it: his coat and shoes by the door, his desk covered in his favourite books. His design bureau in Podlipki has mushroomed into a huge enterprise – RSC Energia, the largest space company in Russia, responsible for the launch of the Russian space station Mir in 1986 and other successes. Podlipki itself has been suitably renamed 'Korolev'. At its heart a large statue of Korolev now presides over the central square. His posthumous fame would have astonished him, casting its admiring light on a life he considered already rich, lived as it was, so close to his dream.

ACKNOWLEDGEMENTS

In writing this book I am indebted to many specialists for their generous assistance with my research. I would particularly like to thank the following consultants who made *Space Race* possible. Historian of science Dr Asif Siddiqi at the Fordham University in New York provided invaluable guidance on Russian history and technology. I also appreciated numerous discussions with Dr Simon Prince, City University, London, whose understanding and enthusiasm for the technology was an inspiration. For her knowledge of Soviet history and her insights into the characters and the period, I am indebted to Svetlana Palmer, who also acted as Associate Producer on the BBC series. Space historian and author Piers Bizony gave up valued time to read and comment on the manuscript, as did Dr Gary Sheffield at King's College London who advised on several historical chapters.

During compilation of material for the book and BBC TV series, many veterans of the space programme in Russia and America made a wealth of information available to us and it would not be possible to thank them all. Specialist interviews and archives are cited in the references and I am grateful to assistant producers Miriam Jones and John O'Mahony without whose detailed investigation into many primary sources and manuscripts this research could not have been completed. The BBC team received excellent support from a large number of specialist archives in Britain, America and Russia and these are credited in the bibliography. At the BBC, I would also like to thank executive producer Jill Fullerton Smith and creative director John Lynch, who made it possible for me to write this book.

At Fourth Estate, I am indebted to Mitzi Angel for her advice and skilled editorial judgement at each stage in the production of the manuscript. It was a pleasure, as always, to work with Nicholas Pearson and Courtney Hodell. Many thanks, too, to Lizzie Dipple and Rachel Smyth; and to Catherine Heaney and Juliet Davis for all their help organizing the pictures. At Curtis Brown, Peter Robinson's advice and encouragement on the project over many months proved invaluable.

Finally, I would particularly like to thank Martin Surr and Julia Lilley for sharing many conversations with me about the characters and the history and whose generous support and great faith in the project made this book possible.

BIBLIOGRAPHY

PART ONE

Agoston, Tom, *Teufel oder Technocrat? Hitler's graue Eminenz* (*Demon or Technocrat? Hitler's Eminence Grise*, Hans Kammler biography). Berlin: E. S. Mittler & Sohn Gmbh, 1993

Albring, Werner, *Gorodomlya. Deutsche Raketenforschen im Russland* (*Gorodomlya: German Rocket Research in Russia*). Hamburg: Lucterhand Literaturverlag, 1991

Andrew, Christopher, and Gordievsky, Oleg, *KGB: The Inside Story of its Foreign Operations from Lenin to Gorbachev*. London: Hodder & Stoughton, 1990

Bower, Tom, *The Paperclip Conspiracy*. London: Michael Joseph, 1987

Braucke, Käte, *Im goldenen Käfig. Unfreiwilligen in Russland, 1946–1952* (*In a Gilded Cage: In Russia Against their Will, 1946–1952*). Frankfurt-am-Main, 1989

Chertok, Boris E., *Rakety I Lyudi* (*Rockets and People*), Volumes 2, 3 and 4. Moscow: Mashinostroyenniye, 1994

Crouch, Tom D., *Aiming for the Stars: The Dreamers and Doers of the Space Age*. Washington, DC: Smithsonian Institution Press, 1999

Dornberger, Walter, *V-2*. New York: Viking Press, 1954

Garlinski, Josef, *Hitler's Last Weapons*. London: Julian Friedman Publishers Ltd, 1978

Gröttrup, Irmgardt, *The Rocket Wife*. London: André Deutsch, 1959

Gröttrup, Ursula, Family archive: letters and diaries of Helmut and Irmgardt Gröttrup (unpublished)

Heather, David M., *Wernher von Braun*. New York: G. P. Putnam's Sons, 1967

Hunt, Linda, *Secret Agenda: The United States Government, Nazi Scientists, and Project Paperclip, 1945–1990*. New York: St Martin's Press, 1991

Huzel, Dieter K., *Peenemünde to Canaveral*. Englewood Cliffs, NJ: Prentice-Hall, 1962

Irving, David, *The Mare's Nest*. London: William Kimber, 1964

Jones, R. V., *Most Secret War*. Ware, Herts: Wordsworth Editions, 1978

King, Benjamin, and Kutta, Timothy, *Impact: The History of Germany's V-Weapons in World War II*. Staplehurst: Spellmount, 1998

Knight, Amy W., *The KGB: Police and Politics in the Soviet Union*. London: Allen & Unwin, 1988

Lang, Daniel, *From Hiroshima to the Moon*. New York: Dell Publishing, 1951

Magnus, Kurt, *Raketensklaven, Deutsche Forscher hinter rotem Stacheldraht* (*Rocket Slaves: German Researchers behind Communist Barbed Wire*). Stuttgart: Deutsche Verlags-Anstalt, 1993

McGovern, James, *Crossbow and Overcast*. London: Hutchinson, 1965

Mozzhorin, Yuri, et al., *Nachalo Kosmicheskoi Ery* (*Beginnings of the Space Era: An Oral History of the Soviet Space Programme*). Moscow: RNITsKD, 1993

Murphy, David, E., Kondrashev, Sergei, A., and Bailey, George, *Battleground Berlin: CIA vs KGB in the Cold War*. New Haven, Conn., and London: Yale University Press, 1997

Naimark, Norman, M., *The Russians in Germany*. Cambridge, Mass.: Belknap Press of Harvard University Press, 1994

Neufeld, Michael J., *The Rocket and the Reich*. New York: Free Press, 1995

Ordway III, Frederick I., and Sharpe, Mitchell R., *The Rocket Team*. New York: Thomas Y. Crowell Publishers, 1979

Williams, Andrew, *D-Day to Berlin*. London: Hodder & Stoughton, 2004

PART TWO

Antonov, Vladimir, *Taynye informatory Kremlia biograficheskiye ocherki o razvedchikakh* (*Kremlin's Secret Informers*). Moscow: Geia Iterum, 2000

Conquest, Robert, *Kolyma: The Arctic Death Camps*. Oxford: Oxford University Press, 1979

Daniloff, Nicholas, *The Kremlin and the Cosmos*. New York: Alfred Knopf, 1972

Golovanov, Yaroslav, *Korolev, Fakti I Mify* (*Korolev: Facts and Myths*). Moscow: Nauka, 1994

—, Notebooks (unpublished)

Harford, James, *Korolev: How One Man Masterminded the Soviet Drive to Beat America to the Moon*. New York: John Wiley & Sons Inc., 1997

Ishlinsky, A., *Akademik S. P. Korolev: uchenyy, inzhener, chelovek* (*Academician S. P. Korolev: Scientist, Engineer, Person*). Moscow: Nauka, 1986

Johnson, Nicholas L., *The Soviet Reach for the Moon*. New York: Cosmos Books, 1995

Keldysh, Mstislav, and Vetrov, Georgi, *Tvorcheskoye naslediye Akademika Sergeya Pavlovicha Koroleva* (*Creative Legacy of Academician Sergei Pavlovich Korolev: Selected Works and Documents*). Moscow: Nauka, 1980

Khrushchev, Nikita, *Khrushchev Remembers: The Last Testament*. London: André Deutsch, 1971

Koroleva, Natalia, *Otets*, Volumes 1 and 2. Moscow: Nauka, 2002

Michel, Jean, *Dora*. New York: Holt, Rinehart and Winston, 1979

Mozzhorin, Yuri, et al., *Dorogi v Kosmos* (*Roads to Space*), Moscow: MAI, 1992

—, *Nachalo Kosmicheskoi Ery* (*Beginnings of the Space Era: An Oral History of the Soviet Space Programme*). Moscow: RNITsKD, 1993

Rakhmanin, V. F. (ed.), *Odnazhdy I Navsegda: Dokumenty I Lyudi o Sozdatele Raketnykh Dvigateley I Kosmicheskikh System Akademike Valentine Petroviche Glushko* (*Once and Forever: Documents and People involved in the Creation of the Rocket Engines and Space Systems of Academician Valentin Petrovich Glushko*). Moscow: Mashinostroyennie, 1998

Rauschenbakh, Boris.V. (ed.), *S. P. Korolev i ego delo: svet i teni v istorii kosmonavtiki, izbrannye trudy i dokumenty* (*S. P. Korolev and his Work: Light and Shadows from the History of Space Research*), compiled by G. S. Vetrov. Moscow: Nauka, 1998

Romanov, Aleksandr (pseudonym), *Nights are Longest There: Smersh from the Inside*, translated from the Russian MS by Gerald Brooke. London: Hutchinson, 1972

—, *Korolev*. Moscow: Molodaya Gvardiya, 1996

Ronblom, Hans, *Wennerstroem the Spy*. London: Hodder & Stoughton, 1966

Sebag Montefiore, Simon, *Stalin: The Court of the Red Tsar*. London: Weidenfeld & Nicolson, 2003

Seth, Ronald, *The Executioners: The Story of Smersh*. London: Cassell, 1967

Siddiqi, Asif, *Sputnik and the Soviet Space Challenge*. Gainesville, Fla.: University Press of Florida, 2004

Smithsonian Institution, *A History of the Third Armoured Division (Spearhead), April 1941–July 1958*. 1999

Stuhlinger, Ernst, and Ordway III, Frederick I., *Wernher von Braun: Crusader for Space*. Malabar, Fla.: Krieger Publishing Co., 1994

Sudoplatov, Pavel, *Spetsoperatsii* (*Special Operations*). Moscow, Olma-Press, 1997

—, *Razvedka i Kreml* (*Intelligence and the Kremlin*). Moscow: Too Geia, 1996

Von Braun, Wernher, and Ordway III, Frederick I., *History of Rocketry and Space Travel*. New York: Thomas Y. Crowell Publishers, 1966

Whiteside, Thomas, *An Agent in Place: The Wennerstroem Affair*. London: Heinemann, 1967

PART THREE

Beschloss, Michael, *Mayday: Eisenhower, Khrushchev and the U-2 Affair*. London: Faber & Faber, 1986

Andrew, Christopher, and Gordievsky, Oleg, *KGB: The Inside Story of its Foreign Operations from Lenin to Gorbachev*. London: Hodder & Stoughton, 1990

Haynes, John, and Klehr, Earl and Harvey, *Venona: Decoding Soviet Espionage in America*. New Haven, Conn.: Yale University Press, 1999

Ivanov, Alexander (aka Oleg Ivanovsky), *Pervye stypeni* (*First Steps*). Moscow: Molodaya Gvardiya, 1980

Kamanin, Nikolai, *Skrytyi Kosmos* (*Hidden Space: The Kamanin Diaries*). Moscow: Infortekst, 1995

Khrushchev, Nikita, *Khrushchev Remembers: The Last Testament*. London: André Deutsch, 1971

Khrushchev, Sergei, *Krizisy I Rakety: Vzglyad Iznutri* (*Crises and Missile: A View from the Inside*). Moscow: Novosti, 1994

—, *Nikita Khrushchev and the Creation of a Superpower*. Philadelphia, Pa: Pennsylvania State University Press, 2000

Medaris, John B., *Countdown for Decision*. New York: G. P. Putnam's Sons, 1960

Mozzhorin, Yuri, et al., *Nachalo Kosmicheskoi Ery* (*Beginnings of the Space Era: An Oral History of the Soviet Space Programme*). Moscow: RNITsKD, 1993

Mitrokhin, Vasily, and Andrew, Christopher, *The Mitrokhin Archives*. London: Hodder & Stoughton, 2003

Oberg, James, *Red Star in Orbit*. New York, Random House, 1981

Rakhmanin, V. F. (ed.), *Odnazhdy I Navsegda: Dokumenty I Lyudi o Sozdatele Raketnykh Dvigateley I Kosmicheskikh System Akademike Valentine Petroviche Glushko* (*Once and Forever: Documents and People Involved in the Creation of Rocket Engines and Space Systems of Academician Valentin Petrovich Glushko*). Moscow: Mashinostroyennie, 1998

Rauschenbakh, B. V. (ed.), *S. P. Korolev i ego delo: svet i teni v istorii kosmonavtiki. izbrannye trudy i dokumenty* (*S. P. Korolev and his Work: Light and Shadows from the History of Space Research*), compiled by G. S. Vetrov. Moscow: Nauka, 1998

Rebrov, Mikhail, *Sergei Pavlovich Korolev, Zhizn i neobyknovennaia sudba* (*Sergei Pavlovich Korolev: His Life and Extraordinary Fate*). Moscow: Olma-Press, 2000

Rhea, J. (ed.), 'Roads to Space: An Oral History of the Soviet Space Program,' compiled by the Russian Scientific Research Centre for Space Documentation. Translated by Peter Berlin, *Aviation Week*, 1995

Richelson, Jeffrey, A *Century of Spies: Intelligence in the Twentieth Century*. Oxford: Oxford University Press, 1997

PART FOUR

Baikonur, Chudo 20 veka, Vospominania veteranov ob akademike Mikhaile Kuzmiche Iangele I kosmodrome (*Baikonur, Miracle of the 20th Century, Collection of Accounts by Baikonur veterans about Academician Mikhail Kuzhmich Yangel and the Cosmodrome*). Moscow: Sovremennyi pisatel, 1995

Baker, David, *The History of Manned Space Flight*. New York: Crown Publishers, 1982

Chertok, Boris, *Rakety I Lyudi* (*Rockets and People*), Volumes 2, 3 and 4. Moscow: Mashinostroyennie, 1994

Doran, Jamie, and Bizony, Piers, *Starma: The Truth Behind the Legend of Yuri Gagarin*. London: Bloomsbury, 1998

Dunar, Andrew J., and Waring, Stephen P., *Power to Explore: A History of Marshall Space Flight Center 1960–1990*. Washington, DC: NASA, 2000

Feoktistov, Konstantin, *Sem' Shagov v Niebo* (*Seven Steps into the Sky*). Moscow: Molodaya Gvardia, 1984

—, *Traektoria zhizni. Mezhdu vchera i zavtra* (*Trajectory of Life: Between Yesterday and Today*. Moscow: Vagrius, 2000

Gallai, Mark, *S. Chelovekom na Bortu* (*With Man on Board*). Moscow: Moskva, 1985

Gagarin, Valentin, *My Brother Yuri. Pages from the Life of the First Cosmonaut*. Moscow: Progress Publishers, 1973

Gagarin, Yuri, *Road to the Stars*. Moscow: Foreign Languages Publishing House, 1962

—, and Lebedev, Vladimir, *Space Psychology*, Moscow: Mir Publishers, 1970

Gerchik, K. (ed.), *Nezabyvayemyi Baikonur* (*Unforgettable Baikonur: Accounts by Veterans of Baikonur*). Moscow: Interregional Committee of Baikonur Veterans, 1998

Glenn, John, *John Glenn: A Memoir*. New York: Bantam Books, 1999

Golovanov Archive Materials: a compilation of newspaper articles and Golovanov's own articles from the Soviet press

Ivanov, Alexander (aka Ivanovsky, Oleg), *Pervye stypeni* (*First Steps*). Moscow: Molodaya Gvardiya, 1980

Johnson, Nicholas L., *The Soviet Reach for the Moon*. Washington, DC: Cosmos Books, 1994

Khrushchev, Sergei, *Nikita Khrushchev and the Creation of a Superpower*. Philadelphia, Pa: Pennsylvania State University Press, 2000

Koroleva, Natalya, *Otets*, Volume 2. Moscow: Nauka, 2002

Kraft, Chris, *Flight: My Life in Mission Control*. New York: Dutton, 2001

Kranz, Gene, *Failure is not an Option*. New York: Penguin, 2000

Mozzhorin, Yuri, et al., *Nachalo kosmicheskoi ery* (*Beginnings of the Space Era: An Oral History of the Soviet Space Programme*). Moscow: RNITsKD, 1993

Pellegrino, Charles R., and Stoff, Joshua, *Chariots for Apollo*. New York: Atheneum, 1985

Raushenbakh, B. V. (ed.), *S. P. Korolev I ego delo: svet i teni v istorii kosmonavtiki, izbrannye trudy i dokumenty* (*S. P. Korolev and his Work: Light and Shadows from the History of Space Research*), compiled by G. S. Vetrov. Moscow: Nauka, 1998

Rauschenbakh, B. V., Sokolskiy, V. N., and Feoktistov, K. P. (eds), *Materialy po istorii kosmicheskogo korablia 'Vostok'* (*Materials on the History of the Vostok Spaceship*). Moscow: Nauka, 1991

Rhea, John (ed.), 'Roads to Space: An Oral History of the Soviet Space Program,' compiled by the Russian Scientific Research Centre for Space Documentation. Translated by Peter Berlin, *Aviation Week*, 1995

Romanov, Aleksandr, *Korolev*. Moscow: Molodaya Guardiya, 1996

Schefter, James, *The Race*. New York: Doubleday, 1999

Scott, David, and Leonov, Alexei, *Two Sides of the Moon*. London: Simon & Schuster, 2004

Shayler, David J., *Gemini*. Chichester, Sussex: Praxis Publishing, 2001

Shelton, William, *Soviet Space Exploration*. New York: Washington Square Press, 1968

Shepard, Alan, and Slayton, Deke, *Moonshot*. London: Virgin Books, 1994

Siddiqi, Asif, *Sputnik and the Soviet Space Challenge*. Gainsville, Fla.: University Press of Florida, 2003

Thompson, Neal, *Light this Candle: The Life and Times of Alan Shepard*. New York: Random House, 2004

Tolubko, Vladimir, *Nedelin. Pervyi gglavkom strategicheskikh* (*Nedelin: Biography of the First Head of Strategic Rocket Forces*). Moscow: Molodaya Gvardia, 1979

Wolfe, Tom, *The Right Stuff*. New York: Farrar, Straus & Giroux, 1979

PART FIVE

Armstrong, Neil, Collins, Michael, and Aldrin, Jr., Edwin. E., *First on the Moon*. Boston, Mass.: Little, Brown, 1970

Baker, David, *The History of Manned Space Flight*. London: New Cavendish Books, 1981

Brooks, Courtney G., *Chariots for Apollo*, NASA SP-4205. Washington, DC: NASA, 1979

Burrows, William E., *This New Ocean*. New York: Random House, 1998

Golovanov's archive: Alexei Leonov compiled interviews to Russian newspapers

Irwin, James B., *Destination Moon*. Portland, Oregon: Multnomah, 1989

Khrushchev, Sergei, *Nikita Khrushchev and the Creation of a Superpower*. Philadelphia, Pa: Pennsylvania State University Press, 2000

Lovell, James, *Lost Moon*. Boston, Mass.: Houghton Mifflin, 1994

Michel, Jean, *Dora*. New York: Holt, Rinehart and Winston, 1979

Mishin, Vasily, *Ot sozdania ballisticheskikh raket k raketno-kosmicheskomu mashinostrojeniyu* (*From the Development of Ballistic Rockets to Space-rocket Building*). Moscow: Informatsionno-Izdatelsky Tsentr Znanie, 1998

—, 'Pochemu my ne sletali na lunu' ('Why We Never Flew to the Moon') (an article in *Znaniye: Seriya kosmnautika, astronomiya* no.12, December 1990)

Murray, Charles, and Cox, Catherine Bly, *Apollo: The Race to the Moon*. London: Secker & Warburg, 1989

Pervushin, Anton, *Bitva za zvezdy*. Moscow: AST, 2003

Piszkiewicz, Dennis, *The Nazi Rocketeers: Dreams of Space and Crimes of War*. New York: Praeger Publishers, 1995

Rakhmanin, V. F. (ed.) *Odnazhdy i navsegda: Dokumenty i lyudi o sozdatele raketnykh dvigateley i kosmicheskikh system akademike Valentine Petroviche Glushko* (*Once and Forever: Documents and People involved in the Creation of Rocket Engines and Space Systems of Academician Valentin Petrovich Glushko*). Moscow: Mashinostroyennie, 1998

Rauschenbakh, B. V. (ed.), *S. P. Korolev i ego delo: svet i teni v istorii kosmonavtiki. izbrannye trudy i dokumenty* (*S. P. Korolev and his Work: Light and Shadows from the History of Space Research*), compiled by G. S. Vetrov. Moscow: Nauka, 1998

Rebrov, Mikhail, 'The Last Argument: A Study of the Designer in Black and White'. *Krasnaya zvezda* newspaper, 1995

—, *Sergei Pavlovich Korolev, Zhizn i neobyknovennaia sudba* (*Sergei Pavlovich Korolev: His Life and Extraordinary Fate*). Moscow: Olma-Press, 2000

Scott, David, and Leonov, Alexei, *Two Sides of the Moon*. London: Simon & Schuster, 2004

Siddiqi, Asif, *The Soviet Space Race with Apollo*. Gainsville, Fla.: University Press of Florida, 2003

Slayton Donald K., *Deke!* New York: Tom Doherty Associates, 1994

Young, John W., *America, Russia and the Cold War*. New York: Longman, 1993

RUSSIAN ARCHIVES

Original exclusive sources:

Private archive of Yaroslav Golovanov, Korolev's main biographer, including unpublished letters and private documents of S. P. Korolev

Notebooks of Yaroslav Golovanov, including unpublished interviews, love letters and the only complete transcript of the KGB file of S. P. Korolev

Original Transcript of Gagarin's flight, never published in its entirety, held at RGANTD, (Russian State Archive of Scientific and Technical Documentation)

Diary of Mikhail Tikhonravov, held privately

Archive of Ketovania Ivanovna S., lover of S. P. Korolev, held at the Academy of Sciences, Moscow

Vasily Mishin Papers, held privately

Letters of S. P. Korolev (limited access), held at the Korolev House Museum, Moscow

US ARCHIVES

Air University, Maxwell Air Force Base: *The Story of Peenemünde, or What Might Have Been: Peenemünde East Through the Eyes of 500 Detained at Garmisch*; microfilm A: 5734

Library of Congress: von Braun's scrapbooks, speeches, correspondence
National Air and Space Museum: oral history interviews with German scientists
NARA (National Archives and Records Administration), Washington, DC: records of the Army Guided Missile Group; records of the Office of the Secretary of Defense; intelligence records on von Braun; FBI files on von Braun
US Space and Rocket Center: oral history interviews with German scientists and US Ordnance officers

BBC PRODUCTION INTERVIEWS

Boris Chertok, Moscow, March 2004; Oleg Gazenko, Moscow, March 2004; Yuri Glushko (son of Valentin Glushko), Moscow, March 2004; Oleg Ivanovsky, Moscow, April 2004; Konstantin Feoktistov, Moscow, April 2004; Natalya Koroleva (daughter of S. P. Korolev), Moscow, December 2003, January and April 2004; Tamara Titova (widow of Gherman Titov), Moscow, April 2004; Viacheslav Rakhmanin (Valentin Glushko's biographer) Moscow, March 2004; Cosmonaut Pavel Popovich, Moscow, December 2003; Cosmonaut/ Engineer Georgy Grechko, Moscow, December 2003; Engineer Sergei Kryukov, Moscow, December 2003; Engineer Valentin Syromyatnikov, Moscow, December 2003; Nina and Vera Mishin, Moscow, March 2004; Ursula Gröttrup, Hamburg, February 2004
Svetlana Palmer interviews with Gherman Titov, Khionia Kraskina and Boris Chertok, 1997
Channel One Moscow interview with Vasily Mishin, 1999

PERIODICALS AND NEWSPAPER ARTICLES

New York Times Magazine, an anthology of articles, 1926–60
'German Scientists', *El Paso Herald Post*, 20 December 1946
'German Scientists', *El Paso Times*, 1 July 1947
'German Pupils', *El Paso Herald Post*, 5 August 1947
'Man Will Conquer Space Soon', series in *Collier's* starting 22 March 1952
'Space Pioneer: Will Man Outgrow the Earth?', *Time*, 8 December 1952
'Space Expert Pledges Allegiance to the US', *Nashville Banner*, 15 April 1955

New York Times, 5 October 1957
'Von Braun Assesses Missile Program', *Washington Post*, 10 November 1957
'Experts to Discuss Space Law Issues', *San Francisco Chronicle*, 10 November 1957
'Army Let in on Satellite Effort', *San Francisco Chronicle*, 12 November 1957
'First Sputnik, then the World', *New York Mirror*, 24 November 1957
'All-round Push to Dominate Space Urged', *Sunday Star*, 15 December 1957
Pravda, Professor K. Sergeev, December 1957
Time magazine, 20 April 1959
New York Times, 22 November 1960
New York Times, April 1961
Life magazine, October 1961
Pravda, Professor K. Sergeev, 31 December 1961
Von Braun, Wernher, 'Dr … Questions About Space', *Popular Science*, Volume 182, No. 1, January 1963
—, Wernher, 'Can We Ever Go to the Stars?', *Popular Science*, Vol 183, No. 1, July 1963 *New York Times*, November 1963
'Soviets Lift Edge of Rocket Shroud', *New York Times*, 7 November 1965
'Russians' Space Ship Designer Korolev Dies', *Washington Post*, 16 January 1966
'S. P. Korolev is Dead at 59 …', *New York Times*, 16 January 1966
'Soviet Shift Hints New Space Chief', *New York Times*, 26 June 1966
'Soviet Space Lag Laid to Economy', *New York Times*, 23 December 1966
'Russia Seems to be Cranking Up', *Washington Post*, 15 January 1967
Richard Porter, GE Public Relations magazine (internal distribution) article on Wernher von Braun

INTERNET

V-2 rocket statistics; Mittelwerk statistics:
http://www.v2rocket.com/start/makeup/design.html
V-2 casualties statistics:
http://gi.grolier.com/wwii/wwii_16.html
Eyewitness accounts from the 104th Division:
http://www.104infdiv.org/nordhausen_survivor.htm

http://members.aol.com/Galione3/timberwolf415b003.htm

TELEVISION DOCUMENTARIES

BBC documentary series: *The Cold War* BBC/CNN 1998

BBC documentaries: *Reputations: Wernher von Braun*, 17 July 1999; *The Paperclip Conspiracy*, 20 February 1987; BBC Red Star in Orbit series, 1984

INDEX

P.S.

Ideas, interviews & features …

About the author

About the book

Read on

The Canvas Emerging

Louise Tucker talks to Deborah Cadbury

Like many children, Korolev and von Braun both wanted to work with rockets when they grew up. What did you dream of doing?
I think those fortunate enough to harness their childhood dreams and achieve them with outstanding success as adults do seem touched by a fate beyond the ordinary, although sometimes a high price is paid for this advantage. In my own childhood, little emphasis was put on pursuing a career and it was only when I reached university that I began to give this serious thought. I realised that I wanted to get involved in some way with stories, characters and plot – and became involved in creative writing. Then I discovered that the BBC ran a training course for documentary makers and suddenly doors started to open.

Was your family background an influence on your career in any way?
I had a very happy childhood and creativity lay at the heart of it. My father was always most content in his garden, seen against an expanse of lawn or among the apple trees or half hidden by the herbaceous bed. My mother loved painting and interior design and I can see her now, paintbrush in hand, listening to music, the vibrant world of the canvas emerging. We had a lot of fun as a family. I have an enduring image from my childhood of French doors flung wide open onto a summer garden with a profusion of colour and scents and the sound of music

coming from somewhere. I think the richness of the experience has helped to give me reserves and resilience which might not have happened from a narrower environment.

As somebody who has combined two careers for many years, did you relate to the engineers struggling to achieve their goals?
Many of the characters I have written about, not just in *Space Race*, but also *Seven Wonders* and even *Dinosaur Hunters*, seem to be driven to the point of obsession, sometimes even to the point of death. I find this theme compelling and I think the truth is I really like a good flawed hero, someone who sets out to achieve the superhuman. It's even better for me if their goal is not just a selfish ambition but something I can really believe in, something that would benefit everyone. And I like to know that it's all true; that this really happened. The beauty about all this as a writer is that you don't have to actually live through the nightmare of turning dreams into reality – you have a ringside seat on all the excitement from the comfort and safety of your study. Yes, I can definitely relate to their struggles – as I think anyone who has ever had a dream would.

What or who inspired you to write this book?
The characters themselves. I was amazed when I first learned of Korolev: a man who set out to reach for the stars, yet his remarkable successes were hidden from ▸

' Many of the characters I have written about seem to be driven to the point of obsession, sometimes even to the point of death. I find this theme compelling and I think the truth is I really like a flawed hero, someone who sets out to achieve the superhuman. '

Author photograph © Jerry Bauer

LIFE
at a Glance

BORN

August 1955 in London

EDUCATED

Guildford Grammar School; University of Sussex; University of Oxford

CAREER

BBC researcher (trainee) 1978–9; researcher 1979–82; TV producer 1983 onwards; author

PREVIOUS BOOKS

The Feminisation of Nature, The Dinosaur Hunters, The Lost King of France, Seven Wonders of the Industrial World

The Canvas Emerging *(continued)*

◀ view. Here was the man behind the headlines that had caught my imagination as a child: the first satellite in space, the first man in space, the first blurry images of the moon itself. It seemed amazing that one man had been the driving force – whilst compelled to live in complete secrecy constantly shadowed by his KGB 'minder'. There was a kind of justice in bringing him out of obscurity, letting him take a bow. I began to piece together the plot and realised he was in competition with a man he never met from another continent, who stood to beat him to every key goal. It was a highly unusual tale of rivalry, with two talented visionaries and two great superpowers locked in a symbolic race. To add to the intrigue, von Braun was a wonderfully complex character; there was evidence that cast him in a glamorous, heroic light and then equally compelling evidence that cast him as a monster. As I started to plot the characters, I realised this was a story I would write.

Britain courteously shared its information with the Americans only to lose out completely. How would space history have changed if the UK had been a little less polite and secured the blueprints and engineers for itself?

It was very entertaining while making the TV series of *Space Race* with American, Russian and German co-producing partners that each country was eager to point up the idea that they were the *real* winner of the race. Without von Braun would the Americans have had the rocket to take them to the moon? The Russians spurred every key step of the race, faltering only as the final goalpost was in sight, with the tragic early death of their charismatic leader. And the Americans, of course, did take that famous 'giant leap for mankind'. The idea that this could have been a British giant leap, if only our intelligence teams had not been so gentlemanly in sharing all their records, is stretching things a bit – but still – nice thought …

How does producing a TV series affect the writing of the book and vice versa?
The book and the film are very different products and yet the work that is required can be handled in a way that complements both. This is particularly true of the research. A book requires exhaustive and searching enquiry and in 100,000 words it is usually possible to do justice to the complexity and subtleties of the story. I like to take the time to pore over the primary sources and see what the characters wrote in their own words and how they saw their predicament. It is important to find the characters' own voices – and this invariably complements the film.

The drama documentary form brings its own discipline. Inevitably there are issues in working out how to turn complex history into a film of less than 5,000 words. This can be as basic as where do you start the story – and where end it? How many characters do we need and can we focus the action through just one lead? How much context do we need at each step to enable the audience to appreciate what is at stake for the characters? What's great when working ▶

TOP TEN BOOKS

A Man in Full
Tom Wolfe

Pride and Prejudice
Jane Austen

Citizens
Simon Schama

Schindler's Ark
Thomas Keneally

The Crimson Petal and the White
Michel Faber

The Sea, the Sea
Iris Murdoch

Half a Life
V.S. Naipaul

Arthur and George
Julian Barnes

Star of the Sea
Joseph O'Connor

Fingersmith
Sarah Waters

The Canvas Emerging *(continued)*

◄ with scriptwriters and directors on a film is that you can brainstorm how to handle 'flaws' in the narrative from a storytelling point of view. I always enjoy the teamwork after the intensity of researching and writing alone and count myself lucky to be a part of both worlds.

Is it difficult to find the motivation to write when you have a full-time job as well?
Sometimes – but the biggest driver is the deadline. If you know the series is due for transmission in the autumn, the publisher has to have the text by April – and if you miss that deadline, you've blown it.

❝ I had not taken writing as an option seriously until Dava Sobel's *Longitude*, but… ❞

Science and history have been at the forefront of successful publishing for many years. Have any books particularly influenced your own writing?
Probably the greatest practical influence was Dava Sobel's *Longitude.* I had not taken writing as an option seriously until then, but this showed there was a huge appetite for a well-written personal true story from history or science. I was thrilled because this was just what I wanted to do and suddenly the market opened up.

Chronologically you've moved forwards. Now that you've reached the twentieth century, what comes next?
I've written a great deal about strong – some would say dominating – male characters and am longing to have a good female lead. I think that will be my next project and I don't mind which era; it just has to be a great true story. ■

A Writing Life

When do you write?
When I get a chance.

Where do you write?
In the studio at the top of our house – and I often review the text the next day in a coffee shop.

Why do you write?
Why doesn't come into it. It just happens.

Pen or computer?
Computer. I love the computer because it's so informal. You can test out an idea and then change your mind or move it around so easily. It makes the whole process so fluid. I normally do several full passes of the whole manuscript before I'm ready to show my editor. Years ago, I did my first story on a typewriter – and I can't think how I managed.

Silence or music?
Silence for drafting. For reviewing and editing text it does not matter.

What started you writing?
I started seriously when I received an approach from a publisher who wanted to turn a film I had done for *Horizon* into a book.

How do you start a book?
For me story is key. I like to block out the whole plot – to see the entire narrative shape on a few pages. Then I still won't start ►

‘For me story is key. I like to block out the whole plot – to see the entire narrative shape on a few pages.’

A Writing Life *(continued)*

◂ writing until I have plotted the storyline in detail across each chapter, exploring the relationship between the characters and the action to create the most satisfying narrative shape.

And finish?
The last full stop is wonderful. We always have a family celebration.

Do you have any writing rituals or superstitions?
No.

Which living writer do you most admire?
Tom Wolfe. ■

Making *Space Race*

by Deborah Cadbury

FOR ME *Space Race* very nearly began and ended in Ilford – when the police rang with an urgent message. It was just a few days before three months of filming was due to start in Romania. The cast was flying into Bucharest. This in itself was quite a feat. One of our Russian starlets had insisted that the entire schedule was changed so that she could travel to location from Moscow by train not plane – complete with bodyguard and boyfriend. Nonetheless, the idiosyncrasies of the cast had been accommodated, locations were booked, props and sets designed, and the directors, Christopher Spencer and Mark Everest, were already in Romania with a crew – when one quiet Sunday night, our production manager, Victoria Gregory, received a call: the lorry carrying all the props for filming had vanished in Ilford – presumed stolen.

At first I could not quite take in this bizarre information. The props lorry contained countless Nazi uniforms, period Soviet and NASA space suits, a replica of the first Russian satellite and oddest of all, a life-size V-2 rocket – in pieces but when assembled all of three storeys high. It was hard to see what any thief could possibly gain from this odd – and very distinctive – assortment of items. At the same time, as series producer, it was beginning to sink in just what a nightmare we faced.

No filming could start without the props and costumes. It had taken the design team the best part of three months to build the ►

Making *Space Race* *(continued)*

◄ V-2. If they had to start again, we would have to stand down the cast and crew. Yet they had been scheduled and booked in a mammoth feat of organisation led by line producer Jules Hussey. Now they would have to be re-hired. However I wrestled with the circumstances, the BBC could not afford to pay for the filming twice. It began to look as though the whole enterprise would have to be shelved. And for me that would be a great loss because we had unearthed a remarkable story.

In over a year of research we had very good access to Russian archives and had been able to explore the Soviet side of the story in full for the first time. I had long been fascinated by the mastermind behind the Russian space programme, the shadowy 'Chief Designer' Sergei Korolev. Such was the fear Korolev would be assassinated by Western intelligence that his real name was a closely guarded secret during his lifetime. It was as though he did not exist. He was rarely seen in public, his name never appeared in official records and he could only publish occasionally under a pseudonym. Korolev's life seemed to epitomise the extremes and random punishments of Soviet rule – a dramatic personal story largely unknown to the West.

Associate producer Svetlana Palmer and assistant producer John O'Mahony had gone to great lengths to track down Korolev's former colleagues and friends to investigate his true story. John was literally knocking on doors in the outskirts of Moscow when he had a major breakthrough. Korolev had an

> 'I was mesmerized when the first man was launched into space, the first probes reached the moon and the first hazy images of its bleak surface reached the earth.'

official biographer, Yaroslav Golovanov, who had been permitted to publish in Russia after his death. Golovanov was survived by his son, who took John into his father's study. There, piled high, was a roomful of files, letters, personal details of Korolev, much of which had never seen the light of day.

I invariably get very excited when we can track down original primary sources. I knew Korolev had been denounced by his own colleagues during Stalin's purges and sentenced, without a trial, to ten years' hard labour in the Gulag where he nearly lost his life. We were able to obtain letters he wrote to Stalin from the notorious Kolyma camp in Siberia protesting his innocence and begging for his release, heartbreaking letters to his wife and even the confession he was forced to make when arrested by the NKVD.

But there were strange anomalies in his story. Despite the extremes of suffering in his youth, Korolev rose to serve Stalin with a fervour and commitment that makes little sense to Western eyes. Broken from the Gulag, and physically weakened, he emerged more driven than ever to put all his energy into something he could believe in – and in the process he led Russia to a series of spectacular firsts. These were great moments that I could recall from my own childhood. Although too young for Sputnik, I was mesmerised when the first man was launched into space, the first probes reached the moon and the first hazy images of its bleak surface reached the earth. It seemed an era of infinite possibility and to find the truth about the man who, through almost superhuman determination, created the Russian space programme against a ▶

‘Such was the fear Korolev would be assassinated by Western intelligence that his real name was a closely guarded secret during his lifetime. It was as though he did not exist’

Making *Space Race* *(continued)*

◄ background of enormous difficulty was very moving.

Yet this is also an unusual tale of scientific rivalry. Piecing together von Braun's character from the threads of evidence was endlessly fascinating; amongst the team we debated each new twist that came to light. There was no doubt that he had been well aware of the horrific conditions of the concentration camp slaves building his V-2; he even helped to recruit skilled labour. How did this fit with the charismatic leader who emerged later in America? As word spread of our production, some of his former colleagues rang us up to explain just how much he was liked and respected. All this was hard to square with his past.

As the characters began to take shape, we were able to explore the true story of the space race in very personal terms, arguably for the first time, mapping it against the bigger picture of superpower rivalry and the Cold War. Above all with new records, we could show how close run the race was – right up until that final mind-bending voyage to the moon.

The weeks before the shoot are invariably the most frantic – and this production was no exception. I've learned never to be surprised by anything – but the theft of the V-2 rocket was a bolt from the blue. After several agonising days of sightings and subplots worthy of the actual story, the props lorry was found, its strange cargo barely disturbed. The only satisfaction I could gain was imagining the look on the thief's face when – having thought he had

‘Above all with new records, we could show how close run the race was – right up until that final mind-bending voyage to the moon.’

stolen something really valuable – he prised open the door to find he had acquired a mock-up of a V-2 rocket. The police escorted it from an industrial estate in Ilford, then began our own personal race to complete the series. ■

Have You Read?

Other Books by Deborah Cadbury

The Feminisation of Nature
As sperm counts fall, and reproductive cancers increase, evidence suggests that some species are becoming 'feminised'. The main cause of this is humanity's increased exposure to chemicals found in everyday life, chemicals which can mimic hormones. In this, her first book, Deborah Cadbury examines recent changes in human reproductive health and their ominous implications for the future.

The Dinosaur Hunters: A True Story of Scientific Rivalry and the Discovery of the Prehistoric World
Another tale of bitter rivalry, this time between Gideon Mantell, who became obsessed with the lost world of the reptiles, and Richard Owen, who claimed Mantell's work as his own, naming the extinct creatures and securing international acclaim for himself.

The Lost King of France: The Tragic Story of Marie-Antoinette's Favourite Son
Aged four, Louis XVII was heir to the most powerful throne in Europe. Yet within five years, he was to lose everything. During the French Revolution, his family was imprisoned and, following the execution of his parents, the Revolutionary leaders declared Louis dead. But was he really?

Seven Wonders of the Industrial World
Seven Wonders tells the stories behind world-changing machines and monuments – and the people like Isambard Kingdom Brunel and Joseph Bazalgette who made them. ■

If You Loved This,
You Might Like…

Moondust: In Search of the Men Who Fell to Earth **by Andrew Smith**
Between 1969 and 1972 twelve men made the journey to the moon. Only nine of them, at the time of writing this book, were left and Andrew Smith set out to interview them all to find out what it was like to reach the heavens, and what it was like to come back down to earth.

The Right Stuff **by Tom Wolfe**
Winner of the American Book Award for non-fiction, Wolfe's narrative details the stellar achievements and the human frailties of the first US astronauts.

Who Really Won the Space Race? Uncovering the Conspiracy That Kept America Second to the Russians **by Thom Burnett**
As *Space Race* makes clear, Wernher von Braun was prevented from working on a rocket for many years. Burnett focuses on the conspiracy that stopped him and explains how it led directly to the election of John F. Kennedy.

Star-crossed Orbits: Inside the US–Russian Space Alliance **by James Oberg**
An examination of more recent space history, the development of the International Space Station. Oberg used to work for NASA, is considered an expert on the Russian space programme and thus has a unique perspective on the relationship between the US and Russia and detailed knowledge of what each country has brought to the alliance. ■

Find Out More

Many websites offer insights into the history of space exploration and its future. The Smithsonian National Air and Space Museum site **www.nasm.si.edu** offers information on exhibitions, lecture series and what's on at its planetarium but is mostly of interest to those visiting the museum. NASA's site **www.nasa.gov** has shuttle launch news, a picture gallery, podcasts of 'This Week at NASA' and videos of launches (such as the Mars Reconnaissance Orbiter). The best treats are an earth observatory, pictures of lakes, mountains and glaciers as seen from space, and a piece of software called 'Worldwind' that allows virtual touring of (and zooming in to) the earth from space. Finally, for those unlikely ever to reach the heavens, **http://hubblesite.org/** (NB no www in this address) offers a tour around them, as seen by the Hubble telescope. Try some of the virtual tours through galaxies and nebulas. ■